The Fundamentals of Modern Astrophysics

Mikhail Ya. Marov

The Fundamentals
of Modern Astrophysics

A Survey of the Cosmos from the Home
Planet to Space Frontiers

 Springer

Mikhail Ya. Marov
Vernadksy Institute
Academician, Russian Academy of Sciences
Moscow, Russia

ISBN 978-1-4614-8729-6   ISBN 978-1-4614-8730-2 (eBook)
DOI 10.1007/978-1-4614-8730-2
Springer New York Heidelberg Dordrecht London

Library of Congress Control Number: 2014949925

Cover photo of the RCW49 Nebula courtesy of NASA/JPL-Caltech/University of Wisconsin-Madison.

Printed on acid-free paper

Springer is part of Springer Science+Business Media (www.springer.com)

This Springer book is published in collaboration with the International Space University. At its central campus in Strasbourg, France, and at various locations around the world, the ISU provides graduate-level training to the future leaders of the global space community. The university offers a 2-month Space Studies Program, a 5-week Southern Hemisphere Program, a 1-year Executive MBA and a 1-year Master's program related to space science, space engineering, systems engineering, space policy and law, business and management, and space and society.

These programs give international graduate students and young space professionals the opportunity to learn while solving complex problems in an intercultural environment. Since its founding in 1987, the International Space University has graduated more than 3,000 students from 100 countries, creating an international network of professionals and leaders. ISU faculty and lecturers from around the world have published hundreds of books and articles on space exploration, applications, science and development.

Preface

Space exploration engendered enormous progress in astronomy and astrophysics. Astronomy was no longer limited by observations in the very narrow optical window accessible from the Earth's surface; instead, space observations in all wavelengths of the electromagnetic spectrum, from the shortest gamma rays to radio waves, opened unique opportunities to carry out astronomical observations and delicate measurements of distant objects beyond the shroud of our atmosphere. This, in turn, allowed us to view our space environment in unprecedented detail and essentially opened up our universe. Astrophysics is a powerful tool for obtaining deep insights into the universe, revealing and interpreting the physical mechanisms and physicochemical processes underlying the astronomical objects and phenomena that we observe with modern science.

This book is addressed to laymen who would like to obtain more insights into our space environment—which stretch from home planet Earth well beyond the solar system to the edge of the universe (which is associated with its origin). The book will also be useful to students who are interested in space and astronomy. The author invites all to join him for this fascinating journey.

The book has its origins in the International Space University (ISU) program.

The International Space University was founded in 1987 as the new world educational body to teach space science and applications. The first session was held in 1988 and I joined this facility a year later. This involvement became a significant part of my personal career. For a quarter of a century I delivered lectures and/or lecture courses on physics and astrophysics to the ISU students, mainly at the summer sessions in different parts of the world. Throughout these years I chaired or co-chaired Space Physical Sciences Department and/or team projects and actively participated in the major ISU events. In 1991 I served as Dean of Faculty. I have been elected several terms in the Academic Council, served a few years in the Board of Trusties, and contributed to the format development of the ISU Master of Space Sciences program and permanent campus selection for which Strasbourg became the place. I benefited from close relationships with student communities and faculty involved in the ISU programs and preserved close and warm relationship with the ISU alumni and colleagues all over the world.

Certainly, my lecture courses were refined from session to session throughout the years running together with an experience I gained on how to teach in the very specific ISU environment different from what I used to pursue with my university lecture courses in Russia and USA. In this case, the bottom line was the necessity to accommodate the Interdisciplinary, Intercultural, and International (3I) approach ISU accepts as the basic philosophy, and therefore to satisfy both professionals (say, specialized in physics or at least in engineering) and non-professionals (say, specialized in policy and law or in social sciences) having different background and cultural roots. These were my lessons learned which I took from the students' response and appreciation. Anyway, based on this multiyear experience, I compiled a lecture set, mainly in electronic/slides format I used in my presentations at the core lectures curriculum. ISU faculty and my colleague Joe Pelton (one of those who supported three MIT graduate students—ISU founders John Hawley, Peter Diamandis, and Bob Richards—with their idea) suggested me to collect and edit the material and prepare the written version of the lecture course on space physics and astronomy and publish it as a short book of the special series covering the ISU program and even extended. The title suggested *The Fundamentals of Modern Astrophysics: A Survey of the Cosmos from the Home Planet to Space Frontiers* fully corresponds to the idea and the book contents.

I should confess that the work on the manuscript turned out not as easy as I initially thought, and required me to apply more efforts to incorporate both fundamentals and modern views in the fast progressing field rather than to simply collect my lectures. Moreover, the contents are broader and include more details not covered in the lectures. I attempted to combine a strict approach and accuracy with accessibility of understanding when discussing quite complicated topics to give an opportunity to comprehend the main sense and, at the same time, to keep it interesting. I therefore essentially avoided mathematics and used simple formulas only when they promote a better understanding of the described physical phenomena. At the same time, I tried to keep the main definitions and to give a brief history of great advancement in the discussed fields, when it seemed appropriate. Many important breakthroughs accomplished by ground based astronomical facilities and space born instruments are addressed and some of them are reviewed in more detail.

The book consists of 11 chapters and basically covers all the main fields/branches in contemporary astrophysics.

Naturally, we start with the space environment closest to our home planet Earth: the solar system, a very confined place in boundless outer space. The solar system is discussed in the first four chapters. In Chap. 1, the general properties of the solar system are described as well as its family members—terrestrial and giant planets, their satellites (moons) and rings, and the numerous small bodies, including asteroids, comets, meteoroids, and interplanetary dust. A special focus is given to the remarkable dynamical properties of the solar system bodies, including orbital and rotational motion, different types of resonances, and small body migration with implications for the planets' evolution.

Chapter 2 is dedicated to the terrestrial (inner) planets (Mercury, Venus, Earth, and Mars), which occupy the region closest to the Sun and are composed of rock.

The manifold features of their surface landforms, geology, and interiors, and their main atmospheric properties are discussed. The core of the chapter is our Earth—the planet uniquely possessing life—and its natural satellite, the Moon. The completely different natural conditions of our neighbor planets Venus and Mars serve as two extreme models of the Earth's evolution. The problems of evolution of Venus and Mars are specifically addressed.

In Chap. 3, the gaseous and icy giant planets (Jupiter, Saturn, Uranus, and Neptune) with their numerous satellites and rings are discussed, including their interiors, structure, and composition, their general patterns of atmospheric dynamics, and their common features of formation and evolution. The unique nature and properties of their numerous satellites and rings are especially emphasized, and in particular the Galilean satellites of Jupiter, Europa, and Ganymede, and the Saturnian satellites Titan and Enceladus as potential abodes for biota.

Chapter 4 deals with the properties of a vast family of small bodies: the asteroids and comets, which are primarily located in the main asteroid belt and the Kuiper belt and in the Oort cloud, respectively, as well as smaller size dodies—meteoroids and interstellar dust. These small bodies are of particular interest first of all as the remnant bodies of which the solar system originated, which therefore preserve in their composition the most pristine matter. Examples of such matter are the various classes of meteorites, which are regarded as fragments of the collisional processes of asteroids. Catastrophic collisions are emphasized as being responsible for the dramatic changes of planetary environments throughout the history of the solar system.

Chapter 5 focuses on the properties of our star—the Sun—and its plasma environment called the heliosphere. The composition, structure, and peculiarities of the main zones of the Sun are outlined with an emphasis on the photosphere, chromosphere, and corona. The generation of the solar wind and its properties, the 11-year solar activity cycle and related events, as well as the different phenomena caused by disturbances in the chromosphere and corona are discussed. This is followed by an overview of the structure of the heliosphere expanding to the solar system outer boundary where plasma interaction with the interstellar medium occurs, including important phenomena within the heliosphere such as the solar wind interactions with the planets and small bodies. Particular attention is given to the Earth's upper atmosphere and magnetosphere, which protect the biosphere from harmful electromagnetic and corpuscular solar radiation.

From the solar system we move further away to the other stars filling the night skies. They are the most familiar example of the fascinating beauty of our space environment and vast universe. Chapter 6 discusses stars—their complex nature and their lifetimes from birth to ultimate death. Special attention is given to the evolution of stars with different masses by tracing them in the Hertzsprung-Russell diagram and reviewing nuclear fusion as an energy source. The final stages of a star's lifetime are specifically addressed, including the red giants/white dwarfs for low mass stars, and the neutron stars/black holes left behind after high mass stellar supernovae explosions. The fascinating properties of black holes are discussed in terms of relativistic physics and general relativity effects.

Stars harbor planets, and hence it is natural to address the common processes of star-disk-planetary system formation, the peculiarities of the physicochemical mechanisms involved, and the subsequent system evolution. Chapter 7 discusses extrasolar planets, whose discovery ranks among the greatest breakthroughs in astrophysics. The study of extrasolar planets (or exoplanets) has become a "hot topic" of modern science. The different discovery techniques for exoplanets and their unusual properties, such as the hot massive super-Jupiters responsible for peculiar exoplanetary system configurations completely different from that of our solar system, are discussed. Recent findings on Earth-type and especially Earth-like planets manifested a great progress in the field and opened new horizons in planetary science. The continuous search for Earth-like planets in the years to come is highlighted as a challenging goal of both astrophysics and astrobiology.

In Chap. 8, the general problems of the origin and evolution of planetary systems (and in particular, the solar system) are considered based on the fundamental concepts of star-planet genesis including important theoretical and observational constraints. This field of astrophysics is called planetary cosmogony. Basically, the scenario includes a sequence of events beginning with protoplanetary gas-dust accretion disc around a solar-type star formation from a primordial protoplanetary nebula, follow-up processes of its fragmentation into original dusty-gaseous clusters, and collisional interactions of these clusters leading eventually to intermediate solid bodies (planetesimals) and, ultimately, planets and a planetary system.

As a logical extension of this topic, we address the potential origin of life on a planet located in the so-called habitable zone in a star's vicinity, where suitable climatic conditions could be set up and supported. In Chap. 9, we briefly discuss these intriguing problems dealing with another interdisciplinary hot topic of modern science: astrobiology. This field is rooted in an astrophysics and biology synergy and is intimately related to planetary sciences, particularly planetary systems formation and evolution. Astrobiology is also closely linked to basic philosophical concepts.

In Chap. 10 we discuss our space environment at large, beginning from our home—the solar system and our own Milky Way galaxy—and reaching to the extreme boundaries of the universe filled with stars and galaxies. The hierarchical system of the revealed structures of progressively growing size, the clusters and superclusters of galaxies forming the cosmic web, is regarded as the remnant of density fluctuations in matter condensing out of the expanding universe after its origin associated with the Big Bang scenario.

Finally, Chap. 11 addresses the problem of cosmology: the study of the origin, evolution, and ultimate fate of the universe. The Big Bang model of the universe's origin and evidence in its support are thoroughly discussed. The supporting factors include the continuing observed expansion of the universe according to Hubble's law; detection of the cosmic background microwave (CMB) radiation at 2.7 K, which is addressed as the remnant of the original explosion; and the present abundance of light elements (hydrogen, deuterium, helium, and lithium) which were synthesized soon after the explosion. Scenarios of the ultimate fate of the universe are based on its total estimated mass with the involvement of dark matter and dark energy, the latter being considered an analog of Einstein's cosmological constant now associated with anti-gravity force and the monstrous vacuum energy. It is emphasized that modern cosmology is viewed as the synergy of micro- and macrophysics and is intimately related to the physics of elementary particles and the four fundamental interactions in nature: strong, weak, electromagnetic, and gravitational, with involvement of the idea of supersymmetry unifying the constants of these interactions under very high energy. Our current understanding of both the early and evolving universe is based on the Standard Model (most recently supported by the Higgs boson discovery) and Superstring (M) theory, the latter underlying the diversity of elementary particles, quantum mechanics, and gravity and integrating the nature of matter. Modern theory assumes that an infinite multitude of universes exists, our own being one of them. The Multiverse concept also assumes that universes are continually born and decay and that they may be coupled through worm holes along a hidden dimension and to hyperspace.

Basically, these 11 chapters contain most essential things known in modern astrophysics. They briefly summarize the main concepts in the field and prospects for future theoretical and experimental studies. I attempted to discuss these topics in as accessible a way as possible, and this is why no special mathematical treatment and particular references were involved. For those who would like to know more and wish to pursue an in-depth study of these topics, a list of Additional Reading is attached.

Moscow, Russia   Mikhail Marov
Nov 2013

Acknowledgements

I would like first of all to acknowledge the ISU community where I taught physics and astrophysics in the international multicultural environment, which became a basis of this book. I am grateful to my wife Olga – professional astronomer – for her valuable assistance and advice on the book contents and great patience throughout the time of my working on the manuscript. I acknowledge O. Devina for the valuable technical assistance. I am thankful to Dr. Joe Pelton who suggested me to write this book, and Springer Editorial for assistance with publication, specifically to Maury Solomon and Nora Rawn for finding the most suitable format to publish the book and thorough efforts in editing.

Short Biography

Mikhail Marov Professor, Academician (Full Member) of the Russian Academy of Sciences and of the International Academy of Astronautics. Born in 1933 in Moscow, graduated from the Moscow Technical University in 1958. He received his Ph.D. in 1964 and full Doctorate degree in Physics and Mathematics in 1970. He was elected in the Russian (former Soviet) Academy of Sciences in 1990. His principal scientific interests are focused on the fundamental problems of mechanics, hydrodynamics, gas kinetics, astrophysics and space physics, with application to solar system studies and planetary cosmogony along with experimental studies of planets.

Marov has been deeply involved in many major endeavors of the Soviet/Russian space program beginning from the first space flights to the Moon and planets up to the present. He worked as Project Scientist and/or Principal Investigator on the VENERA and MARS lander series and made first several in situ measurements in Venus and Mars atmospheres. He has authored above 250 publications in refereed journals and has also published 16 books and monographs. He has occupied a number of distinguished positions in several Soviet/Russian and International scientific organizations and has also served as an Editor for the distinguished International magazines. Currently, he is Chief-in-Editor of the Solar System Research magazine. Since 1989 Mikhail Marov was deeply involved in the International Space University for which he has taught 25 summer sessions as faculty and Co-Chair of the Space Physical Sciences department. He also served as a member of the ISU Academic Council and a Trustee. He received two distinguished National (Lenin and State) awards and the International Galabert Award for Astronautics, International Academy of Astronautics' award for the outstanding book in fundamental science, Alvin Seiff Award for pioneering space studies of planets and COSPAR Nordberg Medal for outstanding achievements in space science and applications.

Contents

Chapter 1
The Solar System

The Solar System Family: An Overview

Our solar system is our closest space neighborhood and surrounds its central star—the Sun. The solar system is just a fleck in the grander cosmos. Our Milky Way galaxy contains hundreds of billions of solar systems, yet it is only one in a sea of galaxies. The solar system (which is naturally most important to us being the place of our habitat) is just a tiny part of the universe, which since ancient times was understood to mean the entirety of existence.

Let us address first of all the subject of distances in outer space. In other words, how do we scale the universe? Astronomers use a unit of measurement called the light-year (ly), which is the distance that light propagating in a vacuum with velocity 300,000 km/s covers during 1 year (1 ly=~ 10^{13} km). A more common unit, however, is the parsec (pc), which is measured as the distance from which a semi-major axis of the Earth's orbit around the Sun is seen for an angle 1 arcsec. This is called a parallax, and hence the word "parsec" is an abbreviation of parallax-second. Note that 1 pc=3.26 ly. Commonly, enormous distances in the universe are measured in thousands (Kpc), millions (Mpc), and billions (Gpc) of parsecs.

The Earth is located approximately 150 million kilometers (km) from the Sun. This most important distance in the solar system is called one astronomical unit (AU).[1] Light covers this distance in about 8 min, and one parsec comprises 205,000 AU. The outermost planet of the solar system, Neptune, is located at ~ 30 AU, or 4.5 billion km, and light from the Sun reaches it in 4.17 h. The size of the solar system is measured as the distance at which the Sun's gravity becomes nearly equal to that from the closest stars. This distance is roughly in coincidence

---

[1] This originally defined rough distance has been transformed from a confusing calculation based on different measurements (from spacecraft, radars, and lasers) into a single number of unprecedented accuracy. The new standard, adopted in 2012 at the International Astronomical Union's XXVIII General Assembly in Beijing, China, is an amazing value: 149,597,870,700 meters.

© Springer Science+Business Media New York 2015
M.Ya. Marov, *The Fundamentals of Modern Astrophysics*,
DOI 10.1007/978-1-4614-8730-2_1

with the outer boundary of the Oort Cloud occupied by the main population of the solar system's comets, which is located at $\sim 10^5$ AU (1.5 ly or ~ 0.5 pc). However, the nearest stars, Proxima Centauri and Alpha Centauri, are located at distances of 4.24 and 4.30 ly, respectively, from the Sun ($\sim$40 trillion km, that is, 270,000 times farther from Earth than the Sun). It would take about 100,000 years for a spacecraft leaving the Earth with an escape velocity of $V = 12$ km/s to reach them. There are 65 stars similar to the Sun located at only 17 ly from us. Compared to our Milky Way galaxy (which is about 100,000 ly or 30 kpc in diameter), the size of the solar system ($\sim$1.5 ly) is negligible ($\sim 10^{-5}$ the size of the Milky Way).

The solar system family (Fig. 1.1) includes the eight major planets with their satellites, numerous small bodies (asteroids, comets, meteoroids), and interplanetary dust. Two groups of the major planets are distinguished: the inner (terrestrial) planets and the outer (giant) planets, all located within 30 AU from the Sun (Fig. 1.2). The terrestrial solid planets are Mercury, Venus, Earth, and Mars, located between 0.4 and 1.5 AU from the Sun; the giant gaseous planets are Jupiter, Saturn, Uranus, and Neptune, located between 5 and 30 AU from the Sun. The orbital positions of the planets are shown in Fig. 1.3. Nearly all the satellites belong to the giant planets, which also maintain systems of rings. The exceptions are the Earth's Moon, and the two tiny satellites of Mars, Phobos and Deimos. Every planet exhibits unique natural properties. The terrestrial planets have experienced the most dramatic changes throughout their evolution, while the giant planets have been preserved essentially unmodified since their origin.

Fig. 1.1 Population of the solar system: the major planets and representatives of small bodies—comets and asteroids (Courtesy of NASA)

Fig. 1.2 Two groups of the major planets: terrestrial planets (*top*) and giant planets (*bottom*) (Courtesy of NASA)

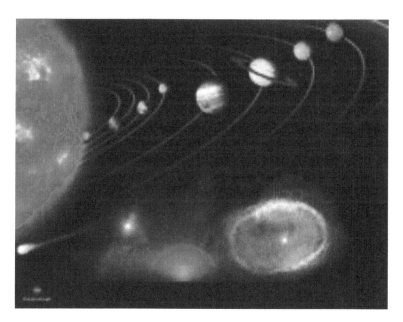

Fig. 1.3 Schematic view of the major planets' orbits around the Sun. At the bottom: planetary nebula to be left behind after the Sun (a G2-type star with the lifetime ~10 Gyr) will exhaust its nuclear fuel in ~5 Gyr from now (see Chap. 6) Adapted from Enciclopedia of Astronomy and Astrophysics (EAA, 2001)

The main characteristics of the planets and satellites are summarized in Tables 1.1 and 1.2, respectively. Both orbital and physical properties are included. Table 1.3 also lists the later-discovered satellites of the giant planets (some very small in size and some closer to the planets). Currently, 67 satellites of Jupiter, 62 of Saturn, 27 of Uranus, and 14 of Neptune are known. As we can see, the remarkable difference between the two groups of planets is their bulk density: the terrestrial planets are much denser than the giants. This is due to the great difference in composition of the rocky inner planets and the mostly gaseous outer planets. Earth and Venus are comparable in size, Mars is about half their size, and Mercury is the smallest planet. The largest planet in the solar system is Jupiter, which exceeds the Earth by more than ten times in size and more than three hundred times in mass, though its density is nearly four times less than that of the Earth. Saturn's density is even less than that of water, and Uranus and Neptune are denser than Jupiter and Saturn because a larger fraction of hydrogen-bearing compounds such as water, methane, and ammonia ices compose their mantles. Let us note also that the giant planets rotate around their axes much faster than do Earth and Mars, while Mercury and especially Venus rotate very slowly.

Beside the major planets, the solar system is populated with numerous bodies of smaller size—asteroids (also called minor planets), comets, meteoroids, and inter-planetary dust—collectively called small bodies (Fig. 1.4). Because some of the asteroids are quite large in size, in 2006 the International Astronomical Union distinguished a group of planets called *dwarf planets,* whose size well exceeds 500 km but are much smaller than the major planets. The only asteroid so desig-nated is Ceres, the largest rock-ice body (976 km across) in the inner solar system, in the main asteroid belt between 2.8 and 3.3 AU in distance. The next asteroids in size are Pallas (544 km) and Vesta (525 km), which are classed as minor planets. Vesta is the brightest asteroid visible from Earth.[2] The recognized other dwarf plan-ets are located in the trans-Neptunian Kuiper Belt populated by comets and aster-oids. The largest of these bodies were distinguished as icy dwarf planets, or plutoids, to acknowledge Pluto, which was demoted from the former ninth most distant major planet of the solar system to be classified as a dwarf planet. The largest of the icy dwarf planets is believed to be Eris, though a final verdict awaits the mission of the New Horizons spacecraft. It is estimated to be 2,326 ($\pm$12) km in diameter and 27 % more massive than Pluto, or about 0.27 % of the Earth's mass. Icy dwarf planets occupy quite a narrow region near the inner boundary of the Kuiper Belt at about 40 AU; its outer boundary extends to nearly 100 AU. It is possible to evaluate the total population of small bodies only statistically; the number of asteroids larger than one kilometer alone exceeds hundreds of thousands and the number grows exponentially towards smaller sizes up to meteoroids of less than a few meters-cantimeters. Comets occupy a vast region well beyond the Kuiper Belt stretching to the periphery of the solar system.

Planetary science involves a vast complex of problems related to a general view of the solar system and the nature of its bodies, first of all, planets and their satel-lites. These problems include system stability and orbital and rotational dynamics of

---

[2] NASA's Dawn spacecraft entered orbit around Vesta in 2011 and explored it for about a year, then left Vesta heading for Ceres to meet it in 2015.

Table 1.1 Principal planetary data

Planet	Symbol	Radius (km)	Mass ($M_E = 1$)	Density g/cm³	Albedo (Bond) A^1	a (AU)	e	i (deg)
Mercury	☿	$2,439.7 \pm 1.0$	0.055	5.427	0.068	0.387 098 80	0.205 631 75	7.004 99
Venus	♀	$6,051.8 \pm 1.0$	0.815	5.243	0.90	0.723 332 01	0.006 771 77	3.394 47
Earth	♁	$6,378^2 \times 6,357$	$5.9\,72 \times 10^{24}$ kg	5.515	0.367	1.000 000 83	0.016 708 617	0.0
Mars	♂	$3,396^2 \times 3,376$	0.107	3.933	0.150	1.523 689 46	0.093 400 62	1.849 73
Jupiter	♃	$71,492^2 \times 66,854$	317.8	1.326	0.52	5.202 758 4	0.048 495	1.303 3
Saturn	♄	$60,268^2 \times 54,364$	95.152	0.687	0.47	9.542 824 4	0.055 509	2.488 9
Uranus	♅	$25,559^2 \times 24,973$	14.536	1.318	0.51	19.192 06	0.046 30	0.773
Neptune	♆	$24,764^2 \times 24,342$	17.15	1.638	0.41	30.068 93	0.008 99	1.770

[1] Integral spherical (Bond) albedo
[2] Equatorial and polar radii

Table 1.2 Data for principal planetary satellites (Adapted from I. de Pater and J. Lissauer, 2010)

Planet and satellite	Radius (km)	Mass (10^{23} g)	Density (g cm^{-3})	Geom. albedo	a (10^3 km)	Orbital period (days)	Eccentricity e	Inclination i (deg)
Earth								
Moon	1,737.53 ± 0.03	734.9	3.34	0.12	384.40	27.321661	0.054900	5.15
Mars								
MI Phobos	(13.1×11.1×9.3)(±0.1)	1.08(±0.01)×10^{-4}	1.90 ± 0.08	0.06	9.3772	0.318910	0.0151	1.082
MII Deimos	(7.8×6.0×5.1)(±0.2)	1.80(±0.15)×10^{-5}	1.76 ± 0.30	0.07	23.4632	1.262441	0.00033	1.791
Jupiter								
JXVI Metis	(30×20×17)(±2)			0.06	127.98	0.29478	0.0012	0.02
JXV Adrastea	(10×8×7)(±2)			0.1	128.98	0.29826	0.0018	0.054
JV Amalthea	(125×73×64)(±2)			0.09	181.37	0.49818	0.0031	0.388
JXIV Thebe	(58×49×42)(±2)			0.05	221.90	0.6745	0.0177	1.070
JI Io	1,821.3 ± 0.2	893.3 ± 1.5	3.53 ± 0.006	0.61	421.77	1.769138	0.0041	0.040
JII Europa	1,565 ± 8	479.7 ± 1.5	3.02 ± 0.04	0.64	671.08	3.551810	0.0101	0.470
JIII Ganymede	2,634 ± 10	1,482 ± 1	1.94 ± 0.02	0.42	1,070.4	7.154553	0.0015	0.195
JIV Callisto	2,403 ± 5	1,076 ± 1	1.85 ± 0.004	0.20	127.98	0.29478	0.0012	0.02
JXIII Leda					11,160	241	0.148	27
JVI Himalia	85 ± 10	0.042 ± 0.006			11,460	251	0.163	175.3
JX Lysithea					11,720	259	0.107	29
JVII Elara	40 ± 10				11,737	260	0.207	28
JXII Ananke					21,280	610	0.169	147
JXI Carme					23,400	702	0.207	163
JVIII Pasiphae					23,620	708	0.378	148
JIX Sinope					23,940	725	0.275	153
Saturn								
SXVIII Pan	17×16×10	5×10^{-5}	0.41 ± 0.15	0.5	133.584	0.57505	0.00001	0.0001
SXXXV Daphnis	(4.5×4.3×3.1)(±0.8)	8×10^{-7}	0.34 ± 0.21		136.51	0.59408	0.00003	0.004

SXV Atlas	21×18×9	7×10^{-5}	0.46±0.1	0.9	137.670	0.60169	0.0012	0.01
SXVI Prometheus	66×40×31	0.00157	0.47±0.07	0.6	139.380	0.612986	0.0022	0.007
SXVII Pandora	58×40×32	0.00136	0.50±0.09	0.9	141.710	0.628804	0.0042	0.051
SXI Epimetheus	(59×58×53)(±2)	0.0053	0.69±0.13	0.8	151.47	0.694590	0.010	0.35
SX Janus	(97.4×96.9×76.2)(±2)	0.019	0.63±0.06	0.8	151.47	0.694590	0.007	0.16
SI Mimas	109×196×191	0.38	1.15	0.5	185.52	0.9424218	0.0202	1.53
SXXXII Methone					194.23	1.00958	0.000	0.02
SXLIX Anthe					197.7	1.037	0.02	0.02
SXXXIII Pallene	(2.6×2.2×1.8)(±0.3)				212.28	1.1537	0.004	0.18
SII Enceladus	249.1±0.3	0.65	1.61	1.0	238.02	1.370218	0.0045	0.02
SIII Tethys	533±2	6.27	0.97	0.9	294.66	1.887802	0.0000	1.09
SXIV Calypso	15×11.5×7			0.6	294.66	1.887802	0.0005	1.50
SXIII Telesto	15.7×11.7×10.4			0.5	294.66	1.887802	0.0002	1.18
SIV Dione	561.7±0.9	11.0	1.46	0.7	377.71	2.736915	0.0022	0.02
SXII Helene	19.4×18.5×12.3			0.7	377.71	2.736915	0.005	0.2
SXXXIV Polydeuces	(1.5×1.2×1.0)(±0.4)				377.71	2.736915	0.019	0.18
SV Rhea	764±2	23.1	1.23	0.7	527.04	4.517500	0.001	0.35
SVI Titan	2,575±2	1,345.7	1.88	0.21	1,221.85	15.945421	0.0292	0.33
SVII Hyperion	(180×133×103)(±4)	0.054	0.6	0.2–0.3	1,481.1	21.276609	0.1042	0.43
SVIII Iapetus	736±3	18.1±1.5	1.09	0.05–0.5	3,561.3	79.330183	0.0283	7.52
SIX Phoebe	(115×110×105)(±10)	0.082	1.63	0.08	12,952	550.48	0.164	175.3
SXX Paaliaq					15,198	687	0.36	45
SXXXVI Albiorix					16,394	783	0.48	34
SXXIX Siamaq					18,195	896	0.3	46

(continued)

Table 1.2 (continued)

Planet and satellite	Radius (km)	Mass (10^{23} g)	Density (g cm^{-3})	Geom. albedo	a (10^3 km)	Orbital period (days)	Eccentricity e	Inclination i (deg)
Uranus								
UVI Cordelia	13±2			0.07	49.752	0.335033	0.000	0.1
UVII Ophelia	16±2			0.07	53.764	0.376409	0.010	0.1
UVIII Bianca	22±3			0.07	59.166	0.434577	0.0003	0.18
UIX Cressida	33±4			0.07	61.767	0.463570	0.0002	0.04
UX Desdemona	29±3			0.07	62.658	0.473651	0.0003	0.10
UXI Juliet	42±5			0.07	64.358	0.493066	0.0001	0.05
UXII Portia	55±6			0.07	66.097	0.513196	0.0005	0.03
UXIII Rosalind	29±4			0.07	69.927	0.558459	0.0006	0.09
UXXVII Cupid					74.4	0.612825	~0	~0
UXIV Belinda	34±4			0.07	75.256	0.623525	0.000	0.0
UXXV Perdita					76.417	0.638019	0.003	~0
UXV Puck	77±3			0.07	86.004	0.761832	0.0004	0.3
UXXVI Mab					97.736	0.922958	0.0025	0.13
UV Miranda	240(0.6)×234.2(0.9)×232.9(1.2)	0.659±0.075	1.20±0.14	0.27	129.8	1.413	0.0027	4.22
UI Ariel	581.1(0.9)×577.9(0.6)×577.7(1.0)	13.53±1.20	1.67±0.15	0.34	191.2	2.520	0.0034	0.31
UII Umbriel	584.7±2.8	11.72±1.35	1.40±0.16	0.18	266.0	4.144	0.0050	0.36
UIII Titania	788.9±1.8	35.27±0.90	1.71±0.05	0.27	435.8	8.706	0.0022	0.10
UIV Oberon	761.4±2.6	30.14±0.75	1.63±0.05	0.24	582.6	13.463	0.0008	0.10
UXVI Caliban					7,231	580	0.16	141
UXX Stephano					8,004	677	0.23	144
UXVII Sycorax					12,179	1.288	0.52	159
UXVIII Prospero					16,256	1,978	0.44	152
UXIX Setebos					17,418	2.225	0.59	158

Neptune

NIII Naiad					48.227	0.294396	0.00	4.74
NIV Thalassa					50.075	0.311485	0.00	0.21
NV Despina	74±10			0.06	52.526	0.334655	0.00	0.07
NVI Galatea	79±12			0.06	61.953	0.428745	0.00	0.05
NVII Larissa	104×89(±7)			0.06	73.548	0.554654	0.00	0.20
NVIII Proteus	218×208×201			0.06	117.647	1.122315	0.00	0.55
NI Triton	1,352.6±2.4	214.7±0.7	2.054±0.032	0.7	354.76	5.876854	0.00	156.834
NII Nereid	170±2.5			0.2	5,513.4	360.13619	0.751	7.23
NIX Halimede					15,686	1,875	0.57	134
NXI Sao					22,452	2,919	0.30	48
NXII Laomedeia					22,580	2,982	0.48	35
NXIII Neso					46,570	8,863	0.53	132
NX Psamathe					46,738	9,136	0.45	137

Table 1.3 New discovered satellites of the planets (Adapted from Wikipedia)

Numeral	Name	Mean radius (km)	Semi-major axis (km)	Sidereal period (d) (r=retrograde)	Discovery (year)
Jupiter					
XVII	Callirrhoe	4.3	24,103,000	758.77 (r)	2000
XVIII	Themisto	4.0	7,284,000	130.02	1975/2000
XIX	Megaclite	2.7	23,493,000	752.86 (r)	2000
XX	Taygete	2.5	23,280,000	732.41 (r)	2000
XXI	Chaldene	1.9	23,100,000	723.72 (r)	2000
XXII	Harpalyke	2.2	20,858,000	623.32 (r)	2000
XXIII	Kalyke	2.6	23,483,000	742.06 (r)	2000
XXIV	Iocaste	2.6	21,060,000	631.60 (r)	2000
XXV	Erinome	1.6	23,196,000	728.46 (r)	2000
XXVI	Isonoe	1.9	23,155,000	726.23 (r)	2000
XXVII	Praxidike	3.4	20,908,000	625.39 (r)	2000
XXVIII	Autonoe	2.0	24,046,000	760.95 (r)	2001
XXIX	Thyone	2.0	20,939,000	627.21 (r)	2001
XXX	Hermippe	2.0	21,131,000	633.9 (r)	2001
XXXI	Aitne	1.5	23,229,000	730.18 (r)	2001
XXXII	Eurydome	1.5	22,865,000	717.33 (r)	2001
XXXIII	Euanthe	1.5	20,797,000	620.49 (r)	2001
XXXIV	Euporie	1.0	19,304,000	550.74 (r)	2001
XXXV	Orthosie	1.0	20,720,000	622.56 (r)	2001
XXXVI	Sponde	1.0	23,487,000	748.34 (r)	2001
XXXVII	Kale	1.0	23,217,000	729.47 (r)	2001
XXXVIII	Pasithee	1.0	23,004,000	719.44 (r)	2001
XXXIX	Hegemone	1.5	23,577,000	739.88 (r)	2003
XL	Mneme	1.0	21,035,000	620.04 (r)	2003
XLI	Aoede	2.0	23,980,000	761.50 (r)	2003
XLII	Thelxinoe	1.0	21,164,000	628.09 (r)	2003
XLIII	Arche	1.5	23,355,000	731.95 (r)	2002
XLIV	Kallichore	1.0	23,288,000	728.73 (r)	2003
XLV	Helike	2.0	21,069,000	626.32 (r)	2003
XLVI	Carpo	1.5	17,058,000	456.30	2003
XLVII	Eukelade	2.0	23,328,000	730.47 (r)	2003
XLVIII	Cyllene	1.0	23,809,000	752 (r)	2003
XLIX	Kore	1.0	24,543,000	779.17 (r)	2003
L	Herse	1.0	22,983,000	714.51 (r)	2003
—	S/2000 J 11	2.0	12,570,000	287.93	2001
—	S/2003 J 2	1.0	28,455,000	981.55 (r)	2003
—	S/2003 J 3	1.0	20,224,000	583.88 (r)	2003
—	S/2003 J 4	1.0	23,933,000	755.26 (r)	2003
—	S/2003 J 5	2.0	23,498,000	738.74 (r)	2003
—	S/2003 J 9	0.5	23,388,000	733.30 (r)	2003

(continued)

Table 1.3 (continued)

Numeral	Name	Mean radius (km)	Semi-major axis (km)	Sidereal period (d) (r=retrograde)	Discovery (year)
—	S/2003 J 10	1.0	23,044,000	716.25 (r)	2003
—	S/2003 J 12	0.5	17,833,000	489.72 (r)	2003
—	S/2003 J 15	1.0	22,630,000	689.77 (r)	2003
—	S/2003 J 16	1.0	20,956,000	616.33 (r)	2003
—	S/2003 J 18	1.0	20,426,000	596.58 (r)	2003
—	S/2003 J 19	1.0	23,535,000	740.43 (r)	2003
—	S/2003 J 23	1.0	23,566,000	732.45 (r)	2004
—	S/2010 J 1	1.0	23,314,335	723.2 (r)	2010
—	S/2010 J 2	0.5	20,307,150	588.1 (r)	2010
—	S/2011 J 1	0.5	20,155,290	580.7 (r)	2011
—	S/2011 J 2	0.5	23,329,710	726.8 (r)	2011
Saturn					
XIX	Ymir	9	23,040,000	1,315.14 (r)	2000
XX	Paaliaq	11	15,200,000	686.95	2000
XXII	Ijiraq	6	11,124,000	451.42	2000
XXIII	Suttungr	3.5	19,459,000	1,016.67 (r)	2000
XXIV	Kiviuq	8	11,110,000	449.22	2000
XXV	Mundilfari	3.5	18,628,000	952.77 (r)	2000
XXVI	Albiorix	16	16,182,000	783.45	2000
XXVII	Skathi	4	15,540,000	728.20 (r)	2000
XXVIII	Erriapus	5	17,343,000	871.19	2000
XXX	Thrymr	3.5	20,314,000	1,094.11 (r)	2000
XXXI	Narvi	3.5	19,007,000	1,003.86 (r)	2003
XXXVI	Aegir	3	20,751,000	1,117.52 (r)	2004
XXXVII	Bebhionn	3	17,119,000	834.84	2004
XXXVIII	Bergelmir	3	19,336,000	1,005.74 (r)	2004
XXXIX	Bestla	3.5	20,192,000	1,088.72 (r)	2004
XL	Farbauti	2.5	20,377,000	1,085.55 (r)	2004
XLI	Fenrir	2	22,454,000	1,260.35 (r)	2004
XLII	Fornjot	3	25,146,000	1,494.2 (r)	2004
XLIII	Hati	3	19,846,000	1,038.61 (r)	2004
XLIV	Hyrrokkin	4	18,437,000	931.86 (r)	2006
XLV	Kari	3.5	22,089,000	1,230.97 (r)	2006
XLVI	Loge	3	23,058,000	1,311.36 (r)	2006
XLVII	Skoll	3	17,665,000	878.29 (r)	2006
XLVIII	Surtur	3	22,704,000	1,297.36 (r)	2006
XLIX	Anthe	1	197,700	1.0365	2007
L	Jarnsaxa	3	18,811,000	964.74 (r)	2006
LI	Greip	3	18,206,000	921.19 (r)	2006
LII	Tarqeq	3.5	18,009,000	887.48	2007
LIII	Aegaeon	0.25	167,500	0.808	2008

(continued)

Table 1.3 (continued)

Numeral	Name	Mean radius (km)	Semi-major axis (km)	Sidereal period (d) (r=retrograde)	Discovery (year)
—	S/2004 S 7	3	20,999,000	1,140.24 (r)	2004
—	S/2004 S 12	2.5	19,878,000	1,046.19 (r)	2004
—	S/2004 S 13	3	18,404,000	933.48 (r)	2004
—	S/2004 S 17	2	19,447,000	1,014.70 (r)	2004
—	S/2006 S 1	3	18,790,000	963.37 (r)	2006
—	S/2006 S 3	3	22,096,000	1,227.21 (r)	2006
—	S/2007 S 2	3	16,725,000	808.08 (r)	2007
—	S/2007 S 3	3	18,975,000	977.8 (r)	2007
—	S/2009 S 1	0.15	117,000	0.471	2009
Uranus					
XV	Puck	81 ± 2	86,000	0.762	1985
XXI	Trinculo	5	8,504,000	749.24 (r)	2001
XXII	Francisco	6	4,276,000	266.56 (r)	2001
XXIII	Margaret	5.5	14,345,000	1,687.01	2003
XXIV	Ferdinand	6	20,901,000	2,887.21 (r)	2001
XXV	Perdita	10	76,417	0.638	1986
XXVII	Cupid	5	74,392	0.613	2003
Neptune					
—	S/2004 N 1	8–10	105,283	0.9362	2013

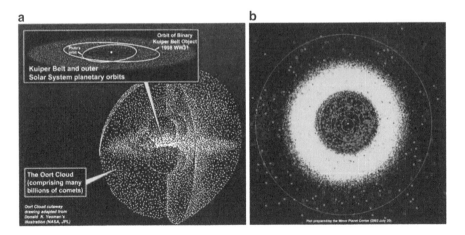

Fig. 1.4 (**a**) The Oort cloud and the Kuiper Belt. The Kuiper Belt is located at the outskirt of our planetary system (40–100 AU). It lies deep inside the Oort cloud whose outer boundary is at a distance of 10^4–10^5 AU; (**b**) The Main Asteroid Belt between the orbits of Mars and Jupiter (2.8–3.2 AU) is indicated by the *green dots*. There are groups of Earth-approaching asteroids (Amur, Apollo, Aton) also called near-Earth objects (NEOs; *red dots*) inside the Belt (Courtesy of D. Yeomans and US Planetary Society)

different bodies, mechanisms of the development of various natural conditions on the planets and satellites, the relationships of planets with small bodies, and interactions of planets with the solar wind plasma, etc. All these problems are intrinsically related with the fundamental concept of the solar system origin and evolution (planetary cosmogony) as well as with the origin of life and its proliferation. The main goal of planetary science is to reconstruct the key physicochemical processes that ensured a unique path of the solar system formation and evolution. In contrast to numerous stars observed at various stages of their evolution, we had until recently only one example of planetary systems: our own solar system. The situation dramatically changed starting in the mid-1990s when the first planets around other stars (exoplanets) were discovered, and since then the number of discovered exoplanets has grown dramatically (see Chap. 7). This has allowed us to apply a statistical approach in analyzing planetary populations, as astronomers have been doing customarily with stars.

Here we address the solar system as a whole, including its orbital and rotational dynamics and resonances, and its migrations and collisions.

Orbital and Rotational Dynamics

The dynamics of planetary systems and primarily of the solar system bodies is one of the most important fields of dynamical astronomy. Its study over several millennia, from Claudius Ptolemy's *Almagest* to the revolutionary change of the world view proposed by Nicolaus Copernicus, laid the foundations of celestial mechanics that were originated by Isaac Newton, Johannes Kepler, and Pierre-Simon Laplace. As we said, only at the very end of the past century were planetary systems around other stars discovered. This extended immeasurably the views on the properties and diversity of the immediate neighborhoods of the stellar population.

The fundamental Newtonian law of gravity governs the motion of the planets around the Sun and of satellites around their planets, whereas Kepler's three laws (Fig. 1.5) define the shape and regularities of their orbits. Newton's law states that the gravitational force F is defined as

$$F = GM_O m / r^2$$

where M_O is the mass of the Sun, m is the mass of a planet, r is the distance between a planet and the Sun, and $G = 6.67 \times 10^{-8}$ cm^3 g^{-1} s^{-2} is the gravitational constant.

Kepler's three laws for the planetary orbits are as follows:

I. A planet's orbit is an ellipse, with the Sun at one focus of the ellipse;
II. A radius vector drawn from the Sun to a planet sweeps out equal areas in equal time intervals;
III. The squares of the periods of revolution P are proportional to the cubes of the semimajor axes, a, of their orbits: $(P_1/P_2)^2 = (a_1/a_2)^3$.

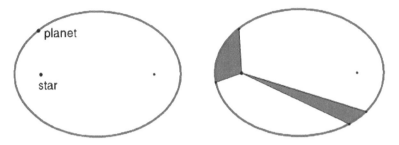

Fig. 1.5 Illustration of the first and second Kepler's laws (Courtesy of Wikipedia)

These laws nicely describe the orbital motions of the whole family of planets, though small perturbations are accounted for in the exact evaluation of planetary motions. The small bodies—comets and asteroids—experience much stronger perturbations, both periodic and secular. Newton's and Kepler's laws were thoroughly used in searching for new planets, and the discovery of Neptune and Pluto established the great success of celestial mechanics. Once these planets were found (although accidentally in the case of Pluto), the possibility of others was more readily considered. In the late 1700s, this was also inspired by the Titius-Bode empirical rule, which at that time seemed to predict the distance of a nonexistent planet at 2.8 AU. This rule, however, has no strict physical and cosmochemical support.

The whole population of celestial bodies in the solar system—the major planets and their satellites, minor planets (asteroids), comets, meteoroids, and meteoric dust—are in a state of permanent dynamical interaction. Mutual gravitational attractions of planetary bodies (tidal interactions) result in perturbations in the orbital and rotational motions. The most explicit manifestations of this effect are the tides on Earth due to the gravity of the Moon and the Sun. Tidal interactions cause planetary orbits to deviate from the Keplerian ellipse, and this gives rise to perturbed motion. In turn, the rotational motions of the planets and their satellites appear to imply an evolutionary history since the time of formation and, in particular, an efficiency of collisional interactions of primordial bodies.

Rotation and the interior structure of a planet are closely related to the generation of the planetary magnetic field. Data on the magnetic field strength of the planets are summarized in Table 1.4. On Earth, the strength of the magnetic field has experienced periodic variations with a mean period of about one million years, and also chaotic changes, as was found from observations of terrestrial magnetized rocks, volcanic lavas, and sediments in the ocean and other water reservoirs. The geomagnetic dipole direction responsible for the change of magnetic poles has varied with a much lower period of an order of thousands to tens of thousands of years (irregular magnetic field inversions). The same seems to be the case for the other planets. The mechanism and physical nature of the origin and maintenance of planetary magnetism are not fully understood. The most plausible mechanism is that of a hydromagnetic dynamo in the planet's interior involving convective heat and mass transfer in the highly conductive mantle. Analytical and numerical models have

Table 1.4 Magnetic fields of planets

Strength of magnetic field at the equator	
Oersted, or 10^5 gamma*	
Mercury	0.0035
Venus	0
Earth	0.35
Mars	0
Jupiter	4.28
Saturn	0.21
Uranus	0.25
Neptune	0.30

*Magnetic field strength H (measured in oersted, or gamma) is related to magnetic induction B (measured in gauss, or tesla T ($1 \text{ T} = 10^4$ gauss) as $B = \mu H$, where μ is magnetic permeability (the degree of magnetization that a material obtains in response to an applied magnetic field). For the real media μ is scaled relative to μ_o in vacuum (free space)

been developed that support the idea of an energy supply-rotation coupling domain in the dipole magnetic field generation. The absence of an intrinsic magnetic field on the very slowly rotating Venus could serve as indirect evidence in support of the hydromagnetic dynamo model.

In celestial mechanics, the *n*-body problem, allowing different assumptions to describe gravitational interactions and to reveal perturbations, is the most relevant one. In a more common limited case, the general three-body problem, including regular and chaotic motions, periodic orbits, and collisions, etc., is pertinent as an appropriate approach. Practically, the plane restricted three-body problem serves to evaluate orbital motions of bodies due to close gravity interactions. It is applied, for example, in calculating the disturbing motion of spacecraft approaching a planet under the influence of the Sun, the planet, and its satellite(s), if any. Here one deals with the study of periodic solutions including degenerate solutions, where the mass of one of the two bodies rotating in circular orbits around their common center of mass approaches zero.

From constant-pattern solutions of the restricted three-body problem, the Lagrange points, or L-points, emerge. These libration points, discovered theoretically by the distinguished mathematician Joseph-Louis Lagrange in the eighteenth century, occupy the five positions (points L1–L5) in an orbital configuration where a small object affected only by gravity can form a constant-shape pattern with two larger bodies, such as a star and planet or a planet and its satellite (see Fig. 1.6). The Lagrange points mark positions where the combined gravitational pull of the two large masses provides precisely the centripetal force[3] required to orbit with them.

---

[3] Centripetal force (from the Latin words *centrum* "center" and *petere* "to seek") is a force that makes a body follow a curved path: its direction is always orthogonal to the velocity of the body.

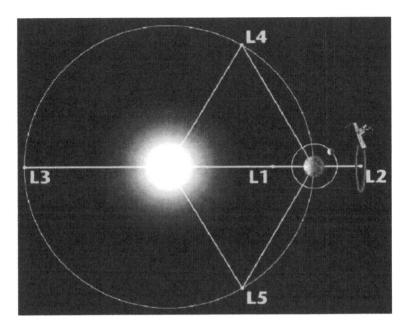

Fig. 1.6 Lagrange points L1–L5 in the central body—satellite system. Basically, all Lagrange points are slightly unstable and a spacecraft's inertia to move away from the barycenter is balanced by the attraction of gravity toward the barycenter. At the L1, L2, and L3 only periodic ("halo") orbits around these points are allowed. At L4 and L5 (sometimes called also triangular Lagrange points or Trojan points) an object appears constantly to hover or orbit around them. At these points heading and trailing satellite bodies are grouped (e.g. Trojans of Jupiter). This particular diagram for the Sun-Earth system with satellite in L2 point shown (Adapted from Wikipedia)

In other words, given two massive bodies in orbit around their common center of mass, there are five positions in space where a third body, of comparatively negligible mass, could be placed so as to maintain its position relative to the two massive bodies. This means that the gravitational fields of two massive bodies combined with the satellite's acceleration are in balance at the Lagrange points, allowing the smaller third body to be relatively stationary with respect to the first two. Typical examples are artificial satellites placed in the Sun-Earth or Earth-Moon systems. Let us note that all Lagrange points are nominally unstable, and a spacecraft's inertia to move away from the barycenter is balanced by the attraction of gravity toward the barycenter. Hence, at L1, L2, and L3 only periodic ("halo") orbits around these points are allowed. L2 and L3 are slightly unstable because small changes in position upset the balance between gravity and inertia, allowing one or the other force to dominate, so that the spacecraft either flies off into space or spirals in toward the barycenter. A satellite at the L1 point would have the same angular velocity of the Earth with respect to the Sun and hence would maintain the same position with respect to the Sun as seen from Earth. Stability at the L4 and L5 points (located along the satellite's orbit and forming an angle 60° from the center of the planet between each point and the fixed straight line) is explained by equilibrium between gravitation and an object's velocity on either a tighter or a wider orbit. The net result is that the object appears constantly to hover or orbit around the L4 or L5 point.[4]

Thus, the thoroughly developed, widely used, exact and approximate analytical methods of celestial mechanics allow one to find solutions to the problems of determining the orbits and their evolution. They also allow one to establish stability and instability regions of the solutions for various classes of motions. However, the methods of direct numerical integration have gained the widest acceptance in the last few decades. These methods, along with the numerical-analytical ones, have turned out to be most efficient, in particular, for the investigation of order and chaos in the solar system. They led to the conclusion that the orbits of planets with low eccentricities and inclinations are only slightly chaotic and do not have any noticeable secular component on time scales comparable to the age of the solar system.

Resonances

A very important feature of orbital and rotational dynamics is that of the different types of resonances caused by gravitational interaction between the solar system bodies—planets and their satellites with the involvement of numerous small bodies. Numerical experiments have found that three-body resonances (and accompanying subresonances) in the Jupiter-Saturn-Uranus system influence some irregular (chaotic) orbital motions of the whole solar system and, hence, may decrease the time of its general stability (the Lyapunov time) by nearly three orders of magnitude. But much more explicit are the resonances and commensurabilities existing in the orbital and rotational motions of planets, satellites, and asteroids. The populations of asteroids and comets change continuously under the gravitational attraction of planets. In particular, this is responsible for the predominant streams of meteorites originating in the resonant zones of the main belt and the Kuiper Belt that make a laboratory study of extraterrestrial matter accessible.

An example of resonance experienced by the solar system bodies is the synchronous rotation of our Moon around the Earth; because of this we can see only the Moon's near side. Similarly, nearly all other satellites rotate synchronously around their home planets. For Mercury, spin-orbit resonance causes its sidereal (relative to stars) orbital period (58.6 days) to be two-thirds of the rotational period (88 days). This determines an unusual year-day relationship for the planet: one Mercurian solar day is equivalent to three sidereal days, or two Mercurian years. A characteristic commensurability is observed in the motions of Neptune and Pluto, meaning that a collision between these bodies is ruled out, although accurate calculations of

---

[4] *Wikipedia* gives a nice analogy of the resulting Lagrange points stability. The L1, L2, and L3 positions are as stable as a ball balanced on the tip of a wedge would be: any disturbance will toss it out of equilibrium. The L4 and L5 positions are as stable as a ball at the bottom of a bowl would be: small perturbations will move it out of place, but it will drift back toward the center of the bowl.

Pluto's orbit have revealed a chaotic component in its motion. However, we cannot reconstruct the evolutionary histories leading to the current dynamical configurations of the planets and their systems of satellites, in particular, the possibilities of rather tight constraints on the primordial rotation states of Mercury and Venus and the stabilizing mechanism for the latter's retrograde spin or the problem of the origin of the moons of Earth and Mars, just to mention a few.

Whereas the possibility of close encounters between planets capable of affecting their motion is ruled out due to their mutual attractions, the comets and asteroids undergo strong perturbations, especially those that are subjected to tidal effects from planets. Indeed, a large number of small bodies are captured into resonant orbits with the nearest planets. The Kirkwood gaps in the main asteroid belt between the orbits of Mars and Jupiter in the region from 2.7 to 3.2 AU attributable to the resonances of the orbital periods of asteroids with Jupiter's period (4:1; 3:1; 5:2; 2:1; 3:2) serve as a remarkable example. Another example is that of the resonances of the mean motion (commensurabilities of periods) with Neptune (4:3; 3:2; 2:1) and the secular resonances (precession of orbits) due to the commensurabilities of the longitude of ascending node and the argument of perihelion for trans-Neptunian bodies in the Kuiper Belt. Interestingly, the orbits of these bodies are stable (outside the resonances) on a time scale of $t\sim 10^8$ years, but the "accumulation of instability" and the sharp growth in eccentricity due to the gravitational influence of Neptune lead to their degree of dissipation from the Kuiper Belt.

Due to the presence of secular perturbations, the region between the "inner" and "classical" belts in the Kuiper Belt (~40–43 AU) is distinguished by the greatest instability. The bodies from this zone migrate into inner parts of the solar system and are initially captured predominantly into Jupiter-crossing orbits (JCOs), while a certain fraction of them subsequently migrate toward the Sun, replenishing the main asteroid belt and three groups of asteroids (Amor, Apollo, Aten) that cross the orbits of terrestrial planets and are generically called near-Earth objects (NEOs). The orbits of these objects, especially the Earth-approaching Apollo asteroids and the Aten asteroids that travel inside the Earth's orbit, are subjected to the greatest chaotization and represent the main hazard of colliding with the Earth.

We must emphasize that the orbits of comets are subjected to the strongest chaotization caused by both gravity interactions with planets and forces of nongravitational origin. This makes it difficult to predict their motions and accurately determine their ephemerides. This is why comets pose an even more serious threat to our civilization. Close encounters with planets have the greatest influence on the dynamics of comets, which can lead to their transition to hyperbolic orbits and their "evaporation" into the interstellar medium. On the other hand, comets of extrasolar origin may come inside the Oort Cloud. Bodies at the edge of the solar system can be thrown inward under the action of tidal perturbations from the nearest stars, thereby providing a connection with the Galaxy. In particular, they give us an opportunity to "probe" material of the protosolar nebula. Thus, the interactions of comets with planets open up possibilities for a better understanding of the role of these primitive bodies in planetary cosmogony and galactic evolution.

Migration and Collisions

Gravitational interactions underlie the mechanism of small body migration, leading to close encounters and numerous collisions with the planets throughout solar system history. One important result of these processes is clearly demonstrated by heavy cratering ("scars") of the surfaces of planets and their satellites (especially those where the traces of impact bombardment were not wiped out by subsequent erosion processes through the influence of the atmosphere and hydrosphere), and these collisions probably served as one of the most important factors of planetary evolution. The Moon, Mercury, the satellites of the giant planets, and even the asteroids themselves, whose surfaces are literally pockmarked with craters, are most characteristic here. In particular, large basins (maria) more than 1,000 km in diameter on the Moon were produced by the impacts of asteroids several hundred kilometers across. At the same time, the lunar crater size spectrum extends from centimeters to thousands of kilometers. The lower threshold of crater sizes on Earth, Mars, and particularly Venus, with atmospheres corresponding to the size cutoff threshold for bodies capable of penetrating to the surface, is considerably higher.

Bodies $\leq$ 10 m in size are destroyed on entering the Earth's atmosphere; only their fragments reach the surface in the form of meteorites. However, there is ample evidence for collisions of even large asteroids and comets with the Earth over at least the last several hundred million years of its history, which led to global cataclysms. For example, in the *Chicxulub* event, an asteroid ~ 10 km in size collided with the Earth. It left a crater about 170 km in diameter on the Yucatan Peninsula in Mexico, at the turn of the Cretaceous and Neocene periods ~ 65 million years ago, presumably with catastrophic global consequences for the terrestrial biosphere. Another example is the local-scale catastrophe that occurred near the Tunguska River in Siberia in 1908, caused probably by the collision with the Earth of a comet fragment whose size is estimated to be ~ 60 m. Smaller (about 20 m) meteoroid Chelyabinsk impacted Earth over heavy populated area of Russian Siberia in February 2013 and effected some destructions. The fall of 22 nearly 1-km fragments from Comet P/Shoemaker-Levy-9, torn apart by tidal forces during its close encounter with Jupiter in 1994, serves as additional convincing evidence that such events occur in the solar system more or less regularly.

At the crossroads of astronomy and geophysics, the importance of migration and impact processes in the evolution of the inner planets has become more well understood. They appeared to exert a significant influence on the formation of their natural conditions, in particular, the setup of the atmosphere and hydrosphere. Indeed, bodies migrating from an ancient main asteroid belt and/or Kuiper Belt could be responsible for supplying volatiles (first of all, water) at the early stage of evolution (especially in the period of Late Heavy Bombardment (LHB) about four billion years ago), thus compensating for loss of volatiles in the high-temperature region of the terrestrial planets' formation. The idea of matter transport inside and outside the solar system through migration-collision processes is intrinsically related with the intriguing question of the origin of life. In this way, comets, apart from their primary role in planetary cosmogony, also attract increasing interest as possible carriers of primordial life forms. We shall return to this problem in Chap. 4.

Chapter 2
The Terrestrial Planets

General View

Unlike the gaseous giant planets, which are preserved essentially unmodified in structure and composition since their origin 4.55 billion years ago, the terrestrial (inner) planets (Fig. 2.1) experienced dramatic changes in the course of their evolution. This is evident in their interior structure, geology, surface landforms, and atmospheres (Fig. 2.2a). The evolution of the inner planets was mainly controlled by both endogenous and exogenous factors involving the original storage of radionuclides, impacts, and distance from the Sun. Endogenous factors were mostly driven by heavy bombardment by asteroid-size bodies in an early epoch dating back to 4.0 billion years ago, and internal heat due to long-lived radio isotope decay, their storage being strongly dependent on the size/mass of the planet. Impacts scarred the surface and left behind numerous craters, while internal heating was responsible for tectonic processes and widespread volcanism. Distance from the Sun determines the incident radiation input to the planet and, hence, its thermal balance. Heat release from the interior (internal flux) is negligible, in contrast to the case of the outer giant planets where it exceeds the solar incident flux. Exogenous factors significantly contributed to the planetary evolution through migration of small bodies from the outer solar system regions and collisional processes.

Planetary geology is closely related to differentiation of the planetary interiors into shells (core, mantle, and crust), accompanied by widespread tectonics and volcanism, as is clearly manifested by different patterns of the surface landforms. All terrestrial and gaseous-icy planets have differentiated interiors, as is supported by the measured quadrupole moments of their gravitational fields and the respectively deduced dimensionless moment of inertia, $I = C/MR^2$ (Fig. 2.2b). Here C is the moment of inertia about polar axis, M and R are mass and radius of the planet, respectively. Note that for an ideal sphere of uniformly distributed density, $I = 0.4$; the less uniform the mass distribution (massive heavy core and lighter mantle in the interior), the lower the I value. This is why for the only partially differentiated

© Springer Science+Business Media New York 2015
M.Ya. Marov, *The Fundamentals of Modern Astrophysics*,
DOI 10.1007/978-1-4614-8730-2_2

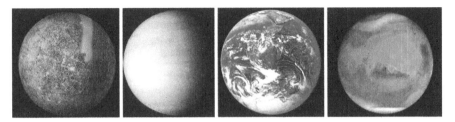

Fig. 2.1 Terrestrial planets (in order from the Sun) Mercury, Venus, Earth, Mars (not in scale) (Author's mosaics of NASA images)

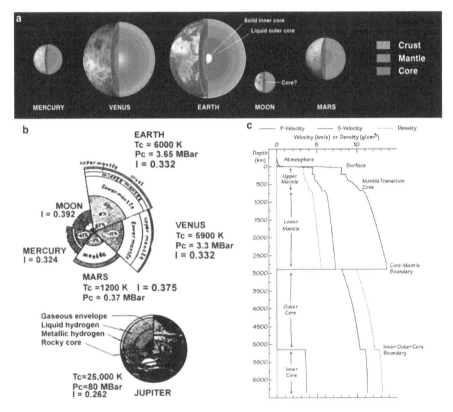

Fig. 2.2 (**a**) Internal structure of the terrestrial planets and the Moon. The order in the arrangement of main regions (core, mantle, crust) is a consequence of the differentiation of their constituent matter into shells, their extension depending on the size (mass) of the planet, the abundances of major components, and the condensation temperature in the formation zone. Mass of the body predetermines the core state and the crust (lithosphere) thickness (Adapted from Wikipedia). (**b**) Parameters of the interiors (temperature T, pressure P, and dimensionless moment of inertia I) of the terrestrial planets and the Moon as compared to those of Jupiter (Credit: the Author). (**c**) Velocity waves propagation and density variations within Earth based on seismic observations. The main regions of Earth and important boundaries are labeled. This model was developed in the early 1980s and is called PREM for Preliminary Earth Reference Model (Adapted from Wikipedia)

Moon with a small core I = 0.392, whereas the minimum I values (less than 0.3) are pertinent to the gaseous giants Jupiter and Saturn with heavy iron-silicate cores and light hydrogen-helium envelopes.

Impacted structures are most clearly seen on the atmosphereless bodies such as Mercury and the Moon. Heavily cratered terrains are also preserved on the Mars surface, though in the presence of its atmosphere, ancient craters were eroded by the processes of weathering. Unlike the main mechanism of global plate tectonics on Earth, the geologies of Mars and Venus are different. On Mars, there are great ancient shield volcanoes elevated up to 26 km above the mean surface level, which (despite the relatively small size of the planet) are among the highest in the solar system. Another remarkable geologic pattern is Valles Marineris—an enormous feature 100 km wide and 8 km deep that extends along the equator for more than 3,000 km. It is poorly associated, however, with the global tectonics. On Venus, geologic structures have been revealed only with radar techniques because its very thick atmosphere and clouds fully obscure the surface when it is observed in optical wavelengths. There are no obvious tectonic features on the planet. Instead, numerous volcanoes were mostly pertinent to the planet's thermal evolution. Volcanic activity appears to have terminated quite recently, less than 100 million years ago, though some planetary geologists believe that some limited activity is preserved today.

Thermal evolution is assumed to be partially responsible for the formation of an atmosphere of secondary origin on a terrestrial planet after the primary atmosphere (presumably retained in the process of the planet's accumulation) was lost. Atmospheres exist on all terrestrial planets, but Mercury has only an extremely rarefied gas envelope equivalent to the Earth's exosphere. The main properties of the atmospheres of these planets are summarized in Table 2.1.

Table 2.1 Properties of atmospheres of the terrestrial planets

Planet	Mercury	Venus	Earth	Mars
Chemical composition (percent by volume)	$He \leq 20$	$CO_2 = 95$	$N_2 = 78$	$CO_2 = 95$
	$H_2 \leq 18$	$N_2 = 3-5$	$O_2 = 21$	$O_2 = 2-3$
	$Ne \leq 40-60$	$Ar = 0.01$	$Ar = 0.93$	$Ar = 1-2$
	$Ar \leq 2$	$H_2O = 0.01-0.1$	$H_2O = 0.01-3$	$H_2O = 10^{-3}-10^{-1}$
	$CO_2 \leq 2$	$CO = 3 \cdot 10^{-3}$	$CO_2 = 0.03$	$CO = 4 \cdot 10^{-3}$
		$HCl = 4 \cdot 10^{-5}$	$CO = 10^{-5}$	$O_2 = 0.1-0.4$
		$HF = 10^{-6}$	$CH_4 = 10^{-4}$	$Ne = < 10^{-3}$
		$O_2 < 2 \cdot 10^{-4}$	$H_2 = 5 \cdot 10^{-5}$	$Kr = < 2 \cdot 10^{-3}$
		$SO_2 = 10^{-5}$	$Ne = 2 \cdot 10^{-3}$	$Xe = < 5 \cdot 10^{-3}$
		$H_2S = 8 \cdot 10^{-3}$	$He = 5 \cdot 10^{-4}$	
		$Kr = 4 \cdot 10^{-5}$	$Kr = 10^{-4}$	
		$Xe = 10^{-6}-10^{-5}$	$Xe = 10^{-6}$	
Mean molecular mass		43.2	28.97	43.5
Surface temperature				
T_{max}, K	700	735	310	270
T_{min}, K	110	735	240	148
Mean surface pressure P, atm	10^{-15}	92	1	$6 \cdot 10^{-3}$
Mean surface density, ρ, g/cm^3	10^{-17}	$61 \cdot 10^{-3}$	$1.27 \cdot 10^{-3}$	$1.2 \cdot 10^{-5}$

We see that the atmospheres of the neighboring planets, Venus and Mars, dramatically differ from that of Earth: the pressure at the Venusian surface reaches 92 atm and the temperature is 735 K, whereas at the surface of Mars the average pressure is only 0.006 atm and the average temperature is about 220 K. The composition of the atmospheres of both planets is mostly carbon dioxide with relatively small admixtures of nitrogen and argon and a negligible mixing ratio of water vapor and oxygen (on Mars). In contrast to Venus, which has no seasonal variations because of the very small obliquity of its equator to the ecliptic, Mars, whose obliquity is nearly similar to that of the Earth's, exhibits pronounced seasonal variations resulting in temperature contrasts between summer and winter hemispheres exceeding 100 K. At the winter pole the temperature drops below the freezing point for CO_2 and thus "dry ice" deposits cover the Martian polar caps, though their main composition is water ice. Seasonal evacuation and release of carbon dioxide in the polar regions is one of the important drivers of atmospheric planetary circulation on Mars involving both meridional and zonal wind patterns.

However, atmospheric circulation on Venus is mainly characterized by the mechanism of super-rotation, or "merry-go-round" circulation, such that a zonal wind velocity of less than 1 m/s at the surface increases up to nearly 100 m/s near the upper cloud level at about 60 km. Venus's clouds consist of quite concentric sulfuric acid droplets, and this unusual composition complements the picture of the very exotic and hostile environment of our closest planet, which until the middle of the last century was thought to be the Earth's twin.

In terms of natural environment, Mars is another extreme, though more favorable in its climate and therefore much more accessible for future human expansion throughout the solar system. Historically, this planet was regarded as a potential target for finding life beyond Earth, and it is still addressed as a possible site that could harbor life at the microbial level and where extant or extinct life could be found. Indeed, unlike Venus, where an assumed early ocean was lost soon after the runaway greenhouse effect responsible for its contemporary climate conditions developed, ancient Mars appears to have had plenty of water until a catastrophic drought occurred on the planet about 3.6 billion years ago for reasons that are not yet well understood. There is evidence that contemporary Mars preserved a substantial part of its water storage; its original bulk is estimated to be equivalent to nearly 0.5 km of an average ocean deep. It is thought to be stored deep under the surface as permafrost and water lenses. Recent space missions revealed many specific geologic structures which confirm such a scenario, as well as the existence of subsurface water at about 1 m depth unevenly distributed over the globe, mostly in the polar regions. Anyway, the general understanding is that ancient Mars had a much more clement climate when water covered much of its surface, until its quite dense atmosphere was lost and water ice was buried beneath thick sand-dust sediments.

We shall start the inner planets' description with our home planet Earth and its large satellite, the Moon; together they form the Earth-Moon system. Because Earth is our home, we address other terrestrial planets first of all in terms of better understanding Earth as a solar system planet and learning what predetermined the unique path of its evolution. In other words, Earth and the Moon should be perceived in the context of the family of Earth-like planetary bodies which store collectively invaluable data about our past history and may help to predict the future trends. However, Earth and the Moon also serve as an important basis for comprehending the peculiarities of natural conditions which formed on other inner planets. The Moon itself can be regarded as a frame of reference to highlight the key processes of the solar system evolution.

We then discuss in a bit more detail the nature of other terrestrial planets, emphasizing our neighbor planets Venus and Mars, which are located very close (on a space scale) to Earth but evolved along completely different paths. Therefore, we may address them as extreme models of potential unfavorable Earth evolutionary paths, provided that the acting natural feedback mechanisms are slowed down. An important implication is that mankind should bear this in mind and carefully control the growing anthropogenic influence on the environment to prevent the development of risky scenarios.

Earth

Main Properties. Earth is the third planet in order from the Sun and the largest body of the four terrestrial planets. It orbits the Sun at a distance equal to semimajor axis $a = 149.6$ million km (1 AU) with a very small eccentricity, $e = 0.017$, which means that the orbit is nearly circular, although the distance varies as the Earth moves from perihelion in January to aphelion in July. The Earth moves along its orbit with a velocity $V = 29.8$ km/s, and its period of revolution is $P_{orb} = 365.256$ days (1 year). The sidereal period of rotation around its axis (relative to the stars, or 1 sidereal day) is $P_{rot} = 23^h \ 56^m \ 4.99^s$. The inclination of the Earth's equator to the plane of circumsolar orbit (ecliptic) equals $i = 23°27'$, which ensures significant seasonal changes on the planet.

Earth is a rather small body by the cosmic scale. Its equatorial radius is $R_e = 6,378$ km, and its polar radius is $R_p = 6,356$ km; hence, its oblateness is 0.0034. The pear-shaped figure of the Earth is called the geoid. The mass is $M_E = 5.974 \times 10^{24}$ kg, the mean density $r = 5.515$ g/cm^3, and the mean acceleration due to gravity is $g = 9.78$ m/s^2 (it is a bit more at the poles compared to the equator). The Earth's surface is represented by the continents and the oceans, the oceans occupying nearly two-thirds of the whole surface. 94 % of the Earth's bulk composition is O, Mg, Si, and Fe. Together with Al and Ca these are rock-forming elements, whereas C and N are volatile-forming elements. Their abundance is < 1 %. The elemental composition of continents and oceans is quite different. The age of the Earth (dated as the time of the solar system origin based on the dating of calcium-aluminum inclusions (CAIs) in the Allende and Efremovka meteorites; see Chap. 4 for more details) is established to be $4,567.5 \pm 0.5$ million years.

Geology. Earth is a geologically evolved planet. The main geological mechanism is global plate tectonics, which means that its outer shell (lithosphere) is not homogeneous but split into 12 large plates that are laterally mobile (Fig. 2.3). The mechanism

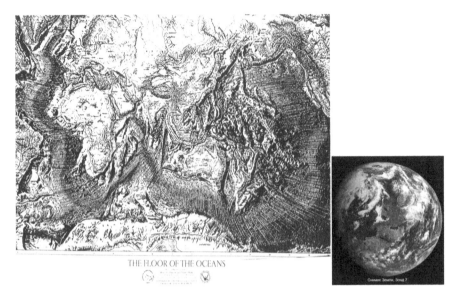

THE FLOOR OF THE OCEANS

Fig. 2.3 The floor of the Earth's oceans. The relief bears traces of lithospheric plate tectonics involving middle-ocean spreading zone and subduction zones at the edges of ocens connected with the most powerful volcanic and erthquakes activity (Credit: B. Heizen. At the right—Image of Earth from the Moon orbit taken by the Soviet Zond 7)

involves a "spreading zone," where hot lava ascends from the upper mantle pushing the lithosphere plates apart and filling "cracks" (rifts) between them, and subduction zones, where plates covered with sediments slowly plunge deep under continents. The seafloor spreading hypothesis was proposed by the American scientist Harry Hess in 1960. The spreading zones coincide with the mid-ocean ridges that run globally at the bottom of all oceans. Both spreading and subduction zones are associated with the sources of the most powerful volcanic activity and earthquakes. There are about 800 active volcanoes on the Earth's surface, and numerous earthquakes occur annually.

Global plate tectonics is responsible for the drift of continents, which continuously retreat from each other, as was first suggested by the German scientist Alfred Wegener as early as 1912, although the idea was not accepted until Hess's hypothesis of spreading zones was known. Reconstruction of the process back in time has caused us to conclude that 300–200 million years ago there was a supercontinent Pangaea which disintegrated into several pieces, giving rise to the now existing continents. In support of this model, one may compare the contours of the eastern part of South America with the western part of Africa and see that they exhibit quite similar configurations. The model was also confirmed by in-depth studies of the bottom of the ocean and magnetic properties of the early emerged lava flows.

The oceans comprise nearly 97 % of all the water storage on Earth (called the hydrosphere) and contain a mass of about 10^{21} kg and cover 361 million km^2. The remaining 3 % is fresh water, found in rivers, lakes, glaciers, and in the atmosphere, as well as in the Arctic and Antarctic polar ice caps. The mean depth of the "world ocean" (if all the oceans were uniformly poured over the Earth's surface) is 3,900 m, while the maximal depth is 11,000 m—this deepest portion is the Mariana Trench in the Pacific.

Interior. The Earth's interior has a rather complicated structure, as revealed by the seismic sounding technique. The propagation rate of longitudinal P and transverse S seismic waves depends on the density and elasticity of rocks, and the waves also experience reflection and diffraction on the boundaries between different layers, while the transverse waves decay in liquids. The technique has distinguished the Earth's main zones, which are shown in Fig. 2.2c. They are the upper firm crust, partially melted upper and lower mantle, and liquid outer and inner rigid cores. The thickness of the crust amounts to 35 km on continents and about half that in the oceans. The region between the crust and upper mantle is the lithosphere, which is about 70 km deep, underlain by a more fluid layer, the asthenosphere, extending to 250 km in depth. The boundary between the crust and upper mantle is known as the Mohorovicic discontinuity, or just the Moho. As we said above, the lithosphere is split into a dozen large plates "floating" on the asthenosphere, ensuring the mechanism of global plate tectonics in action.

The crust is composed of bedrock (basalt and granite) having a mean density of ~ 3,000 kg/m^3, and it makes up less than 1 % of the bulk mass of the Earth. The mantle composes nearly 65 % and the core 34 % of the total mass. The thickness of the mantle is 2,900 km, and it is composed mainly of silicate rocks as well as silicate and magnesium oxides modified under high pressure. This results in a density increase to 5,600 kg/m^3, which is manifested as a sharp increase in seismic wave rates. The strong reflection and decay of these waves below 2,900 km (this level is called the Gutenberg discontinuity) marks the outer boundary of the Earth's liquid core, which is 2,250 km thick and composed of melted iron and nickel. Finally, inside the outer liquid core lies the inner rigid core, which has a radius of 1,220 km.

Let us note that near the Earth's surface the temperature gradient is about 20°/km; the temperature reaches ~ 1,800 K at 100 km and ~ 5,000 K at the mantle-core boundary, where the pressure amounts to $P = 1.3 \times 10^{11}$ Pa (~1.3 million atmospheres, mil. atm). In the center of the Earth $T = 8,000$ K, $P > 3.6 \times 10^{11}$ Pa (~3.6 mil. atm), and the density exceeds 10,000 kg/m^3. In the contemporary epoch, the liquid state of the core and the partially melted mantle are primarily maintained by the original storage of long-lived radionuclides—uranium, thorium, and potassium—in the Earth's interior. The liquid core and the planet's rather fast rotation are thought to be responsible for the significant magnetic field of the Earth, following the mechanism of a magnetic dynamo. The field strength is 0.31 oersted (roughly corresponding to a magnetic induction equal to 3×10^4 nanotesla) at the equator and 0.63 oersted at the poles. In the contemporary epoch, the North geographic pole is located close to the South magnetic pole (the angle between geographic and magnetic field axes is 11.5°). For the last 10 million years there were recorded 16 magnetic field inversions apparently caused by periodic core overturns. The region around Earth stretching out to 5–10 Earth's radii is controlled by its intrinsic magnetic field and is called the magnetosphere. The magnetosphere shields our planet from solar corpuscular radiation (energetic protons and electrons).

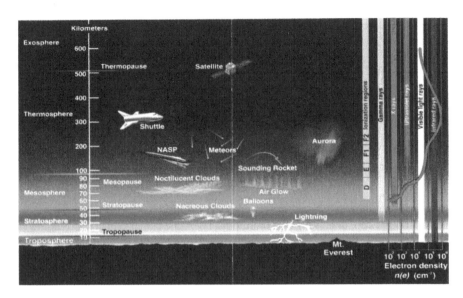

Fig. 2.4 Structure of the Earth's atmosphere from the surface to 600 km. Different regions (troposphere, stratosphere, mesosphere, thermosphere, and exosphere) in the lower, middle, and upper atmosphere are distinguished. Levels of penetration of the solar electromagnetic radiation, specifically that responsible for ionization of atmospheric species and ionosphere formation, are shown (Credit: R. Courtis)

Atmosphere. Earth has a unique atmosphere in the solar system. Unlike the other planets, its atmosphere is composed mainly of nitrogen and oxygen with an irregular temperature stratification. It has several distinguished regions, sometimes called layers (see Fig. 2.4), called the troposphere, stratosphere, mesosphere, thermosphere, and exosphere.

The troposphere stretches from the surface to 12 km (a bit higher at the equator and lower at the poles). The mean temperature at the surface is +12 °C with variations from −85 °C (inner regions of the Antarctic) to +70 °C (Western Sahara). The temperature lapse rate is roughly 6°/km, which nearly corresponds to the wet adiabatic gradient. A relatively small amount of carbon dioxide in the troposphere (0.03 %) is mainly responsible for the greenhouse effect, raising the Earth's effective temperature by about 20 °C. The troposphere has significant though variable humidity depending on location, season, time of day, and height. The mean gas density at the surface under a temperature of 15.0°C and relative humidity $f = 0$ % is $\rho = 1,225$ kg/m^3 (a standard atmosphere).

In the stratosphere (12–50 km above the Earth), the temperature first drops to −50 °C at 25 km and then rises due to solar ultraviolet (UV) radiation absorption in 200–300 nm wavelengths by ozone. This shields the planet from intense electromagnetic radiation and prevents biosphere destruction. Ozone raises the temperature of the upper stratosphere to nearly 0 °C at 50 km. Above this level is the mesosphere, where the temperature drops again, reaching its minimum (about −90 °C) at the level called the mesopause at 85 km. At the beginning of this level direct absorption of extreme ultraviolet (EUV, at wavelengths shorter than 200 nm) and soft X-ray radiation occurs, which is responsible for the photochemical processes of dissociation and ionization of predominantly oxygen and some other species and rarefied atmospheric gas heating. Here the temperature steadily increases up to 800–1,000 K, but it varies from about 500 to 1,500 K between the minimum and maximum phases of the 11-year solar activity cycle. The latter results in atmospheric density variations by more than two orders of magnitude at about 400 km and, hence, dramatically influences the lifetimes of artificial satellites and orbital stations. In the thermosphere is located the ionosphere with its most prominent layers E, F_1, and F_2, where the ion concentration is high enough to reflect radio waves in the short wavelength and thus provide long distance radio communication on Earth. The atmosphere itself is nearly transparent to radio frequencies between 5 MHz and 30 GHz, while the ionosphere blocks signals below this range. The thermosphere smoothly transits at about 500–600 km to the very outer upper atmosphere region called the exosphere, consisting mostly of hydrogen/helium atoms and extending to thousands of kilometers.[1] The transition layer is called the exobase, from which atoms not experiencing collisions when moving upward and having sufficient thermal velocity can dissipate into outer space.

The Earth exhibits quite complicated atmospheric dynamics. Generally, air mass transport on different planets depends on four forces acting on an elementary volume of gas (gravity force, Coriolis force, viscous friction, and pressure gradient). The ratio of the planetary rotation rate (frequency f) and thermal atmospheric relaxation involving these forces mainly defines the planetary circulation patterns. The relative contribution of the Coriolis force to the atmospheric dynamics is determined by the Rossby number $Ro = U/fL$, where U and L are typical horizontal velocity and length scales for synoptic processes. If horizontal pressure gradients are balanced by the Coriolis forces ($Ro \ll 1$), it results in geostrophic types of circulation flows. In the Earth's atmosphere $Ro \approx 10^{-1}$, which means that the forces due to pressure gradients are essentially balanced by the Coriolis forces. Therefore, a geostrophic wind is a typical synoptic characteristic, and its zonal and meridional components can be determined from a known pressure distribution.

The combined effect of the planet's rotation and change of frequency f with latitude generates Rossby planetary waves, which are the main weather generators, cyclones and anticyclones being formed in their troughs and crests. In the Earth's atmosphere, these planetary waves propagating horizontally westward have a period exceeding both the planetary rotation period and the periods of three other types of atmospheric wave motions: tidal, acoustic, and internal gravity waves (IGWs). Short-period IGWs with relatively small amplitudes belong to the category of micro-meteorological atmospheric oscillations and, along with convection, serve as the source of small-scale turbulence. Thus, they energetically (through dissipation) affect the formation of large-scale weather processes.

---

[1] Note that an exosphere is the most abundant type of planetary weak atmosphere contiguous to the surface, and is characteristic of the Moon, Mercury, some satellites of the giant planets, and even some large asteroids.

The current state and ultimate fate of the Earth are entirely dependent on our Sun and its evolution. The Sun ensures all necessary energy resources including those accumulated in the past as organic fuels, and photosynthesis serves as the main biological factor for life persistence. Projecting into the future, we should not only be aware of short-term concerns such as climactic changes, resource exhaustion, and natural environment pollution/damage (temporarily putting aside the world's social problems and relevant conflicts), but we should also deliberate about very far-off perspectives. At its red giant phase the Sun will increase to 250 times its present radius, its high-temperature outer shell expanding beyond the Earth's orbit to swallow our planet. Should it escape incineration in the Sun, all its water will still be boiled away and its atmosphere will escape into space, ending all terrestrial life. However, provided civilization develops smoothly, future generations millions to billions of years from now will find an opportunity to avoid this catastrophic scenario by moving to a home planet that is harbored by another younger star.

The Moon

Main Properties. The Moon (Fig. 2.5) is the brightest object in our skies after the Sun: its maximum stellar magnitude is -12.7^m. The semimajor axis of the Moon's orbit $a = 383,398$ km and the eccentricity $e = 0.055$. The inclination of the Moon's orbit to the ecliptic equals $5°09'$. The sidereal period of revolution around the Earth is 27 days, 7 hours, and 43 minutes, while its synodic period (relative to the Sun corresponding to the lunar phases change) is 29 days, 12 hours, and 44 minutes.

The Earth-Moon system is a unique formation in the solar system with the largest satellite-to-planet mass ratio: 1/81. This means that there is a significant mutual gravity influence, clearly manifested by the lunar tides on Earth, which significantly exceed those caused by the Sun. In turn, Earth locks the Moon's orbit and its intrinsic rotation in resonance, such that the period of its rotation equals the sidereal period. This means that the Moon orbits Earth for exactly the same time it takes to complete one rotation around its axis, and this is why the Moon permanently faces

Fig. 2.5 Near (*left*) and far (*middle*) sides of the Moon and false color image of the Moon Near side showing different rocks composition (Author's mosaics of NASA images)

Earth with the same hemisphere. Nonetheless, because of its quite high eccentricity, the orbital motion of the Moon is irregular—faster near perigee and slower near apogee—whereas the intrinsic rotation of the Moon is regular. Both motions result in librations in longitude, the maximum value reaching 7°54'. This allows us to observe from the Earth's surface, besides the near side, very narrow strips on the far side of the Moon that make available about 59 % of its surface. It is of interest to note that the tidal energy exerted by the Earth on the Moon is responsible for a permanent small increase of the semimajor axis of its orbit, with the implication that the Moon moves steadily away from Earth by about 3 cm per year.

In size the Moon is nearly 3.7 times smaller than Earth, and its figure is very close to a sphere. Its equatorial radius $R = 1,737$ km, and its mass $M = 7.3476 \times 10^{22}$ kg, which is less than that of Earth by a factor of 81.3. Its bulk density, $\rho = 3.35$ g/cm^3, is comparable with the Earth's mantle density, which poses questions about the Moon's interior structure and origin. The acceleration due to gravity on the surface is only $g = 1.63$ m/s^2, which, from an energy viewpoint, makes the Moon a very efficient launch pad. Obviously, the mass of the Moon is too small to retain an atmosphere, whether or not one was formed in the past.

With no atmosphere and with sharp temperature contrasts on the surface amounting to 300° (from −170 °C at night to +130 °C during the day, both day and night lasting nearly half a month), the Moon is an inhospitable world. There is no surface water, although some water ice deposits are thought to be present in the polar regions and some water is assumed to store in the interior.

Surface, Interior, Geochemistry. A relief of the Moon is represented by high elevations and deep depressions (Fig. 2.6). These features still have the historical names given to them by Italian astronomer Giovanni Riccioli in the seventeenth century following the association of bright spots with continents and dark spots with oceans and seas (these were erroneously thought to exist on the Moon). Thus names such as Alps, Apennines, and Caucasus Mountains, Mare Tranquilitatis, Mare Crisium, and Mare Serenitatis are given to prominent morphological features on the Moon's surface. The surface is heavily cratered, and the large craters were also named long ago after great astronomers such as Tycho (Brahe), Ptolemy, Copernicus, and many others. The naming of features (including those on the far side of the Moon) exploded at the beginning of the space era, and currently about 2,000 names and events are commemorated on the Moon's surface.

Basically, craters are impact-formed in origin, and in the absence of an atmosphere they experience rather slight erosion caused only by meteorite bombardment and solar wind direct interaction with the surface. The rate of erosion is therefore small enough so that artifacts can be preserved on the surface for a long time. The impact origin is clearly manifested by such typical features as a ray system around many craters (which could be craters of secondary origin formed at the explosion), central hills, terraces on inner slopes of rims, etc. The range of sizes stretches from centimeters to hundreds of kilometers, the total number of craters with size more than 1 km on the near side of the Moon exceeding 300,000. Heavy cratering confirms the great importance of the above-mentioned collisional processes in the solar system. The Late Heavy Bombardment (LHB) culminated about four billion years

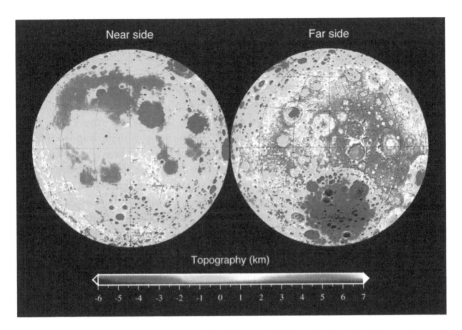

Fig. 2.6 Relief of Near side and Far side of the Moon marked in color from blue (low regions and basins) to *yellow-red* (elevations and mountains). Unlike the near side with many basins (mare) the far side is mostly represented by elevated areas (continents) with exception for the great basin nearby the South pole (Credit: NASA/J. Green)

ago when large primitive chondritic asteroids (remnants of planetesimals and sources of exogenous material) fell on the Moon's surface, leaving behind enormous basins which we now call the maria (seas). There were probably common Earth-Moon crossing impactors at the basin-forming epoch. At that time, the Moon is thought to have had a melted mantle close to the surface that was broken up by powerful impacts followed by cracks and lava extrusions that filled the basins. Indeed, the bottoms of the basins bear traces of lava emergence and flow. These very basins serve as sources of mass concentrations (*mascons*) which are responsible for the pronounced gravity field anomalies revealed by lunar orbiters as early as the 1960s.[2] Great basins are often surrounded by high elevations, probably associated with the catastrophic events. The mountains in the equatorial and middle latitudes reach heights from 2 km (Montes Carpatus) to 6 km (the Apennines). The most irregular relief represented mostly by "continents" is characteristic of the far side of the Moon and especially its polar regions. The Moon's topography generally correlates with the gravity field of the near and far sides of the Moon, inhomogeneities roughly corresponding to gravity anomalies—mascons—in basins (Fig. 2.7).

---

[2] NASA's twin Gravity Recovery And Interior Laboratory (GRAIL) lunar orbiting spacecraft, launched in 2011, halved the average operating altitudes to 23 km (taking them within 8 km of the highest lunar areas) and allowed researchers to obtain remarkable data on the gravitational influence of the Moon's surface and subsurface features.

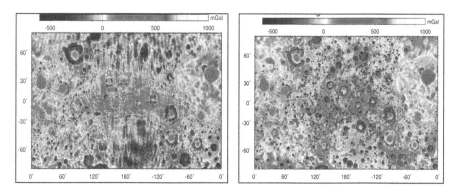

Fig. 2.7 Gravity field of the near (*left*) and far (*right*) sides of Moon in mGals marked in color from low (*blue*) to high (*yellow-orange*) values. Inhomogeneities roughly correspond to the Moon topography. Gravity anomalies in basins at the near side (mass concentrations—mascons) are clearly seen as red spots (Credit: NASA/J. Green)

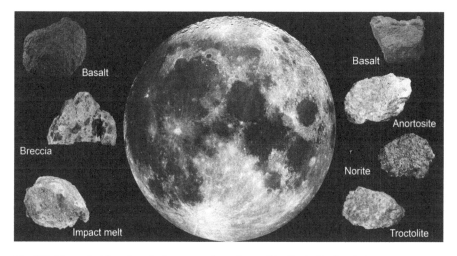

Fig. 2.8 The main Moon's rocks in mare and continents (Credit: A. Basilevsky)

The surface of the Moon is generally of basaltic composition and is intrinsically related with its interior structure and geologic history. Lunar rocks contain the same common rock-forming minerals that are found on Earth, such as olivine, pyroxene, and plagioclase feldspar (anorthosite) (see Fig. 2.8). Plagioclase, an important rock-forming series of silicate minerals within the feldspar family, is a major constituent of the lunar crust, specifically rocks in the highlands. Plagioclase feldspar is composed primarily of iron-poor and calcium-rich anorthosite. Pyroxene and olivine are typically found in the lunar mantle.

The lunar maria are composed predominantly of basalts different from those in the lunar highlands, with a higher abundance of olivine and pyroxene and a lower

Fig. 2.9 South Pole of the Moon, heavily cratered terrain with potential water ice deposits. Telescopic view (Credit: NASA)

one of plagioclase. They are enriched with iron and especially the titanic iron oxide mineral called ilmenite (with an abundance of up to ~15 % compared to ~4 % in terrestrial minerals) and other related minerals. A special group of lunar basalts are the KREEP basalts, which are abnormally rich in potassium (K), rare earth elements (REE), and phosphorus (P).

Unfortunately, geochemical measurements are pertinent only to the upper layer of regolith; therefore, the composition of the lunar surface at a relatively small depth can differ from that of the underlying layer of igneous rocks because of the significant contribution of meteorite bombardment to the surface material. The bombardment left behind a special type of broken mineral fragments—the melt and other impact processes formed a surface layer called regolith.[3] The high abundance in regolith breccias of normally rare elements in basaltic rocks, such as nickel, osmium, and iridium, favors the idea of their contamination over time by chondritic meteorites. The composition of the meteorites is supposed to be quite similar to the primordial composition of terrestrial planets before these elements were sequestered into the planet's cores, and therefore, their presence in the crust is caused by meteorite bombardment.

Of special interest is the region on the far side of the Moon near the South Pole (Fig. 2.9). Here a multilayered structure known as the Aitken basin extends 2,500 km and 13 km in depth, the largest known impact crater in the solar system formed when a large asteroid impacted an estimated 3.9 Byr ago during the Late Heavy Bombardment period. There is convincing evidence that water ice deposits could lie inside this basin in the permanently shadowed regions. Assuming the collision of a

---

[3]Generally, "regolith" is the name for a planetary surface layer that is pulverized by meteor impacts; it is sometimes also called "soil." Regolith breccias are impact-derived melts, compacted and cemented into rocks. Many regolith breccias are present in the samples returned by lunar missions. The finer regolith, the lunar soil of silicon dioxide glass, has a texture like snow and smells like spent gunpowder.

comet with the Moon, the mass of delivered water could be as high as $\sim 3 \times 10^8$ tons. Water ice deposit by comets is one of the plausible sources of water in the lunar soil (in the form of interparticle frost and crystalline water); other sources could be solar wind protons and pristine water exiled from the interior. The Moon could also preserve different volatiles and organics of regolith delivered through impact processes on the surface. Finding these substances is the main goal of future Moon exploration, and is closely related to our most intriguing astrobiological problems.

The lunar landforms and the mechanisms of their formation are intrinsically related to its interior structure, which is still quite poorly known. One of the most intriguing questions is whether the Moon preserved its liquid core and, if it did, what the size of the core is (see Fig. 2.2a, b). Other relevant questions are the nature of magma oceans and primary/secondary crustal formation. According to the modeling and partially supported by Apollo seismic data, the thickness of the upper crust ranges from 60 to 100 km, and the thickness of the upper and middle mantle layers below is about 400 and 600 km, respectively. Together they form a powerful solid lithosphere that fully excludes fissures and lava (if any) emergence, though volcanic activity was widespread on the early Moon. The lower mantle occupies a height range between 1,100 and 1,600 km; thus, the core could be roughly about 300 km in size. The size of the liquid core is also limited by the Moon's well-known dimensionless moment of inertia I. The very low magnetic field strength of the Moon ($\sim 10^{-4}$ that of Earth) could be related to a partly solidified core and/or the Moon's slow rotation.

The absolute age and dating of events in the Moon's history were determined from U-Pb, Rb-Sr, and K-Ar isotopic ratios in lunar rock samples. This led to the conclusion that the Moon was formed 50–70 million of years after the solar system formation, which is dated by the age of refractory CAI inclusions in the meteorites (see Chaps. 4 and 8). The Moon appears to have experienced interior differentiation with core separation within the succeeding 200 million years (Myr). Differentiation and impact processes left behind significant height changes on the lunar surface, reaching 18 km in the polar regions on the far side. The impact energy and the mantle heating by radiogenic isotopes led to the outflow of lava from the mantle to the surface from a depth of more than 100 km and to the filling up of lunar maria on the visible side, where the crust was considerably thinner. Some peculiarities of the lunar morphology are associated with lava cooling, which gave rise to faults, highlands, and valleys. The endogenous processes ceased 3.18 billion years (Gyr) ago, and the Moon was subsequently subjected only to impact bombardment by bodies mostly migrating from the outer solar system. The processes of meteorite pounding of the surface material, which is responsible for the origin of regolith, have a very slow time scale. As a result, the Moon stores the chronicles of events on its surface, including its interaction with the Sun and the interplanetary medium, over several billion years. Interestingly, under the average level of meteorite bombardment and small erosion, the tracks of astronauts and lunar rover wheels will be retained without noticeable changes for at least several million years (Fig. 2.10).

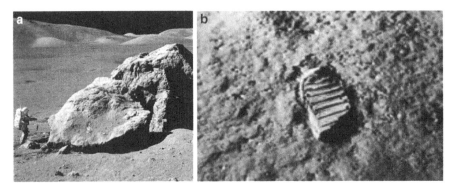

Fig. 2.10 (a) A large piece of rock on the Moon surface inspected by astronaut. (b) Astronaut's footprint on the Moon surface. In a rather weak erosion conditions on the Moon it is estimated to preserve millions years (Courtesy of NASA)

The knowledge we have gained allows us to argue that, as the body nearest to the Earth and a key proxy for the early Earth, the Moon is of primary interest for fundamental planetology and Earth sciences: geophysics, geology, and geochemistry. We pay special attention to the Moon in the context of better understanding the Earth's geological history, because the most ancient rocks are preserved only on the Moon's surface—on Earth they were eroded by the hydrosphere-atmosphere-biosphere system. Thus, we study the Moon as a window into the early Earth. Unfortunately, despite the great progress in our knowledge of the Moon provided by the unprecedented (both in scale and cost) American and Soviet space programs in the 1960s and 1970s, many problems remain unsolved. After a long break, interest in studying the Moon resumed in the 1990s with the flights of the American spacecraft Clementine and Lunar Prospector, followed by the Japanese Kaguya, Chinese Chang E, and Indian Chandrayaan-1 in the early 2000s. They were focused mostly on the study of lunar relief morphology and the gravity field. Chandrayaan-1 also found evidence of H_2O/OH distribution on the lunar surface. Recently, China's Yutu Moon rover and the Chang E-3 lander that carried it landed on the Moon, though they operated for only a short time.

A new breakthrough was accomplished by the launch of the US Lunar Reconnaissance Orbiter (LRO) and LCROSS—the final stage of the rocket which impacted the lunar surface to release volatiles through explosions. These experiments confirmed quite reliably that the surface of the Moon in the polar regions contains a significant abundance of hydrogen-bearing compounds associated with water ice (up to a few percent). Especially convincing results were obtained by the spectroscopic observations of plumes of volatiles released from ejected surface material, which accompanied the LCROSS impact. From an impact area of 30–200 m^2 heated to >950 K, ~300 kg of water ice sublimated, LCROSS detecting 155 ± 12 kg of water, or 5.6 ± 2.9 % by mass of excavated soil. Other trace molecule species such as H_2S, SO_2, CO_2, NH_3, OH, CH_4, C_2H_4, and CH_3OH were identified. As we can see, the relative abundance of water in the subsurface polar region of the

Moon turned out to be unexpectedly large. Moreover, the recent more detailed analysis of the lunar soil samples delivered by Apollo missions brought the earlier missed evidence that glasses embedded in the soil contain a small amount of water as well. Because water traces appear to connect with magma eruption from the interior, one may assume that the water content in the lunar mantle is close to that of Earth. This flurry of recent developments dramatically changed the persistent view of the Moon as an extremely dry planetary body to that of a rather wet one and has intensified interest in its further study.

Origin. The key unresolved problem is that of lunar genesis: the Moon's origin, accretionary history, and tidal evolution. Basically, there is a trade-off between two main hypotheses. The first one is based on the giant impact scenario first proposed by William Hartmann and Donald Davies in 1975: a catastrophic collision with early Earth of a Mars-size body, which struck Earth a glancing blow and sprayed nearby space with disk-shaped debris, and then the reassembly of exploded debris into a consolidated body in a near-Earth orbit. The second idea, proposed by Eric Galimov, which has received blows since the early 1990s, proceeds from the Moon forming jointly with the Earth from partially differentiated protoplanetary disk matter followed by the contraction of two separate gas-dust clumps of nearly similar chemical composition (Fig. 2.11). The former scenario is supported by the low bulk density of the Moon similar to that of the Earth's mantle from which it presumably has formed, while the latter idea is in much better accord with important cosmo-chemical constraints—an identity of certain isotope ratios (the so-called isotope shift) for Earth and the Moon. Basically, both scenarios satisfy important constraint placed by the resulting angular momentum of the Earth-Moon system. However, the most conclusive evidence that the Earth and the Moon were formed from a single gas-dust reservoir is the similarity in the mass-dependent fractionation of oxygen isotopes ^{16}O–^{17}O–^{18}O for both bodies, unlike that for other planets and meteorites (Fig. 2.12). This and other isotopic ratios (such as Hf/W, Ru/Sr, etc.) contradict the giant impact hypothesis when one considers the relative contribution to the Moon's formation of the target material (terrestrial mantle) and a projectile (cosmic body), because the latter would constitute up to ~80 % of the ejected matter, as modeling has shown. In other words, ~80 % of the projectile's material would enter the bulk mass of the Moon. Again, this contradicts the similarities in isotopic composition of both bodies. Moreover, there is no evidence of a separation of light and heavy iso-topes (kinetic isotopic fractionation), which would accompany any high-temperature explosion in the framework of the giant impact hypothesis. Thus, the above evi-dence essentially deprives the giant impact hypothesis of geochemical support.

Unfortunately, geochemical measurements are pertinent only to the upper layer of regolith. Therefore, the composition of the lunar surface at a relatively small depth can differ from the underlying layer of igneous rocks because of the signifi-cant contribution of meteorite bombardment to the surface material. The modeling of the processes involved in both scenarios is also not yet sufficiently advanced. In particular, to remove the major geochemical problems of the giant impact hypoth-esis, a scenario involving an impactor an order lower in mass but with higher speed was suggested, though it results in an angular momentum much higher than what

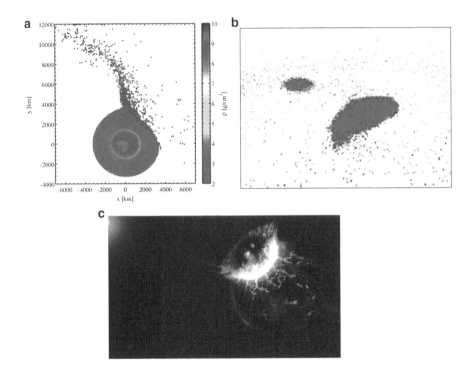

Fig. 2.11 (**a**) Diagram of the Moon formation according to mega impact scenario (Credit: W. Hartmann and D. Davies; A. Cameron and W. Ward). (**b**) Diagram of the common Moon-Earth system origin from the protoplanetary nebula (Credit: E. Galimov and A. Krivtsov). Mega impact hypothesis neatly explains the bulk density and dynamics of today's Earth moon system but not its geochemistry, the latter is in accord with hypothesis of the Moon-Earth common formation. (**c**) Artist's view of mega impact scenario: a planetesimal plowing into the young Earth (Credit: www.sciencemag.org; Science, 2013)

astrophysicists measure today. To avoid this difficulty, an evolution of the former Earth-Moon system toward its current state through a cyclic precession motion which slowed down until it became locked in its now-existing fixed position relative to the Sun—a rhythm known as evection resonance—was invoked. The problem of the Moon's origin remains one of the most challenging in the planetary sciences.

A Perspective. The Moon will probably be the first target along the track of humanity's expansion into outer space. Mastering the Moon and beginning to utilize its resources are on the agendas of space-faring countries. Stored Fe, Al, Si, Ti, and other elements and their compounds in rocks can be used to manufacture materials for building and in-site technology development and, in parallel, to produce air, water, and even fuel from H_2 and O_2 locked in rocks, and possibly to excavate stored water. However, we need to know more, including how lunar potential resources are distributed, the chemical reactivity of the lunar regolith, and which technologies should be applied most efficiently for the treatment/utilization of excavated soil. Solar incident radiation ensures essentially unlimited energy production. Some scientists

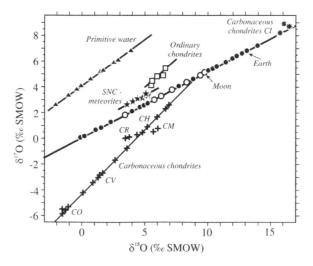

Fig. 2.12 Diagram of oxygen isotopes $\delta^{17}O$ and $\delta^{18}O$ ratio. The diagram characterizes shifts in the oxygen isotopes ratios $^{17}O/^{16}O$ and $^{18}O/^{16}O$ relative to adopted standard SMOW (*Standard Model Ocean Water*). Earth and the Moon matters fit the same line in contrast to different types of Chondrites and SNC meteorites (Courtesy of E. Galimov)

value an opportunity to extract the ^3He isotope deposited on the Moon's surface (whose abundance in ilmenite is estimated to be ~ 10 ppb) and in the far perspective even to deliver it to Earth for use in a "pure" nuclear reaction. In this reaction only protons are generated, which can be easily shielded, rather than energetic neutrons which are difficult to protect against. There is no doubt that deployment of the lunar infrastructure will provide a new stage of space explorations, especially in planetary science and radio astronomy, which could be referred to as "investigations on and from the Moon."

A great challenge for robotic missions and permanent human habitation is related to the Moon's South Pole. This is a target of the Russian projects Luna-Glob and Luna-Resource to be implemented before 2020. (Let us recall that in 1911 the Earth's South Pole was first explored by humans, and 100 years later Antarctica became a continuous habitation site for a few thousand people from about 30 countries doing innovative and critical scientific research.) The most promising lunar areas in terms of planetary science advancement, technology development, and lunar local resource utilization are the earlier mentioned Aitken basin, the largest known impact crater in the solar system, as well as Shackleton, Cabeus, and other permanently shadowed craters where water ice deposits are expected and water for life support and propulsion could be extracted. In turn, Mons Malapert offers a "peak of eternal light" which opens opportunities for permanent site illumination, power generation, and continuous line-of-sight communications with Earth. "One small step" on the Moon decades ago did not quench the human thirst for exploration of this body, or dull the exciting prospects for further expansion in outer space and settlements, probably using the lunar launch pad. There is no doubt that by the

Fig. 2.13 Artist's view of Lunar Base deployment involving a lunar mining facility harvests oxygen and possibly water from regolith as the local resource—a precursor for the further development of the Moon infrastructure and industry with iron, aluminum, magnesium, and titanium production. (**a**) NASA concept (Credit: NASA/Pat Rawlings) (**b**) Roscosmos concept (Credit NPO Lavochkin)

middle of the twenty-first century lunar base deployment (Fig. 2.13) will become a new cornerstone in human space exploration and peaceful use, a worthy achievement for the centennial anniversary of the first satellite launch.

Mercury

Main Properties. Mercury (named in honor of the Roman god of trade, equivalent to Hermes in ancient Greece) is the closest planet to the Sun at less than 0.4 AU. Its semimajor axis *a* is 0.387 AU (57.9 million km), its eccentricity *e* is 0.205, its sidereal period of revolution around the Sun (Mercurian year) is about 88 days, and its inclination to the ecliptic plane is an unusually high 7.0°. Owing to its large eccentricity (and, hence, very elliptical orbit), Mercury approaches the Sun (at perihelion) at the minimal distance of 46 million km and moves off (at aphelion) at the maximal distance of 69.8 million km (nearly twice as far as in perihelion). As we have mentioned, because Mercury is locked in a spin-orbital resonance, it gives rise to a unique mode of Mercury's rotation relative to the stars and revolution around the Sun. As a result, one Mercurian stellar day (58.65 Earth days) equals two-thirds of the Mercurian year and turns out to be much shorter than one Mercurian solar day (176 Earth days). It is assumed that initially the intrinsic rotation of Mercury was faster but that it was slowed down by tides from the Sun which, by degrees, took a significant part of the Mercury's momentum. The orbit of Mercury in the Sun's proximity experiences a quite pronounced precession caused by the effects of general relativity. There were attempts to explain the peculiarities in the motion of Mercury by its origin as a satellite of Venus, but this idea was not confirmed in light of contemporary knowledge, specifically the planets' composition and geology. Obviously, Mercury formed independently in the inner region of the protoplanetary disk, depleted of volatiles and with a larger fraction of iron and other heavy refractory elements.

Mercury (see Fig. 2.1) is the smallest planet in the solar system, having a radius of only $2,439.7 \pm 1.0$ km. This is even smaller than the radii of the largest satellites of Jupiter (Ganymede) and Saturn (Titan). The mass of the planet is 3.3×10^{23} kg and its density is 5.43 g/cm^3, which is only slightly less than the Earth's density. The acceleration due to gravity on Mercury equals 3.70 m/s^2, and the escape velocity is 4.25 km/s. Its visual stellar magnitude varies between—1.9^m and 5.5^m, but it is difficult to observe because of its small angular distance from the Sun (28.3° at maximum elongation). The axis of Mercury's intrinsic rotation is nearly perpendicular to the plane of its orbit, and therefore there are no seasonal changes on the planet. Because it is located close to the Sun, Mercury attracts special interest like many discovered exoplanets in the proximity of their parent stars.

The most complete information about Mercury came via space flights. Mariner 10 was the first spacecraft to fly by the planet, doing so three times in 1974–1975 at a minimum distance of 320 km and sending back to Earth images of about 45 % of Mercury's surface. The second spacecraft, MESSENGER (MErcury Surface, Space Environment, GEochemistry, and Ranging), following three gravity assist maneuvers (one of Earth and two of Venus), made its first flyby of Mercury in 2008, and then after two additional gravity assists of Mercury was placed into orbit around the planet in 2011. Since then it has imaged more than 95 % of the planet's surface with rather high resolution and transmitted other important information as well.

Surface, Interior. The heavily cratered surface of Mercury resembles that of the Moon, though it is more uniform. There are no large basins with the exception of Caloris Planitia, a ring structure 1,550 km across of impact origin filled with lava formed at the hub (Fig. 2.14). Mercury's craters range in size from a few to hundreds of kilometers, the largest being the crater Rembrandt, which is more than 500 km across. Generally, craters are well preserved, giving evidence of no geological activity on the planet for the last 3–4 billion years and very weak erosion. However, evidence of widespread volcanism discovered in Mercury's northern plains implies that volcanic activity was of great importance in the thermal evolution history of the planet, in particular indicating that lava has been extremely hot (Fig. 2.15). Obviously, because of the rather small size of Mercury, the stored

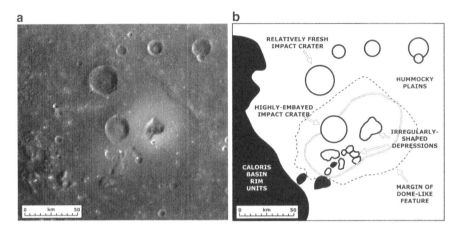

Fig. 2.14 (**a**) Messenger's image of impact craters around volcano in the largest basin on Mercury Caloris Planitia of ~1,600 km across. (**b**) Diagram of the main landforms right of Caloris basin rim units. (Credit: NASA)

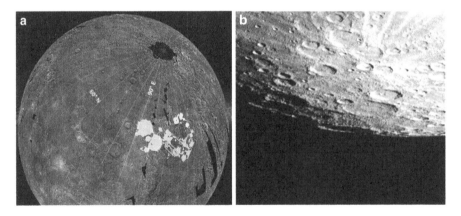

Fig. 2.15 (**a**) Mercury Northern plains where widespread volcanism was discovered. This serve as important implication for volcanic activity in Mercury's thermal evolution history and in particular indicates that some lavas were extremely hot. (**b**) Messenger's view of the South pole of Mercury (Credit: NASA/J. Green)

radionuclides responsible for heating the planet's interior were exhausted during its first 1–2 billion years, and since then Mercury, composed of 83 % iron stored mainly in its core, has steadily cooled and shrunk. This resulted in a dramatic deformation of the crust and a general decrease in the planet's surface area by about 1 % accompanied by the formation of sharp cliffs (lobate scarps) hundreds of kilometers long. For example, the giant lobate scarp Discovery Rupes, 3 km in height, extends for about 350 km.

Measurements of the elementary composition of Mercury's surface with X-ray fluorescent spectrometry showed that, in contrast to the lunar surface which is

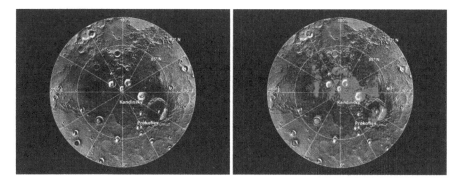

Fig. 2.16 A mosaic of Messenger radar images of Mercury's North (*left*) and South (*right*) polar regions. In the North polar region, all of the polar deposits are located on the floors or walls of impact craters. Deposits farther from the pole are seen to be concentrated on the north-facing sides of craters. In the South polar region, permanently shadowed polar craters were revealed (similar to those on the Moon) shown in *red*. The comparison of Messenger background mosaic with Earth-based radar data (shown in *yellow*) indicates that all of the polar deposits associated with water ice are located in areas of persistent shadow (Credit: NASA/John Hopkins University/Carnegie Institution/Arecibo Observatory)

enriched with plagioclases and feldspars, the surface rocks of Mercury are depleted of calcium, aluminum, titanium, and iron (all iron and siderophiles seem to have sunk down to the core) but contain more magnesium and sulfur. This means that the composition of Mercury's surface occupies an intermediate position between typical granites and ultramafic rocks and is similar to terrestrial komatiites - a type of ultramafic mantle-derived volcanic rock having low silicon, potassium and aluminium, and high to extremely high magnesium content. The most detailed data on the relief and composition of the planet's surface were obtained with the MESSENGER spacecraft. Of special interest are radar images of Mercury's north and south polar regions (Fig. 2.16). It was found that in the north polar region all of the polar deposits are located on the floors or walls of impact craters and deposits farther from the pole are concentrated on the north-facing sides of craters. In the south polar region permanently shadowed polar craters were revealed (similar to those on the Moon). They were compared with Earth-based radar data and provided additional evidence that all of the polar deposits associated with water ice are located in areas of persistent shadow.

Rather unusual is Mercury's interior, which is different from those of all other solar system planets. A major part of its mass occupies an iron-rich core with a radius of 1,800–1,900 km, overlaid by a solid silicate mantle 600 km thick and a 100–300 km solid silicate crust. The core is composed mostly of iron (with some admixture of sulfur) and is assumed to be at least partially liquid, as deduced from radar observations of the planet's specific rotation patterns. A liquid outer core and perhaps a solid inner core can be distinguished, as modeling suggests, in accord with the MESSENGER spacecraft tracking. The core liquid state is probably maintained by the pronounced tidal interaction of the planet with the Sun, stipulated by the high eccentricity of its orbit. Mercury appears to have formed from a high-temperature (refractory) fraction of its protoplanetary disk, and its original matter was rich with iron. Its extended iron-rich core occupying a major part of the body corresponds to the model of a planet having a quite uniform radial density distribution and fully satisfies a dimensionless moment of inertia $I = C/MR^2 = 0.353 \pm 0.017$, and has also been found to have three mass concentrations (mascons) of more than ~100 mGal deep resembling those of the Moon.

Based on the dynamo theory invoked to explain planetary magnetic fields, the circulation in a partially liquid core and its slow rotation together are assumed to be responsible for Mercury's rather weak magnetic field. Its field strength is a hundred times less than the Earth's (350 nanotesla). It has a quite regular dipole structure; its axis is shifted from the rotation axis by only $10°$. Mercury's intrinsic magnetic field is sufficient to form a rather small magnetosphere and specific patterns of interaction with solar wind plasma. In particular, MESSENGER discovered peculiar magnetic eddies hundreds of kilometers across and breaks in the magnetosphere through which solar plasma may reach the surface, possibly caused by frequent reconnections of interplanetary and Mercurial magnetic field lines.

Natural conditions on Mercury's surface are extremely hostile. There is essentially no atmosphere: the surface pressure is 5×10^{11} times less compared to that of Earth (its density of $\sim 10^{-17}$ g/cm^3 is like that of the Earth's exosphere), and its integral spherical albedo A is quite low (0.12). The surface temperature exhibits sharp variations from 90 K at night to 700 K in the day (between $-180\ °C$ and $+430\ °C$), the widest range in the solar system. However, because of the low thermal conductivity of the loose upper layer (regolith), the temperature at about a 1 m depth tends to stabilize to an average value of about $+75\ °C$. The thin atmosphere is composed of oxygen (42 %), sodium (29 %), hydrogen (22 %), and helium (6 %), with admixtures of water, argon, nitrogen, potassium, calcium, and magnesium (altogether within 1 %). The atmosphere's origin is related to particles delivered by the solar wind (mostly hydrogen and helium), supplied by radioactive decay (helium, sodium, potassium), and sputtered by solar wind plasma from the surface (all other species). Interestingly, particles experience less frequent collisions in the very rarefied atmosphere than with the surface, the atmosphere being renewed about every 200 days. As mentioned above the permanently shadowed polar regions could preserve large water ice deposits covered with dust particles. This idea is supported by the higher reflectivity of polar regions revealed by radar observations and the discovery in the atmosphere of such ions as O^+, OH^-, and H_2O^+, which are closely related to water. The discovery of a comet-like tail of Mercury extending for more than 2.5 million km (angular size about $3°$) showed another interesting phenomenon of the planet.

Venus

Main Properties. Venus is the second planet from the Sun (Fig. 2.1), at a mean distance of 0.723 AU (108 million km). Its name honors the goddess of love in the Roman pantheon (Aphrodite is Venus's analog in Greek mythology). The maximum visual stellar magnitude of Venus is −4.9^m (though typically −4.5^m), and it is the third brightest object in our skies (after the Sun and the Moon). The maximum elongation of Venus from the Sun is 47.8°. Because the planet is best observed in different seasons before sunrise or after sunset, it is often called the morning or evening star.

The planet orbits the Sun for 224.7 Earth days (this is the sidereal period relative to the stars, i.e., a Venusian year). The orbit of Venus is nearly circular (its eccentricity is only 0.0068). Its distance from Earth varies from 259 to 40 million km at the superior and inferior conjunctions, respectively. The rotation of the planet around its axis is extremely slow—one revolution in 243.02 Earth days. Venus rotates in a retrograde (clockwise) direction, in contrast to the anticlockwise (prograde) direction of all other planets except Uranus. The inclination of Venus's axis of rotation to the plane of the ecliptic is only 3.4° (2° to the plane of its orbit); therefore, there are essentially no seasonal changes on the planet. Because of its slow rotation, the shape of Venus has no oblateness. The combination of intrinsic rotation and revolution in orbit give rise to Venus's motion relative to the Sun (the synodic period) 584 Earth days. Its motion relative to the Earth is 146 Earth days, i.e., exactly four times less, whereas one solar day on Venus is 116.8 Earth days. Interestingly, in every inferior conjunction Venus faces Earth with the same side. It is not yet clear whether this is caused by the mutual gravitational interaction of the two planets.

Venus is very similar to Earth in size, total mass, and density. Its radius is 6,051.8 km (95 % of Earth's), its mass is 4.87×10^{24} kg (81.5 % of Earth's), and its mean density is 5.24 g/cm^3. Its acceleration due to gravity is 8.87 m/s, and its escape velocity is 10.46 km/s. Like Mercury, Venus has no satellites. Due to its proximity to the Sun, the solar constant at Venus's orbit is approximately twice that of the Earth (S = 2,621 W/m^2), although its integral spherical albedo is also approximately twice as much (A = 0.76). Therefore, the incident flux of solar energy to the rotating planet (insolation) on Venus (157 ± 6 W/m^2) is nearly the same. Let us note that while both planets derive comparable radiant energy from the Sun in the contemporary epoch, the situation was undoubtedly different at the earliest stages of both planets' evolution.

When observed from Earth in visible light, Venus looks like a completely featureless planet covered with a gaseous/hazy atmosphere (see Fig. 2.1); only in near-ultraviolet light are some features revealed (Fig. 2.17). The surface is impossible to see. This is why only very limited data about the nature of the planet were known until late in the space age. Original ideas regarded Venus as our sister planet with a favorable climate and pleasant conditions at the surface, including a wealth of vegetation. However, the reality turned out to be completely different. The first hints that Venus could have unusually high temperatures were revealed by ground observations in radio wavelengths (centimeter-decimeter waves), though their interpretation was not unambiguous. The principal breakthroughs in our knowledge about Venus occurred only with the beginning of space flights. The American Mariner 2 was the first spacecraft to fly by Venus in 1962, and it provided more reliable evidence that the recorded high radio brightness temperature could be caused by a hot surface on Venus. However, estimates of atmospheric pressure and composition remained fully speculative. The most reliable data about parameters of the

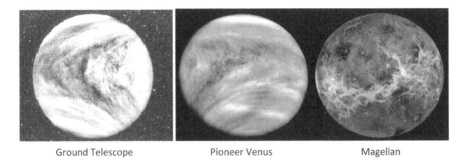

| Ground Telescope | Pioneer Venus | Magellan |

Fig. 2.17 Images of Venus obtained from the Earth in the ultraviolet spectral range (*left*); from Pioneer Venus Orbiter (*middle*); and in the centimeter radio wavelength—mosaic of images of the Venus surface obtained by aperture synthesis radar and returned by the Magellan spacecraft (*right*). While the visible range contains little information (only the isolated bands associated with the inhomogeneous structure of sulfuric-acid clouds can be seen) characteristic ordering structures are observed in the ultraviolet due to the presence of an absorber (probably sulfur particles) at the upper cloud boundary. These structures ("ultraviolet clouds") move over the Venusian disk with a speed of ~ 100 m/s that exceeds the wind speed near the surface by two orders of magnitude due to the superrotation of the atmosphere. The surface of Venus can be seen only in the radio wavelength for which the atmosphere and clouds are transparent. Radio mapping has revealed many relief features and peculiarities of the Venusian surface (Courtesy of NASA)

Venusian atmosphere were derived by direct in situ measurements of temperature, pressure, and composition made by the Soviet Venera landers. Venera 4 in 1967 was the first space vehicle to enter the planet's atmosphere, descending by parachute. It was followed by several more generations of Venera missions of much more complicated scenario and capability performance through Venera 14. The probes were able to survive on the surface in a very hostile environment and make unique scientific observations. Valuable information was also achieved by the American Mariner 10 and Pioneer Venus missions, and later by the European Venus Express. Space exploration revealed the planet to be fully different from our original views— a truly inhospitable world existing just beside us.

Atmosphere. Venus stands out among terrestrial planets of the solar system primarily by its massive gaseous envelope and thermal regime. The most intriguing question is what ultimately caused the formation of a hot atmosphere on this planet. The atmosphere is composed mostly (by ~ 97 %) of carbon dioxide (CO_2) with about 3 % nitrogen (N_2) and only trace contents of water, carbon monoxide CO, sulfur-bearing gases (SO_2, COS, H_2S), hydrogen chloride (HCl), and hydrogen fluoride (HF). There is essentially no oxygen. The temperature on the surface is 735 K (472 °C)—higher than that on Mercury which is located closer to the Sun, and well above the roast setting on your kitchen oven! Some metals are melted under such high temperatures. There are no essential temperature variations between the Northern and Southern Hemispheres because of the great thermal capacity (enthalpy) of the atmospheric gas. The pressure at the surface is 92 atm (like that in the Earth's oceans at nearly 1 km depth!), and the atmospheric density is only about an order of magnitude less than that of water. Diurnal and latitudinal temperature variations become pronounced only in the atmosphere above the cloud deck, in the stratosphere-mesosphere and thermosphere. Venus's clouds are also quite exotic: they are mostly composed of micrometer-sized sulfuric acid droplets and form a three-layered structure ~ 20 km thick between about 49 and 65 km. This and other unique properties fundamentally distinguish Venus from the other two terrestrial planets with atmospheres, Earth and Mars. As the closest analog of Earth, it is worthwhile and instructive to consider Venus as an extreme model of the home planet evolution. While Earth is the garden spot of the solar system, Venus is an awful place to visit.

Measurements of the monochromatic and integrated solar radiation fluxes as a function of solar zenith angle serve as the key limiting factor for thermal balance estimations. Measurements of solar flux attenuation in the Venusian atmosphere and illumination of the surface made with Venera landers 8, 9, and 10 and the Pioneer Venus probe showed that more than 65 % of the solar incident flux is absorbed in the upper cloud layer and the upper cloud haze (in a range of altitude of 60–90 km), about 8 % is absorbed in the middle and lower cloud layers (49–60 km), and approximately 27 % is absorbed by the lower atmosphere and the surface. Averaging the measured incident solar radiation flux over the entire surface of the planet gives rise to a flux per unit surface area of less than 20 W/m². This is only about 10 % of the averaged energy absorbed by Venus (~160 W/m²). Since the surface albedo estimated by analyzing the photometric measurements and panoramas of Venus transmitted by the Venera landers does not exceed 10 %, this implies that almost 90 % of the solar radiation flux reaching the surface is absorbed. No direct solar rays reach the surface due to the large optical thickness of the atmosphere and clouds. Scattered radiation dominates below about 60 km, and the surface is illuminated rather poorly, from about 300 to 3,000 lux (meter-candles) depending on the solar zenith angle. Thus, illumination on the Venus surface in the daytime is like that in deep twilight on Earth. Attenuation of solar light in the short-wavelength range increases as the surface is approached (the blue part of the solar spectrum is "extracted"), and red rays dominate. An observer on the surface would see an orange sky and a reddish landscape.

Surface, Relief, and Geology. Radio waves, in contrast to optical wavelengths, are capable of penetrating through the dense atmosphere and cloud layer. We already mentioned how important the first ground radio measurements were in advancing our knowledge about Venus. Later the radar technique gave us an opportunity to map the Venusian surface with spacecraft (Fig. 2.17). This started with the Venera 15 and 16 orbiters, which mapped the Northern Hemisphere, and the technique was then performed globally and with better resolution by the Magellan spacecraft. These missions allowed us to investigate Venus's surface in great detail, as shown in the topographic map of Fig. 2.18. Numerous landforms preserving certain patterns of Venus's geological evolution which were not quite similar to terrestrial ones were revealed. Generally, on Venus terrains are represented by elevated areas, vast plains, and valleys. A typical landform is the plateau Lakshmi Planum in the near-equatorial region surrounded by numerous ridges (Fig. 2.19). Another example of plains typical

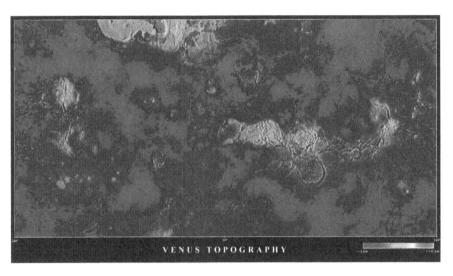

Fig. 2.18 Topographic map of the Venus surface in Mercator projection based on the Magellan radar mapping. *Green* color—elevated areas, *blue* color—low lands. The mostly high areas (*white* to *red*) are Ishtar Terra with Maxwell Mounts in high latitude (*upper left*) and Alta Regio—Aphrodite Terra and Beta Regio in the equatorial regions (Credit: NASA)

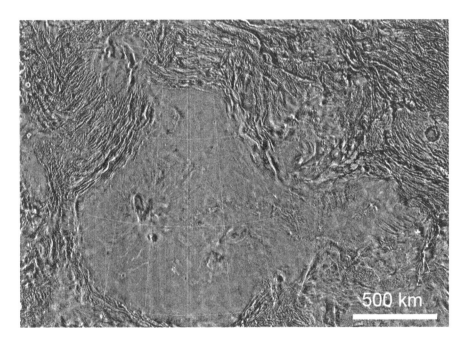

Fig. 2.19 Plateau Lakshmi in the near-equatorial region surrounded by numerous ridges (Venera 15/16 image) (Courtesy of Russian Academy of Sciences)

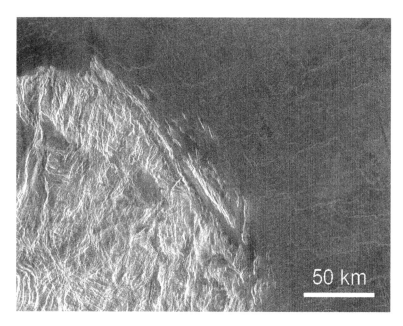

Fig. 2.20 These strongly deformed structures (ridges of compression and furrows of extension) called Tessera are typical for Venus plains (Courtesy of JPL, NASA)

of Venus are strongly deformed structures, ridges of compression and furrows of extension called tessera (Fig. 2.20). The largest highlands in the equatorial and middle latitudes are Aphrodite Terra and Ishtar Terra, which are comparable in size to the Earth's continents. The highest elevation is the mountain range Maxwell Montes at about 11 km in height and about 800 km in diameter (Fig. 2.21) on Ishtar Terra. Interestingly, Maxwell is one of three (historically preserved) male names on the planet, together with Beta Regio and Alpha Regio; by the decision of the International Astronomical Union, entirely female names are attributed for the surface features of Venus. Another elevation of presumably tectonic origin is Beta Regio (Fig. 2.22); however, unlike on Earth, there are no prominent tectonic structures that would dramatically modify the planet's surface.

However, there is strict evidence of widespread volcanism exhibited by large volcanic mountains (Fig. 2.23) and the surface outpouring of volcanic lava called "pancakes" (Fig. 2.24). There are also some local volcano-tectonic features represented by coronas—specific circular or oval-shaped structures of the Venusian surface morphology (Fig. 2.25). Similar to the coronas are arachnoids, peculiar ring structures left behind by ancient geologic processes of not quite clear nature. Volcanic activity apparently ceased on Venus relatively recently (~100 Myr ago). Although Venus possesses a very thick atmosphere, it has been bombarded throughout its history by large projectiles (a few kilometers across) capable of reaching the surface. Evidence of impact craters ranging between about 10 and 50 km in size with characteristic central dome and outbursts of soil of different

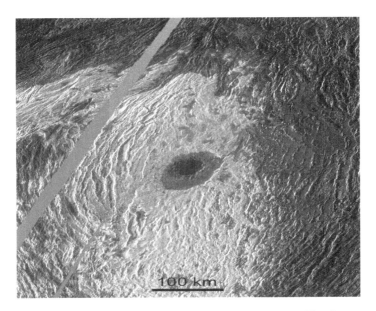

Fig. 2.21 Maxwell Mont of about 11 km in height, the highest on Venus. They have very bright surface above ~5 km in the radio wavelength (a sort of "snow line") that is probably caused of metallic type of substance the mountain is composed of (Courtesy of JPL, NASA)

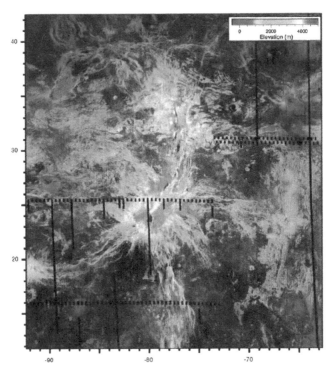

Fig. 2.22 Beta Regio—elevation of tectonic origin dissected by Devana canyon running from south-west to north-east. Scale of elevations is shown at the *upper right*. (Courtesy of NASA)

Fig. 2.23 Volcano Maat in Alta Regio—Aphrodite Terra on Venus in perspective projection. The height of the mountain is ~5 km; the traces of lava flows are clearly seen. The vertical scale is ten times enhanced for clarity. Surface brightness is also artificially exaggerated (Courtesy of NASA)

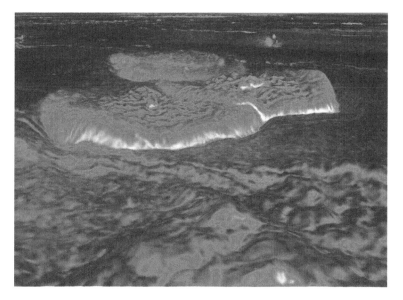

Fig. 2.24 An image of the surface outpouring of volcanic lava ("pancake") in perspective projection (Courtesy of NASA)

configurations formed in the dense atmosphere were revealed (Fig. 2.26). The most spectacular images are those of Venus's surface—color panoramas taken on the very hot surface by the Soviet landers Venera 13 and 14 and returned back to Earth (Fig. 2.27). An inhomogeneous relief of probably volcanic origin illuminated by the scattered reddish-orange solar light is clearly seen in the vicinity of both landers.

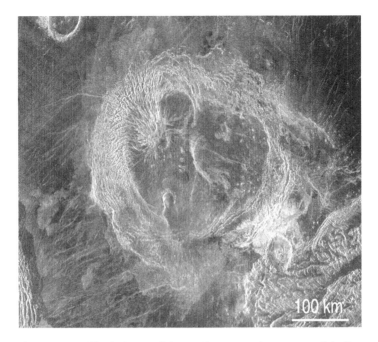

Fig. 2.25 Coronas—specific circle or oval shape volcano-tectonic structures of the Venus surface morphology (Courtesy of NASA)

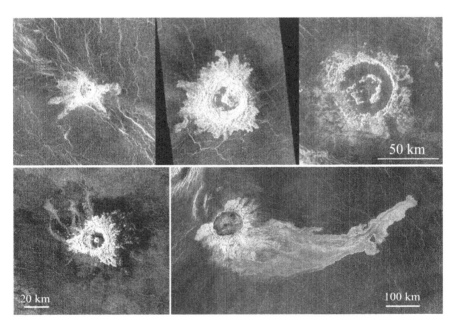

Fig. 2.26 Types of impact craters on the Venus surface with characteristic central dome and outbursts of soil of different configuration formed in the dense atmosphere. The size of craters varies between about 10 and 50 km and is indicative that projectiles of several km size penetrate the very thick Venus atmosphere and reach the surface (Courtesy of NASA)

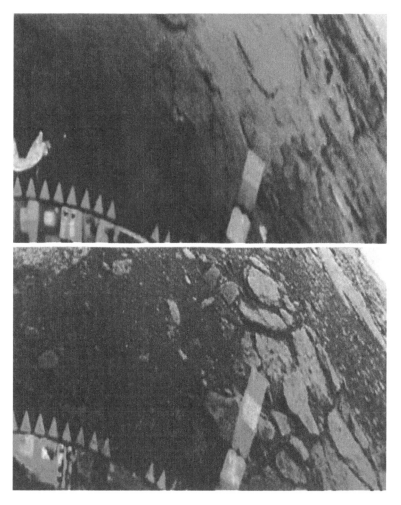

Fig. 2.27 Color panoramas of the Venus surface returned by Venera 13 (*up*) and Venera 14 (*bottom*) landers. Irregular relief is seen with heap of stones of presumably tuff composition. Horizon is at the *upper right* corner. At the *bottom left*—element of lander with tooth-shape aerodynamic ring and test-color plate. *Reddish-orange* color illuminating the surface corresponds to scattered light of the solar spectrum shifted to the longer wavelength (Courtesy of A. Selivanov)

A montage of the Venusian landscape at the Venera 13 lander site in perspective transformation is shown in Fig. 2.28. Note that unlike our blue sky, on Venus one would see skies orange in color.

The interior of Venus is still poorly known, and its model generally resembles that of the Earth. The same three main shells are assumed to exist: an upper crust of about 16 km thick, then a mantle extending to about 3,300 km in depth, and finally a massive central iron core comprising nearly a quarter of the overall mass of the planet (see Fig. 2.2a, b) Because of the very slow intrinsic rotation of Venus, the

Fig. 2.28 Venus landscape at the Venera 13 lander site: montage and perspective transformation by Don Mitchell. The incident solar light penetrating to the surface (~3–5 %) is reflected by very low surface albedo (Courtesy of D. Mitchell)

dynamo mechanism is apparently damped and no electric current is generated in the core, which explains why Venus has no magnetic field.

Heat Regime and Dynamics. A fundamental question is: Why did a neighboring planet very similar to Earth and only 40 million km away take a completely different path in its evolution? This question is addressed in more detail in the last section of this chapter. Here we preliminarily touch upon only some basics.

One may assume that the unusual, in many respects exotic, climate of Venus was predetermined by many factors, primarily by the runaway greenhouse effect that probably resulted from a loss of water from the surface of the planet followed by decomposition of carbonates in surface rocks. We can hardly reproduce this process, but we can study in sufficient detail the physical mechanism of the formed stable thermal equilibrium state on the planet.

The problem of fundamental importance is how a thermal regime on a planet is formed. Let us note that, in contrast to stellar atmospheres, the influx of radiant energy absorbed by a planetary atmosphere, which is a permanently acting factor, leads to heat flux variability with altitude. The interaction of radiation with the gaseous medium characterizes the energy release. Thus, to understand the main features of the thermal regime in Venus's atmosphere, it is necessary to answer the questions of how the outgoing and incident radiation are balanced, in other words, how net heat flux is formed and what interrelationship occurs between the radiative and dynamical heat exchanges. To judge the patterns of radiative heat exchange, greenhouse effect formation, and the role of atmospheric dynamics, it is important first of all to compare the altitude profiles of solar and thermal radiation, each of which is the downflow-upflow difference (the net flux).

Evaluation of the radiative transfer involving optical properties of the main and minor atmospheric components, their abundance, and their height distribution is key for the greenhouse mechanism modeling. Thermal infrared radiation is absorbed by almost all gases of Venus's troposphere. Carbon dioxide release from the crust with involvement of positive feedback leading to irreversible increase of both temperature and pressure is mainly responsible for the runaway greenhouse setup and its escalation. The optical properties of CO_2 are determined by a number of strong vibration-rotation bands and several weaker bands produced by asymmetric isotopes. Because of the great CO_2 content in the path of ray propagation ($\sim 10^9$ atm cm) and under high temperature, the contemporary atmosphere of Venus is opaque in the regions of intense absorption bands and radiation is transferred between them only in a few "windows of transparency." These windows become narrower, however, in the induced CO_2 absorption spectrum under high temperature and pressure because of the formation of far wings of strong bands. The weak bands falling into these windows also reinforce and contribute to the radiation absorption. Nonetheless, minor atmospheric species, primarily water vapor and sulfur-bearing compounds (SO_2 being the most important), whose bands fall in the CO_2 windows of transparency most efficiently influence the radiation transfer there, although their relative content (mixing ratio) amounts to only hundredths or even thousandths of a percent. Let us note that in a homogeneous medium, the mixing ratios are preserved as constant only for the components that are not subjected to chemical or phase transformations in the range of temperatures under consideration, whereas water vapor and sulfur dioxide contents change with altitude, especially in the cloud layer. The clouds themselves additionally contribute to the radiation transfer and net flux evaluation.

Calculations for more than 500 wavelength ranges with a high spectral resolution showed that almost all outgoing atmospheric radiation at the surface temperature of 735 K is concentrated in the infrared spectral range from 1.5 to 1,000 µm, with the intensity peak being at 4 µm and shifting to 8.2 µm near the lower cloud boundary at a temperature of 365 K in response to the Planck radiation law. It turns out that thermal balance is accomplished in the carbon dioxide atmosphere, provided the mixing ratio of water vapor is between 10^{-4} and 10^{-3} throughout the troposphere, which is in accord with the available measured data. The contribution to the opacity of sulfur dioxide is weak and is mostly pronounced in the upper troposphere region, above ~ 40 km.

In addition to the radiative transfer, an important role in the formation of the thermal balance of Venus is the atmospheric dynamics. Obviously, if there is a radiative equilibrium in the atmosphere (as in the terrestrial stratosphere), then the solar radiation energy is compensated for at each level by the outgoing radiation flux. In other words, the rates of solar energy input and atmospheric cooling through infrared radiation are equal in this case. If, alternatively, the radiative-convective equilibrium condition is met (as is usually the case in the troposphere), then the input of solar radiation is compensated for at each level in the atmosphere by the net heat flux through outgoing radiation and convective heat transport. In fact, these alternatives serve as a local approximation to the more complete picture of heat transport including dynamical processes of various spatial scales (e.g., planetary circulation, convection, turbulence).

While the condition for global-scale heat balance on the planet (characterized by the effective temperature near the upper boundary of clouds, $T_e = 228$ K for Venus) is rigorously met, an inequality between the input and output energies in different regions of the planet serves as the source of motions with various spatial scales. In other words, the emergence of thermal inhomogeneities (horizontal temperature gradients due to differential heating) is compensated for by the development of large-scale motions with a broad spectrum of spatial sizes, including the cascade fragmentation of turbulent vortices to sizes at which the dissipation of mechanical energy into heat takes place. This may explain the imbalance between the calculated heat fluxes and the measured altitude profile of the incoming solar radiation, as well as significant latitude dependence of the vertical temperature profiles revealed by microwave sounding of the Venusian stratosphere and difference in heat fluxes according to radiometric measurements from the Pioneer Venus probes.

The atmospheric dynamics results from transformation of a part of the incident energy flux from the Sun. Generally, we can say that planetary dynamics reflect the balance between the rate of potential energy generation through solar radiation and the rate of kinetic energy loss through dissipation. From this viewpoint, the planetary atmosphere is often compared with a heat engine in which the equatorial and polar regions serve as the heater and refrigerator, respectively. Note that the efficiency of such an engine is very low, no more than a few percent.

Thus, the obvious difference in heat influx from the Sun to the equator and poles on Venus is apparently not compensated for by the outgoing radiative heat flux due to the noted change in the contents of H_2O and SO_2 with latitude. In other words, at low latitudes influx of the solar radiation is larger and the radiative heat losses are smaller, while the reverse is true at high latitudes. If we take into account the earlier mentioned fact that the diurnal variations in the thermal regime of the lower atmosphere are negligible, then an explanation of this phenomenon can be found in the mechanism of planetary circulation through which a redistribution of heat and, accordingly, additional heating or cooling of the lower atmosphere occurs. Indeed, the imbalance of the latitude gradient of solar heating and infrared cooling in the lower atmosphere is an important energy factor in the planetary circulation mechanism.

The dynamical transport processes on Venus are very efficient. In contrast to the Earth, the influence of Coriolis forces on the slowly rotating Venus is insignificant, which means that the Rossby number $Ro \gg 1$. Therefore, the cyclostrophic rather than geostrophic balance condition is valid. This is characterized by the fact that a zonal velocity component increasing with altitude is superimposed on a Hadley circulation cell emerging due to the equator-pole temperature gradient. In this case, a significant heat efflux at high latitudes forms due to the existence of downward motions of atmospheric gas accompanied by backward flows in the lower atmosphere in the meridional direction and formation of huge vortex structures at the poles. The emergence of the well-known phenomenon of super-rotation (4-day circulation) in the Venusian atmosphere and the formation of vortices near the poles, with the angular momentum for atmospheric circulation being generally conserved, must be associated with the same mechanism.

Atmospheric super-rotation (also called carousel or merry-go-round circulation) causes the zonal wind velocity in the Venusian atmosphere to increase from ~ 0.5 m/s near the surface to ~100 m/s at an altitude of about 65 km and higher. The latter corresponds to the upper cloud boundary near the tropopause, where one observes characteristic ultraviolet contrasts of variable configurations (mosaics of white-and-black stripes), presumably attributable to the absorption of solar radiation by sulfur allotropes contained in the upper clouds and moving with this velocity. Let us note that one of the key elements of the atmospheric dynamics on Venus is turbulence. In particular, turbulence could play an important role in the early evolutionary stages of Venus if we consider the hypothesis about a primordial ocean that was subsequently lost due to the development of a runaway greenhouse effect.

Our current views on the structure of Venus's atmosphere with the most prominent regions at the different height levels and layers of the sulfuric acid cloud deck with lower and upper haze zones are summarized in the diagram of Fig. 2.29.

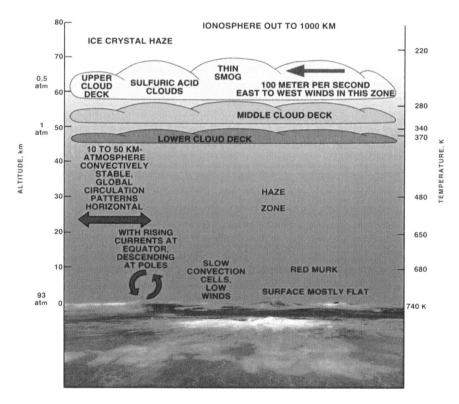

Fig. 2.29 Diagram of the Venus atmosphere structure with the most prominent zones at the different height levels. Temperature distribution along the *right* vertical axis is shown depending on altitude and respective pressure values along the *left* axis (Credit: T. Donahue & C. Russell)

The temperature distribution along the right vertical axis is shown depending on the altitude and respective pressure values along the left axis. The main mechanisms of atmospheric dynamics are indicated, including slow convection cells in the lower, mostly stable troposphere, rising currents at the equator and descending ones at the poles, and general horizontal global circulation patterns. Sustainable atmosphere-lithosphere interaction responsible for cloud formation is maintained by complicated geochemical mechanisms involving several cycles with the key role of sulfur-bearing compounds in the hot carbon dioxide atmosphere.

Mars

Main Properties. Mars is the fourth planet, located at 1.5 AU from the Sun and at a mean distance of about 0.5 AU from Earth, i.e., a bit further away than Venus but from the opposite side (Fig. 2.1). In ancient Roman mythology, Mars is the god of war (the Greek mythological equivalent is Ares), and the planet Mars was named for its reddish ("bloody") color when observed from Earth (see Fig. 2.1). Mars's visual stellar magnitude is rather high (maximum -2.88^m), though less than that of Venus and Jupiter. The semimajor axis of the Mars orbit $a = 1.52$ AU, and the eccentricity $e = 0.09$. This means that Mars orbits the Sun in a very elliptical orbit, approaching the Sun in perihelion by 206.65 million km (1.38 AU) and retreating at a maximum distance in aphelion by 249.23 million km (1.67 AU). The sidereal period of revolution (relative to the stars, or the Martian year) is 686.98 Earth days (1.88 Earth years) and the synodic period (relative to the Earth) is 779.94 Earth days. The period of rotation (which corresponds to an astronomical day) is 24^h 37^m 23^s, while one solar day is a bit longer at 24^h 39^m 35^s. The inclination of Mars's orbit to the ecliptic plane is 1.85°, and the inclination of the axis of intrinsic rotation (angular distance between the axis of rotation and perpendicular to the plane of its orbit) in the contemporary epoch is 25.19°. It is very close to that of the Earth's, and therefore, Mars experiences similar seasonal changes during the year. It is interesting to note that spring and summer in the Southern Hemisphere occurring near perihelion are shorter and hotter than in the Northern Hemisphere occurring near aphelion, and vice versa.

Mars approaches Earth at a minimal distance of 55.76 million km when the Sun and Mars are at opposite positions relative to Earth (they are in opposition repeating every 26 months at different points of the Mars orbit) and removes to 401 million km when the Sun is just between Mars and Earth (Fig. 2.30). Once every 15–17 years opposition occurs near to perihelion and is called the "great opposition," corresponding to the closest Mars approach to Earth when it reaches its maximum angular size (25.1″) and maximum brightness (-2.88^m). The last great opposition was on August 28, 2003, when Mars was at only 0.373 AU from Earth. The next one will occur on July 27, 2018.

Mars is approximately half Earth's size and nearly one-tenth its mass. Its equatorial radius equals 3,396.2 km and its polar radius is 3,376.2 km; hence, Mars' polar

Fig. 2.30 Mars images near the closest approach in 2003 taken by Hubble Space Telescope (Credit: NASA and Dr. Tony Irving of the University of Washington)

oblateness is 0.006 (close to the Earth's). The mass of the planet is 6.42×10^{23} kg, its mean density is 3.93 g/cm^3 (0.71 of the Earth's), its acceleration due to gravity is 3.71 m/s^2, and the escape velocity is 5.03 km/s. The solar constant in the Mars orbit is 2.25 times less than that of Earth, and the mean annual surface temperature is ~220 K (−53 °C).

Historically, Mars was considered as a planet which could harbor life, even intellectual life. This is why the planet has attracted special attention since ancient times, but telescopic observation could reveal only the most prominent details. The distinguished large dark and bright areas, by analogy with the Moon, were called seas and continents, and it was found that more seas are located in the Southern Hemisphere. The concept of life's existence on the planet was supported by the observations of some surface features experiencing seasonal changes, and especially more or less regular linear structures associated with channels (Italian "canali"), thought to be built by Martian civilization. However, these canals turned out to be illusive because of the poor telescopic resolution available. The reality buried hopes of a civilization, but instead opened to our view an extremely interesting world that has evolved completely differently than our home planet. The present-day Mars is a cold desert with a plethora of craters, systems of mountain ridges, plateaus, uplands, and valleys that retained the traces of a paleomagnetic field, ancient widespread volcanism, destroyed igneous rocks, and the influence of atmospheric dynamics on the surface landscapes (weathering).

Enormous progress in the study of the nature of Mars has been achieved only in the last few decades due to space missions. The US Mariner 4 was the first spacecraft to fly by Mars, and it measured parameters of its atmosphere more accurately than was possible from the Earth. The following Mariner 6 and 7 flybys transmitted the first images of the Mars surface. Although the images were of rather poor resolution, the details were discouraging because they resembled those of the Moon's surface. However, images of much better quality from Mariner 9 and also from Mars 5 revealed a completely different planet with manifold landforms and traces of past geological activity, as well as unusual properties of its tenuous and very dynamic atmosphere. The Soviet Mars 3 was the first space vehicle to land on the Martian surface, though with a very short time of signal transmission, while Mars 6 made the first direct in situ measurements of the atmospheric parameters during its descent from about 60 km (well above the height of the Martian tropopause) to the surface. The most successful explorations were the US Viking 1 and 2 missions, both orbiters and landers, in the mid-1970s, which operated for more than one Martian year. They performed many important measurements, including a search for life on Mars, which was unsuccessful. This caused a loss of interest in Mars missions for more than 10 years; they were only resumed with the Soviet Phobos missions at the end of the 1980s. These missions were only partially successful. In the 1990s two new launches to Mars with NASA's Mars Observer and the ROSCOSMOS (RSA) Mars 96 were undertaken, and unfortunately both failed. Similarly, both the NASA Mars Polar Lander and RSA Phobos-Grunt missions launched later were also ill-fated.

Especially great advancements in the study of Mars began at the end of the last century and during the first decade of the 2000s with the successful flights of the US Mars Global Surveyor, Mars Reconnaissance Orbiter, Mars Odyssey, and Phoenix, and the European Mars Express. The first US Martian small rovers Pathfinder, and especially Spirit and Opportunity, significantly contributed to our knowledge about the planet. Recently, the large, much more capable, and well-equipped NASA Curiosity rover landed on Mars in 2012, with the goal of answering many intriguing questions about Mars, including those related to its geological/climate evolution and biology. Curiosity's operation provided important information, specifically about Martian soil composition and water content to highlight the planet's evolution, though no signs of life were detected.

From the viewpoint of evolution, Mars is the antipode of Venus and is another opposite of Earth, although it is more similar in its natural properties to our planet and may have been a close analog of the Earth at early evolutionary stages. The most topical problems that now face researchers of Mars include studying the short-term and long-period variations in climate, the geological and chemical composition of the rocks forming the planetary crust, the seasonal cycles of the water and carbon dioxide contents in the atmosphere, and searching for the distribution of various forms of water both at depth and in the subsurface layer—liquid, bound, and water ice. Basically, the research and development of various climatic, geological, and evolutionary models are focused on the problem of water in the geological past and present of Mars and are associated with the presence of possible traces of primitive extant or extinct life (bacterial fossils).

Fig. 2.31 Images of Mars from spacecraft. (*Left*) Clouds above the huge shield volcanoes in the Tharsis region, relief of the Northern polar region, and Valles Marineris rift zone are distinguished in this image. (*Right*) Mosaic of images in which the main features of the Martian relief, among the highest in the Solar system shield volcanoes with a height up to 26 km above the mean surface level in the northwest and the huge tectonic fracture Valles Marineris extending for more than 3,000 km nearly along the equator (Courtesy of NASA)

Surface, Relief, and Geology. The Mars relief is very complicated: different landforms dominate the Northern and Southern Hemispheres, the former being about 1–2 km lower than the mean level. There are heavily cratered terrains in the higher Southern Hemisphere, presumably 3–4 billion years old, in contrast to the obviously much younger lower plains and unique geological structures in the Northern Hemisphere. Here the most prominent features are the great rift zone Valles Marineris and the elevated regions Tharsis and Elysium with the giant shield volcanoes Olympus Mons, Arsia Mons, and Pavonis Mons. Olympus is the highest volcano in the solar system (Fig. 2.31). Its height is about 27 km above the Mars lowland level, the size of its base is nearly 600 km across, and its caldera is 70 km across and 3 km deep. It dwarfs the Earth's largest mountain, Everest, which is only 8.8 km high. Also, the largest known basin of suspected impact origin in the solar system, about 8,000–10,000 km in size, was discovered in the Northern Hemisphere and may be the cause of the north–south dichotomy. Rift zone Valles Marineris in the equatorial region is a great canyon, probably of tectonic origin, that extends more than 3,000 km and is about 600 km wide and 7–10 km deep with diverse geological structures located within it.

The most detailed mapping of Mars's very complicated relief was taken by the Mars Orbiter Laser Altimeter (MOLA) on the Mars Global Surveyor (MGS) orbiter (Fig. 2.32). The relief stretches from vast lowland plains in the Northern Hemisphere to heavily cratered terrains in the Southern Hemisphere. The Tharsis uplands with several huge shield volcanoes and steep slopes and the adjoining great rift zone Valles Marineris running nearly parallel to the equator southeast of Tharsis are clearly distinguished. They are shown in more detail in Figs. 2.33 and 2.34, respectively. Of special interest are the polar caps of Mars which are intrinsically related with its atmospheric circulation patterns and climate, the latter depending on Mars's

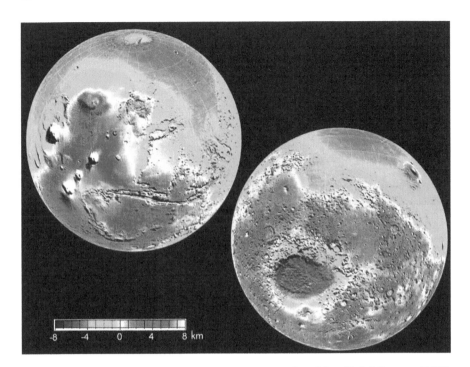

Fig. 2.32 Mars relief according to laser altimetry (MOLA) from Mars Global Surveyor (*MGS*) orbiter. Height scale (from *dark blue* basins to *red-white* elevations/mountains) is shown at the *bottom left*. Very complicated relief is revealed, from vast low land plains in the Northern hemisphere to heavily cratered terrains at the Southern hemisphere with the prominent transition regions. Tharsis upland with several huge shield volcanoes and steep slopes are clearly distinguished at *left* and adjoining great rift zone Valles Marineris running nearly parallel equator southeast of Tharsis (*upper* image). There are enormous basins Hellas of about 2,000 km across (*dark blue*) and Argir of about 900 km across (*white blue*) in the Southern hemisphere, (*bottom* image) (Credit: NASA)

obliquity. The perennial or permanent portion of the north polar cap consists almost entirely of water ice. Both polar caps contain water ice beneath the dry ice cover as much as about 3 km thick with a volume of ice of ~1.6 million km^3 (Fig. 2.35). There are enormous basins, Hellas, about 2,000 km across and Argyre. The northern polar cap accumulates condensed CO_2 as a thin layer of about 1 m thick in the northern winter only. The southern poar cap has a permanent dry ice cover about 8 m thick and much deeper water ice layer beneath it about 900 km across, in the Southern Hemisphere. Remarkably, there exist prominent transition regions between the Northern and Southern Hemispheres (Fig. 2.36).

There is no doubt that the surface and atmosphere of Mars changed dramatically over its geological history as a result of intense impact bombardment and volcanic, tectonic, and erosion processes closely related with water history and climate. Geologists distinguish three main epochs in the Mars evolution: Noachian, covering 4.6–3.7 billion years of Martian history, when the oldest crust formed, heavy impact

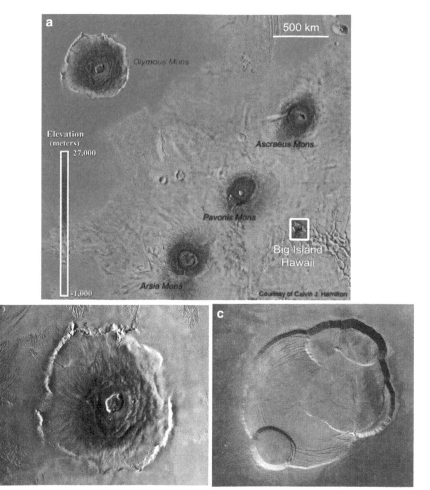

Fig. 2.33 Martian volcanoes. (**a**) Distribution of huge shield volcanoes in Tharsis region. Height is scaled in color at the *left*, horizontal size is shown and compared with Big Island Hawaii. (**b**) The largest Martian volcano Olympus Mons taken by Viking of about 500 km in foot. (**c**) An exquisite image of the 85×60 km summit nested caldera of Olympus Mons taken by Mars Express. Note how few impact craters are present (Credit: NASA)

bombardment ceased by the end of the period, early Mars was warm/wet, and a valley network formed; Hesperian, covering 3.7–2.9 billion years with volcanic activity, the great lava plains and outflow channels formation, as well as an assumed ocean existence; and Amazonian, 2.9 billion years ago to our time, attributed to a lower rate of impact crater formation and younger surface features appearance, in particular because of volcanic eruptions, including continuing Olympus Mons activity, as well as continuing outflow channels and late stage polar cap formation (Fig. 2.37). During this late period, the Martian climate dramatically changed, eventually becoming the contemporary cold, dry planet.

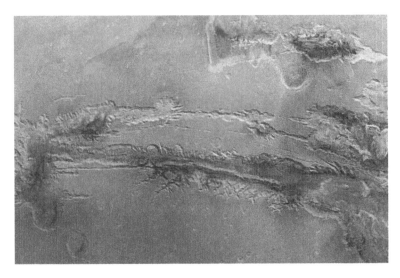

Fig. 2.34 Rift zone of tectonic origin running by more than 3,000 km along equator of maximum width 100 km and depth 8 km (Credit: NASA)

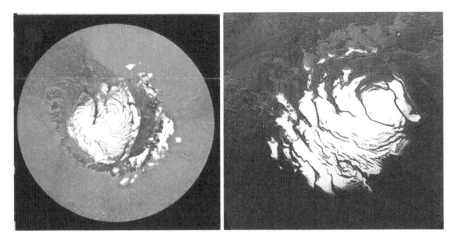

Fig. 2.35 Northern (*left*) and Southern (*right*) Polar Caps on Mars. In the Northern polar cap CO_2 accumulates as a thin layer about 1 m thick in the northern winter only. It has diameter of ~1,000 km during the northern Mars summer above perennial water ice layer having thickness 3 km and volume of ice ~1.6 million km^3. Southern Polar Cap maintains a permanent dry ice cover about 8 m thick. Its diameter ~350 km. (NASA Courtesy)

The surface features are intrinsically related to the interior structure. Obviously, Mars experienced differentiation resulting in the emergence of a liquid core, a partially melted mantle, and a solid crust in a very early epoch, soon after its origin, although differentiation was not as complete as on Earth (see Fig. 2.2a, b).

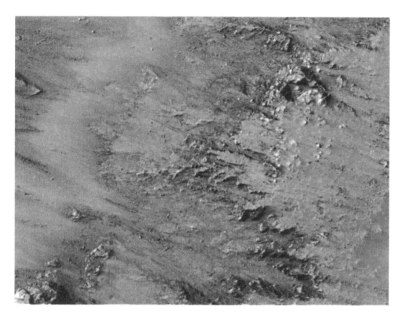

Fig. 2.36 Typical region of sharp transition between elevated terrain and plains at the Southern-Northern hemispheres boundary of Mars (Credit: NASA)

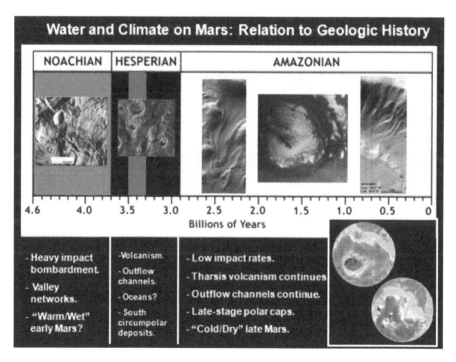

Fig. 2.37 Mars geologic history with implication to water and climate. The three main geologic periods Noachian, Hesperian, and Amazonian are fixed to time scale from 4.6 billion years to now. They are characterized by the most prominent processes which formed Mars' landforms and climate (Credit: J. Head/Brown University)

The widespread volcanism with formation of the great shield volcanoes is attributed to the first billion years. Mars cooled fast enough because of its relatively small size and limited storage of the radiogenic long-lived isotopes uranium, thorium, and potassium. According to the models developed, the contemporary Mars interior includes a thick (50–130 km) crust, a silicate mantle about 1,800 km thick, and a still partially liquid iron-sulfur core having a radius of 1,480 km and comprising 14–17 % of the planet's mass. The inner structure resembles that of Earth with the caveat that phase transitions and emergence of minerals forming under high pressure (for example, the transition of olivine to spinel) occurred in the Martian interior at much deeper levels because of the smaller acceleration due to gravity.

There are indications of an ancient liquid core existence that produced a magnetic field which eventually decayed, but left behind some remnant magnetic anomalies on the Mars surface, mostly in the Southern Hemisphere (paleomagnetism). The field is stripe-like and correlates with the most abundant hematite (iron oxide) contents on the surface (Fig. 2.38). The magnetic field strength at the Martian equator does not exceed 60 nanotesla, which is nearly 500 times weaker than the Earth's field. Because of the weak magnetic field, there are peculiar details of solar wind interaction with Mars's upper atmosphere–ionosphere involving formation of an irregular quasi-magnetosphere around the planet.

Tectonics (though probably unlike Earth's plate tectonics) strongly affected the evolution of ancient Mars. There is progressively growing support for the idea that early Mars during the Noachian-Hesperian epochs was quite different from what we see now. In particular, water can't be retained on the contemporary Mars surface, because it easily evaporates under very low atmospheric pressure. However, an ancient atmosphere comparable in density to the terrestrial one is assumed to have existed, which could provide a sufficiently high surface temperature capable of

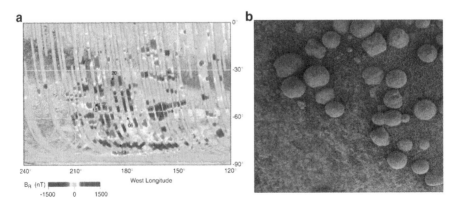

Fig. 2.38 Evidence of liquid core caused ancient magnetic field that decayed, however, later on leaving behind. (**a**) The stripe-like magnetic field remnants on the Mars surface in the Southern hemisphere (paleomagnetism). They correlate with the most abundant hematite (iron oxides) contents on the surface (Data obtained by Mars Global Surveyor). (**b**) Hematite spherules in the Martian soil. (Data of Mars Exploration Rover Opportunity. NASA Courtesy)

retaining liquid water and even its circulation between the surface and atmosphere. In any case, the appearance of a hydrological cycle and a secondary atmosphere 3.9–3.6 Gyr ago is to be associated with tectonic and volcanic processes. The mechanism of atmospheric weathering, involving combined hydrologic and glaciological processes on the surface and meteorological processes in the dense atmosphere (first of all, wind and water), was responsible for severe erosion of craters and modification of the Martian landscape. Obviously, the oldest craters were virtually wiped off the face of Mars. The present-day atmosphere could not generate such effects so destructively. We assume that this took place more than 3.5 Gyr ago, before the catastrophic collapse of the ancient atmosphere.

Water History. Many of the signatures that are an integral feature of Mars geology suggest that water furrowed its surface for several hundred million years of its history, probably most actively in the period from about 3.8 to 3.6 Gyr ago (Figs. 2.39 and 2.40). Numerous examples of geological features in support of this idea are preserved on the Martian surface from that epoch. These include primarily the systems of valleys and gullies resembling the beds of ancient dried-up rivers with numerous tributaries extending for hundreds of kilometers. Some of them can be associated with vigorous water flows produced by the melting of subsurface ice during the denudation of ice lenses, which can be likened to the motions of Antarctic glaciers. The region to the north of Elysium, where water could rise to the surface through faults in the Martian crust, is an example of such structures.

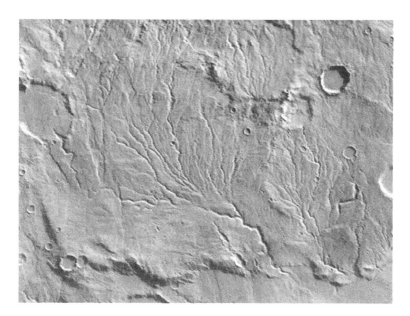

Fig. 2.39 Fluvial processes on the Mars surface—the beds of ancient dried-up rivers with numerous tributaries extending for hundreds of kilometers as a evidence of running water flows on the ancient Mars (Image returned by Viking. Credit: NASA)

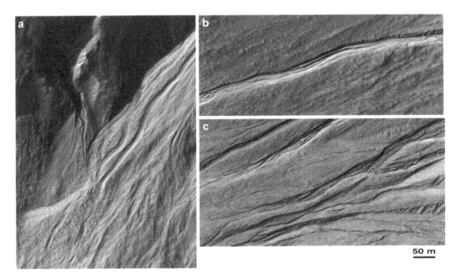

Fig. 2.40 Furrows on the Mars surface left behind by running water—ancient dry river beds (Credit: NASA/J. Green)

There are numerous examples of water erosion on Mars, some of which are shown in Fig. 2.41. These are the fissures—eroded traces of flows on steep slopes which could be produced by the fall of heavy rains or even torrents—or the spectacular Candor region in the central part of Valles Marineris. In some regions an ordered structure formed against a general chaotic background of the relief was detected in regions of ancient tectonic activity. Some regions of chaotic terrain formation are related to water channels, specifically those in river mouths. Of special interest are gullies that could be formed due to the emergence of water on the surface from below (seepage). A correlation was found between the ice-rich latitude-dependent mantle and related gully features including gully alcoves, channels, and fans (Fig. 2.42). These and some other morphological features argue for an occurrence of liquid water on the Martian surface. A close-up view of conglomerates from the Mars rovers indicates flowing water in the planet's history (Fig. 2.43). Surprisingly, a large ice lake inside one of the impact craters was detected on an image taken by the Mars Express spacecraft. This may serve as evidence of water's sporadic appearance even on the present-day planet (Fig. 2.44).

Rather young gullies were also detected on the inner slopes of impact craters or on the walls of deep troughs, mainly on their southern side, whose age is estimated to be not billions but less than millions of years old. These relief features may be associated with the challenging idea about short-term climatic variations owing to periodic changes in obliquity (the position of Mars's axis of rotation in space). They are attributed mainly to the short- and long-period variations of the Martian orbit due to the influence of the strong gravitational field of Jupiter, and are modulated by the periodic precession of the Martian rotation axis (Fig. 2.45). The total variation is in the range from 15° to 35° on a time scale of only a few million years (obliquity

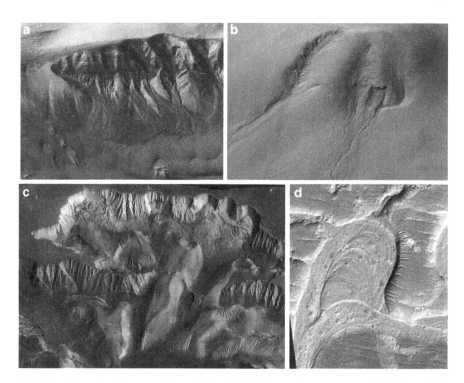

Fig. 2.41 Examples of water erosion on Mars. (**a**) The fissures that were most probably formed by water flows at steep slopes; (**b**) the gullies that could be formed due to the emergence of water on the surface (seepage); (**c**) the Candor region in the central part of Valles Marineris on Mars. Isolated more ordered structures whose formation can obviously be associated with giant water flows. They are distinguished against the background of chaotic morphology attributable to the early tectonic-volcanic processes; (**d**) Remnants of water erosion forming peculiar isolated ordered structures in chaotic terrain whose formation can obviously be associated with giant water flows (Courtesy of NASA)

cycle). Let us also note that Mars has no large satellite like our Moon that would be capable of stabilizing the position of the rotation axis in space (the two small asteroid-like satellites of Mars cannot perform this role). Theory predicts that on even longer time scales, the perturbations can become chaotic in nature, so that the inclination of the rotation axis may change from 0° to 60°. This creates the prerequisites for enormous climate changes due to a change in insolation at the poles (the warmth of polar summers) and in the intensity of the transfer of volatiles (mostly carbon dioxide and water) into the atmosphere from the polar caps, along with deposits of dust, and a change of permafrost boundaries in the summer and winter hemispheres (Fig. 2.46).

Basically, the relief of high-latitude regions is consistent with such a possibility. In particular, it was found that the surface between the poles and the equator is over-laid with sedimentary rocks with a thickness of 4–6 km in the north and 1–2 km in the south, while the surface itself is crisscrossed by cliffs and fractures that, as it

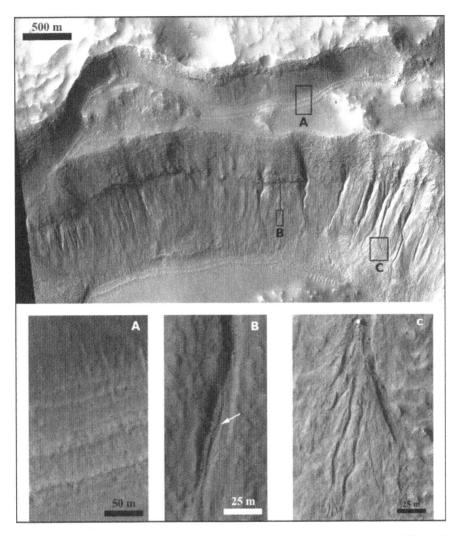

Fig. 2.42 Correlation of ice-rich latitude-dependent mantle and related gully features. (**a**) Exposed layerd mantle. (**b**) Galey alcoves and channels. (**c**) Gally channels and fans (Credit: J. Dickson; S. Schon and J. Head)

were, are wound around the poles. The sedimentary cover itself has a layered structure, backing the assumption about periodic climate changes.

Unfortunately, the huge layers of dust-sand material on the Martian surface hide many of the original structures, including large deposits of subsurface water ice formed after the supposed climate change on ancient Mars. The presence of such deposits at a depth of about 1 m predominantly at high latitudes was confirmed by neutron spectrometry measurements during monitoring from the Mars Odyssey satellite orbit and later from the Mars rovers. The ice content in rocks reaches 50 % by mass.

Fig. 2.43 Conglomerates: Indication of flowing water. Close up view of Curiosity rover (Credit: NASA/J. Green)

The individual relief features, especially those on slopes, were probably formed by the involvement of water flows, and features that are tempting to associate with periodic seepage of subsurface water in the recent past are detected on a number of surface structures. One may assume that it is salt water, which is capable of surviving more easily under Martian conditions. With a high probability, one might expect even larger reserves of water in the form of ice lenses and inter-layers to have been preserved at a depth of several hundred meters, and the possibility that liquid water can be near the lower surface of a thawing ice lens due to the accumulation of heat through the internal heat flux under the very low thermal conductivity of ice cannot be ruled out. General ideas about the contemporary hydrological system on Mars covering both Northern and Southern Hemispheres are shown in Fig. 2.47. The system spans the subsurface region (basement) to megaregolith and permanently frozen (cryosphere) layers below and above the water freezing point with the involvement of the polar caps and troposphere. Obviously, climatic changes due to obliquity variations exert only a small influence on the diagram.

Therefore, it is unlikely that water in rivers could appear on the Mars surface in periods of comparatively short-term climatic warming. This is evidenced by the age of the craters near these relief shapes, which most likely date from the first billion years of the planet's evolution. At the same time, it is hard to reconcile the above data on the layered structure of sedimentary rocks at mid-latitudes with these views, so certain contradictions are retained. Also, they by no means rule out an episodic hydrologic activity on the present-day Mars, along with preserved residual volcanism. The lake appearance within an impact crater (Fig. 2.44) could be connected with such activity. To resolve the problem, mapping of the planet's mineralogy could help, specifically the search for minerals forming only in the presence of water.

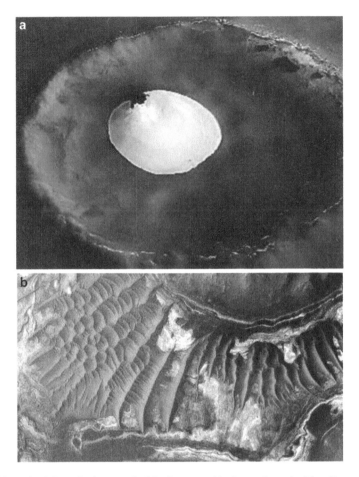

Fig. 2.44 (**a**) Icy lake at the *bottom* of a Martian crater (An image from the Mars Express space-craft Credit: ESA). (**b**) Evidence of filling up Endeavor crater with water, possibly twice in geologic history (Credit: R. Arvidson, Saint Lewis University)

We know that generally basalts form the Martian surface. Our more detailed knowledge of the Martian soil is based mostly on the landers' measurements, which discovered some variations in the surface mineral composition containing silicon, magnesium, aluminum, calcium, sodium, and sulfur compounds. Martian soil is enriched with iron oxides (up to 15–18 %), which is compatible with the basaltic composition of its crust and, as we mentioned, was apparently caused by the less complete Mars differentiation compared to that of Earth. The NASA Viking landers supported the conjecture that the distinctive overall reddish color of Mars was caused by surface dust deposits that strongly resembled mixtures of clay and ferric oxides, like those produced by weathering of terrestrial lavas. Of special interest was the finding by the Opportunity lander of rock samples composed of clay and enriched by aluminum, such as montmorillonite, which could form on ancient Mars

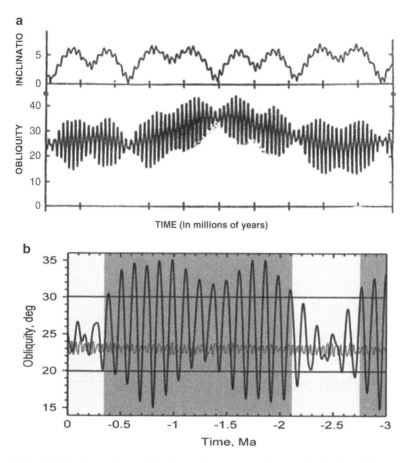

Fig. 2.45 (**a**) Periodic variations of the Mars obliquity caused by combined effect of the changes in the inclination of Martian orbit and precession of the axis of intrinsic rotation. (**b**) Obliquity changes during the last three million years. The current value fits 24° (Credit: J. Laskar)

only in the presence of fresh rather than oceanic salt water. Interestingly, the alkalinity of the Mars surface is close to that of Earth and theoretically could be suitable for cultivation.

The most intriguing idea concerns the existence of an ocean on ancient Mars. Indeed, the relief structures in the Martian northern regions, as revealed by observations from the Mars Global Surveyor satellite, resemble the coastlines of an ancient ocean bounding regions of constant surface height. This can be explained by the uniform precipitation accumulated in large volumes of water on the Martian northern plains. Moreover, there was reported a detection of layered structures (resembling sedimentary rocks on the Earth's ocean bed) with a high concentration of chlorine and bromine salts at Meridiani Planum by the Martian rover Opportunity. Figure 2.48 helps confirm the idea of the existence of an ancient ocean on Mars.

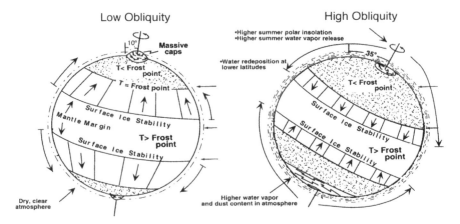

Fig. 2.46 Mars obliquity variations between low and high and consequences. The surface features and Mars climate dramatically change. At very low obliquity (<20°) collapse of CO_2 atmosphere, deposition near poles and perennial accumulation of CO_2 on steep slopes seems occur. At the high obliquity (35–40°) the mean temperature increases because of higher summer polar insolation resulting to a higher summer water vapor/dust release from the pole transited to the atmosphere and higher water deposition at lower latitudes occur. Surface ice stability zone becomes narrower, atmosphere thicker (more opaque) because of higher water and dust content (Credit: M. Kreslavsky and J. Head, Vernadsky Institute and Brown University.)

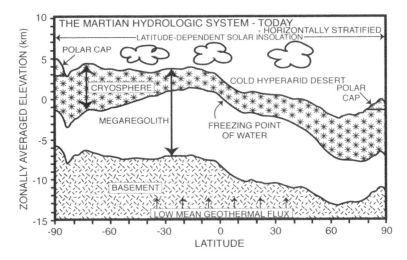

Fig. 2.47 Schematic view of the contemporary hygrological system on Mars covering both Northern and Southern hemispheres and spanning from subsurface region (basement) to megaregolith and cryosphere layers below and higher water freezing point with the involvement of the polar caps and troposphere (Credit: J. Head/Brown University)

Fig. 2.48 These rocks at the Martian Meridiani Planum resemble those on the Earth's ocean floor. High concentrations of chlorine and bromine salts detected in them also relate their emergence with the Mars ancient oceans. The region in the images may be associated with a coastline where cyclic evaporation and/or freezing occurred on shallow water (Images from the Opportunity Mars rover. Credit: NASA)

These deposits can actually be associated with the coastline of an ancient ocean where the cyclic processes of evaporation and/or freezing took place on shallow water, leaving behind layered sedimentary rocks rich in chlorine and bromine salts which are common to the Earth's oceans. The search for such terrains is one of the goals of the rover Curiosity currently operating on Mars, which is capable of tracing water abundance in Mars minerals using the technique of neutron monitoring with a better resolution than that from an orbiter. Of particular interest is the discovery of the rock's enrichment in sodium, giving it a feldspar-rich mineral content, which makes it very similar to some rocks erupted on ocean islands on Earth.

In any case, there is little doubt that Mars had a water ocean in the past. According to estimates by geologists, the total mean depth of the ocean (if it were poured uniformly over the surface) could reach 0.5 km. This value apparently restricts the maximal reserves of water that could be preserved on Mars, minus the losses (~30 %) due to aeronomic processes, as was shown by calculations of the nonthermal escape of hydrogen and oxygen atoms from the planetary atmosphere. As we discussed above, the challenge is to find out not only what caused dramatic climate change but also the mechanism of water loss. A possible aeronomic mechanism involving superthermal atmospheric particle escape is depicted schematically in Fig. 2.49.

A significant breakthrough in the space research of Mars was accomplished with the launch in 2011 of the NASA Curiosity (Mars Science Laboratory) rover (Fig. 2.50), which has been operating at the bottom of large Gale Crater (Fig. 2.51) since August 2012. It advanced both surface and atmospheric studies of the planet, and we will address the main results here in more detail.

The analysis of volcanic rocks, in particular the pyramid-shaped stone called "Jake Matijevic," obtained more insight into the problem of Mars's geologic evolution and the rate of differentiation of its interior. Soil samples were taken and analyzed with the onboard Sample Analysis at Mars (SAM) instrument including a gas chromatograph and a mass spectrometer. The rover's scoop was used to collect dust, dirt, and finely grained soil from a sandy patch known as Rocknest. Some authors

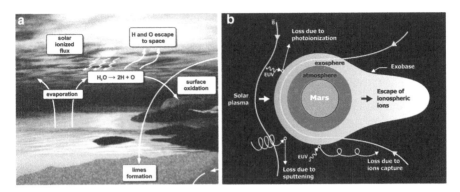

Fig. 2.49 (**a**) Model for the loss of water from the Martian atmosphere caused by the aeronomy processes (superthermal particles formation and escape). This mechanism could be responsible for loss of approximately a third of the Mars ancient ocean over the geological time. (**b**) Diagram of the mechanisms of hydrogen and oxygen dissipation from the atmosphere (as a consequence of the photolysis of water molecules and deactivation of the excited O atoms), with the formation of super-thermal oxygen atoms with an energy sufficient for their escape from Mars (Credit: V. Shematovich and M. Marov)

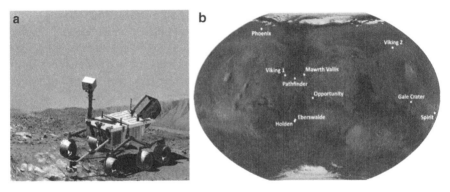

Fig. 2.50 (**a**) Mars Curiosity rover (sketch). (**b**) Landing sites of Mars landers on the Mars surface (marked in *yellow*). Curiosity landing site is in Gale crater (*right edge* on the map) (Credit: NASA)

have said that the grains in the samples, a "garden variety of Martian soil," are "finer than sugar but coarser than something like flour." Gases from a sample taken inside the instrument were released after the soil was heated to 835 °C, and the composition was measured. Besides a significant abundance of carbon dioxide, oxygen, sulfur-bearing compounds, chlorinated methane compounds, and water were found. The chlorine was probably of Martian origin and together with oxygen came from perchlorate, which has previously been detected on Mars. The water abundance turned out to be quite high, about two weight percent, the water molecules being bound to fine-grained soil particles. Because a similar water abundance was detected in several samples taken along the rover route, one may argue for water distribution over the whole planet rather than mostly in high latitudes, as was earlier shown by neutron monitoring. Martian water was found to be enriched with the heavy

Fig. 2.51 (**a**) Gale crater (4.5°S, 137°E, −4.5 km) contains a 5-km sequence of layers that vary from clay-rich materials near the *bottom* to sulfates at higher elevations. Curiosity landing site is shown with a *yellow circle*. (**b**) Curiosity landing site—quite hummocky local terrain—in a perspective view (Credit: NASA/J. Green) (**c**) Picture taken by Curiosity as it departed the latest sample site on April 14, 2014 (Courtesy NASA/W. Huntress)

hydrogen isotope[4] deuterium in proportions analogous to that of the thin atmosphere. Atmospheric origin water in the soil has been assumed.

More complicated is the problem of organic compounds, which are not likely to be preserved in surface soils exposed to harsh radiation and oxidants. Trace levels of organic compounds containing both carbon and chlorine were detected in the experiments with the SAM instrument as well. The results are not clear, because organic compounds could likely be formed during the samples' heating and reaction with terrestrial organics already present in the SAM instrument background. This means that the measured organics are not clearly Martian in origin. Another possibility is that they formed in the processes of interaction of exogenous organics due to comet/asteroid bombardment of early Mars (see Chap. 4) with atomic chlorine released from toxic perchlorates found in Mars's polar regions as early as in 2008 and now discovered in the equatorial region as well. Notice that the general extension of toxic substances on the planet's surface not only prevents us from determining

---

[4] Isotopes are variants of the same chemical element with different numbers of neutrons and therefore different atomic weights.

whether organics are indigenous to Mars, but also seems to complicate prerequisites for the origin of life and may pose a problem for the future crew of the Mars human mission.

Curiosity also has made a comprehensive mineralogical analysis on the Martian surface using a standard laboratory method for identifying minerals on Earth: X-ray analysis. This allowed us to identify 10 distinct minerals, but also to find an unexpectedly large amount of amorphous ingredients, rather than crystalline minerals, in the Rocknest composition. Amorphous materials, similar to glassy substances, are a component of some volcanic deposits on Earth. The results on both crystalline and noncrystalline components in the soil and the chemical compositions of the rocks provide clues to the planet's volcanic history and, in particular, can be used to infer the thermal, pressure, and chemical conditions under which they crystallized.

Atmosphere. Mars has a very thin atmosphere. The average surface pressure (corresponding to the triple point of water) is 6.1 mbar, which is a factor of 160 lower than the Earth's. The pressure varies depending on relief, increasing to 12.4 mbar at the bottom of Hellas basin and dropping to 0.5 mbar at the top of Olympus Mons. The atmosphere is composed mainly of carbon dioxide (95 % by volume). Other components are nitrogen (2.7 %) and argon (1.6 %). Minor species are oxygen (0.13 %), carbon monoxide (0.02 %), and water vapor (0.01 %). This water abundance, confirmed by exploring the water signature found by Curiosity in the ubiquitous Martian dust, is in coincidence with the tiny amount of ambient humidity in the planet's arid atmosphere. The atmospheric water content is negligible (it corresponds to tens or hundreds of micrometers of precipitated water). However, this abundance turns out to be near saturation in the very rarefied atmosphere and may condense out on the surface as hoarfrost. Of special interest in terms of potential Mars biology is the minor species methane. There were reports about finding methane at the upper level of only one hundred million's abundance, though no traces were detected. But one should be aware that principally it could be either organic or inorganic (volcanic) in origin and that its presence would not be key in determining potential Martian biology.

In contrast to Venus, there is no noticeable greenhouse effect on Mars, although even a tenuous atmosphere raises the surface temperature, however by only a few degrees. Seasonal-diurnal temperature variations exceed 150°C, from +20°C in some regions near the equator in summer to −130°C on the winter polar cap, where dry ice (CO_2) condenses. The temperature lapse rate in the troposphere is about −2.5°/km, and above the tropopause (in the stratosphere) the temperature reaches a nearly constant value of −129°C. Because the atmosphere of Mars is so rarefied and permanently contains suspended dust, the colors of Martian skies are different from those on Earth. In contrast to our blue sky, for which Rayleigh scattering is responsible, the daytime sky on Mars is orange-red and becomes dark violet at the horizon. Interestingly, the SAM instrument on the Curiosity rover found that the ratios of isotopes in the soil, including the above-mentioned hydrogen-to-deuterium ratio and carbon isotopic ratios, are similar to those found in the atmosphere, indicating that the surface soil has interacted heavily with the atmosphere. This tends to support the idea that as the dust is moving around the planet, it is reacting with some of the gases from the atmosphere.

The dynamics of the Martian rarefied atmosphere with its low thermal inertia differs significantly from the terrestrial one primarily by the absence of oceans, which are heat accumulators and damp the diurnal-seasonal temperature inhomogeneities. For the fast-rotating Mars ($Ro \ll 1$), the geostrophic balance condition is met. The global circulation model (GCM) predicts a topology of motions in the troposphere and stratosphere generally similar to that of Earth, with a predominance of winds blowing eastward at high latitudes in winter, near the subtropics in summer, and westward at other latitudes. The polar caps also play an important role in planetary circulation. As we said above the polar caps are composed of water ice (a secular component) overlaid by carbon dioxide ice with a lower temperature of phase transition, and so there is freezing (CO_2 collapse over the cap) in winter and melting in summer. The thickness of the permanent north polar cap of water ice amounts to a few km, while the layer of freezing carbon dioxide does not exceed a few meters but may expand to $50°$ latitude in each hemisphere. The seasonal carbon dioxide exchange between the atmosphere and cryosphere in the polar caps serves as the main driving mechanism of air transport in the meridional direction, as illustrated by the diagram in Fig. 2.52. This gives rise to Hadley cell configurations with upflows and downflows, rearranging the system of winds near the surface and at high altitudes in the summer and winter hemispheres and causing seasonal changes of the cloud cover.

Convection compensates for the high static instability of the Martian atmosphere close to saturation even at a very low relative content of water vapor. Nonetheless, even under conditions of low moisture and the dry adiabatic gradient, massive clouds form. The convection excitation efficiency during daytime hours is approximately an order of magnitude higher than that in the Earth's atmosphere, while during the night it is completely blocked due to the formation of an inversion layer with a positive temperature gradient near the surface. Convection also maintains a

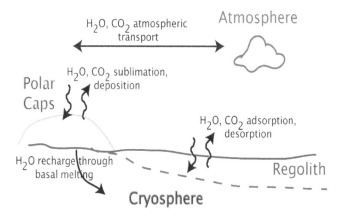

Fig. 2.52 Cryosphere-atmosphere volatiles exchange driving global atmospheric circulation on Mars (Credit: J. Head/Brown University)

constant high content of dust in the Martian troposphere and produces an additional dynamical effect superimposed on the global system of winds and the formation process of a windblown drift of sand and dust. This effect arises from a positive feedback between the dust content and the degree of atmospheric gas heating, which manifests itself as thermally generated diurnal and semidiurnal tides. This peculiar feature of the Martian atmosphere is most pronounced during periodically emerging global dust storms when fine dust rises to an altitude above 30–40 km because of turbulent mixing. Since the troposphere is highly opaque, a reverse greenhouse effect strongly damping circulation transport is produced near the surface. (The discovery of this unique natural phenomenon during a global dust storm on Mars in 1971 served as the impetus for drawing an analogy between it and the possibility of "nuclear winter" on Earth as an inevitable consequence of using atomic weapons and helped to cool down some political ambitions.)

Note also that the circulation in the Martian atmosphere is strongly affected by the surface relief (areography) on which both the observed wind patterns and generation of horizontal waves with various spatial scales depend. In turn, the planetary waves attributable to the baroclinic instability of the atmosphere and also internal gravity waves manifest themselves as irregularities in the temperature and vertical motion profiles in the stratosphere. The observed wave motions in the structure of clouds from the leeward side when flowing around obstacles, suggesting the existence of strong shear flows in the Martian atmosphere, are associated with them as well. The entire near-surface atmosphere turns out to be turbulent due to shear flows, even under conditions of relatively stable stratification.

The turbulence of a dispersed medium whose pattern depends significantly on the dynamical and energetic interaction of the gas and dust phases undoubtedly plays an important role in the complex of processes responsible for the formation, maintenance, and decay of a Martian dust storm, although the details of these mechanisms are not yet completely clear. Nonetheless, some estimates can be made based on the turbulent momentum and heat fluxes deduced from the measured altitude profiles of the mean velocity and temperature at the surface boundary layer, which are, however, modified in the case of turbulent flows with a heavy admixture. This approach allowed the conclusion that the presence of relatively small dust particles in the near-surface flow gives rise to the growing velocity gradient and increases the effect of saltation, transporting much larger amounts of dust particles into the atmosphere. This mechanism could be responsible not only for global dust storms lifting billions of tons of dust into the atmosphere but also for the formation of local eddies (dust devils) one thousand times in excess of what forms on Earth.

Moons. Mars has two small satellites of irregular shape: Phobos ($27 \times 22 \times 18$ km) and Deimos ($15 \times 12 \times 10$ km); see Fig. 2.53. In Greek mythology the names mean "fear" and "horror," acknowledging two sons of Ares who accompanied him in battle. Both satellites are locked in a synchronous rotation that results in their facing Mars with the same side. Their orbits lie close to the plane of the Martian equator. Interestingly, the period of Phobos's revolution around Mars is nearly two times shorter than the period of Mars's intrinsic rotation around its axis; therefore, an

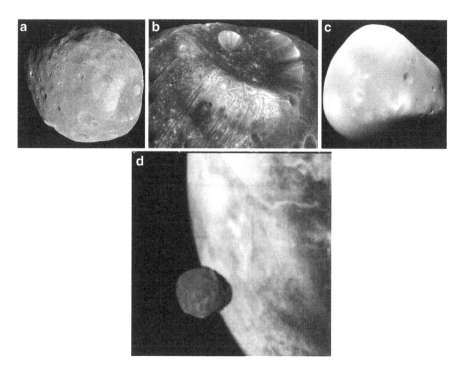

Fig. 2.53 The Martian moons Phobos (**a**, **b**) and Deimos (**c**) (NASA Courtesy). Image of Phobos at the background of Mars taken from the Russian Phobos 88 (**d**) (Credit: the Author). Phobos has accumulated dust and debris from the surface of Mars, knocked into its orbital path by projectiles colliding with the planet. A sample-return mission to Phobos would thus return material both from Phobos and from Mars

observer on the surface of Mars would see Phobos rise and set two times every day, crossing the sky quite quickly during the night. The period of Deimos's revolution is much longer, and both moons can often be observed simultaneously.

The problem of the origin of both Martian satellites remains unresolved. They could be either remnants of primordial bodies from which the planet formed and preserved since then, or asteroids captured by Mars much later. The latter scenario seems more plausible, although it is not clear how asteroid orbits eventually evolved from originally elliptical to nearly circular. Phobos experiences a significant tidal influence from Mars that results in a slow contraction of its orbit, and it will ultimately fall to the planet in a hundred of million years. Changes of the Phobos orbital parameters were already noticed in the middle of the last century, and as an explanation of this fact, the speculative idea was put forward that the satellite is hollow inside and artificial in origin, designed and launched by Martians. This intriguing idea died soon after perturbations from the irregular gravity field of Mars (strong gravity anomalies) were found. Images of both satellites transmitted by space vehicles brought evidence of the numerous impact craters on their surfaces, some of them comparable in size with the moons themselves (e.g., Stickney on Phobos). The bulk density of these satellites is low enough to testify to a rather porous structure of their interiors. Another possibility is that they had ice in their interiors or were once completely fragmented and assembled back like a loose rubble pile. There is a dust torus along the Phobos orbit produced by the meteorite bombardment including that erupted from Mars and eventually gathered as sediment on Phobos. Moreover, Phobos probably has accumulated soil, debris, and rock from the surface of Mars, knocked into its orbital path by projectiles colliding with the planet. This means that the tiny moon has been gathering Martian castoffs for millions of years, and thus a sample return mission to Phobos would be a twofer, returning material both from Phobos and from Mars. Unfortunately, the Russian Phobos-Grunt sample return mission in 2011 was ill-fated, and it is planned to be repeated in collaboration with the European Space Agency (ESA) in the early 2020s.

We see that historically predetermined interest in Mars has been maintained and has even increased in the last decades. The bottom line includes not only the wish to disclose why the nearby planet took a path of evolution different from Earth's but also the still-remaining hope of finding life signatures on past or present Mars, extinct or extant life features. One possibility is that abiotic primitive life origins could be related with groundwater cycles in volcanic regions, schematically shown in Fig. 2.54, similar to the supposed abode of ancient life on Earth. In the last decade of the last century, there was great excitement related to the in-depth study of meteorites collected in the Antarctic, some of them coming from Mars as the result of catastrophic asteroid impacts and blowing out matter from the Martian surface into space millions of years ago. Scattered debris traveled for millions of years in interplanetary space before some of it reached other planets, in particular Earth, and

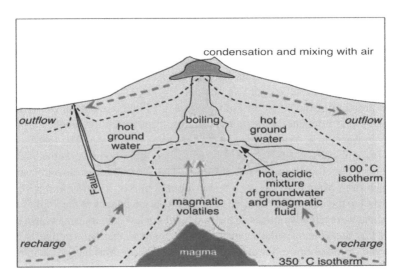

Fig. 2.54 Schematic representation of ground water cycle in volcanic region as abode of abiotic life origin (Credit: J. Head/Brown University)

 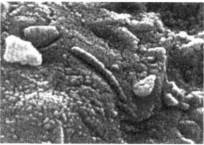

Fig. 2.55 Antarctic meteorite of achondrites class ALH 84001 (mass 1,931 kg) found on Dec. 27, 1984 in Alan Hills mountains in Antarctic (*left*), where traces of bacterial fossils of worm-shape were announced to be found. Later study showed these structures are rather of inorganic origin (Credit: D. McKay)

impacted the surface. The Antarctic ice cover is the best area for meteorites to be preserved and found. Many meteorites came from the Moon, but some were identified as being of Martian origin because of the unique composition and isotopic ratios of trapped gases, similar to those of the Martian atmosphere. They were distinguished by mineral composition as the class of shergottite-nakhlite-chassignite (SNC) meteorites (see Chap. 4 for more details). Note that recently another meteorite also associated by gas composition with Mars origin was found, this time in Morocco, and was called Tissint for the name of the town near the fall site. Of specific interest is the Antarctic meteorite ALH84001 (Fig. 2.55), where traces of bacterial fossils were thought to be found. Unfortunately, the follow-up studies did not confirm this discovery and related it rather with structures of inorganic origin.

Mars not only attracts progressively growing interest from scientists, it also draws public interest. The attention could be increased by the virtual presence of people on Mars. An interesting proposal was made some time ago by the Planetary Society to set up a robotic Mars outpost network with access to the transmitted real-time information/images and data collected through the Internet. The basic idea is to select a few of the most intriguing sites on the Martian surface and to deploy there (quasi) permanent robotic stations including both stationary platforms equipped with communication, navigation, and other supportive systems, and rovers. Balloons could also be launched and controlled from a stationary stations network. All these facilities would comprehensively investigate the surrounding terrain and different patterns of planetary geology, geochemistry, atmosphere-surface interactions, atmospheric dynamics, and weather. Researchers from various countries, college/university students, school children, amateur astronomers, and even the general public could participate in this endeavor, getting real-time access to data acquisition and analysis. Students could be a part of special teams and participate not only in the study of Martian landforms but also in operation planning including rover maneuvering through craters, dry river beds, and basins suspected to be ancient lakes of biological interest. Several rovers could operate in parallel, for example, for the study of geological history at a denudated steep wall of cliffs. Another challenge is the deployment from a stationary station drilling facility to excavate soil patterns from different horizons. All this would imply a nice integration of the basic science and public interest/involvement, because people from all over the world could collaborate and have a virtual presence on the planet, thus ensuring much better public/taxpayer support for planetary exploration.

Outlook. Summarizing this brief overview, we may conclude that, among the terrestrial planets, Mars with its satellites continues to draw great attention as a neighbor world which evolved differently than Earth but is nonetheless still close enough to the home planet by its natural conditions to warrant inspection. The outlook for further Mars study is as follows:

– History of the planet's seemingly once-volcanic and aquatic history, its water and climate, relationships with geological structures in support of evidence of an earlier wet and warm Mars; what caused transitions from global warming to global cooling;
– Role of abundant liquid water and pluvial activity in pervasive alteration of crust: water inflow, rainfall, filling and outflow, open-basin lakes; valley origin and history; ancient ocean at shallows; subsurface permafrost as uppermost of the surface, seasonal changes in hydration, polar caps, glaciers;
– Topology of hydrological system and its alteration throughout the Noachian/ Hesperian/Amazonian periods in Mars history;
– Evolution of the Martian crust and deeper regions, surface morphology in relation with the volcanic diversity, mineralogy, loose rocks, sand and dust; search for punctuated volcanism and hydrothermal sources to alter crust composition; elemental composition of the surface, including hydrogen-bearing compounds, sulfates, chlorides, and metal-bearing clays; search for carbonates;
– History of magnetic field in conjunction with the interior, paleomagnetism;
– Thermal regime/dynamics of the atmosphere (specifically the role of CO_2, H_2O, SO_2, and chlorine compounds) thermal balance near the surface, role of aerosols; minor constituents (in particular, methane) and noble gases, isotopic ratios; estimates of dissipation to outer space of an earlier denser atmosphere;
– Search for organics and biosignatures/life features, extinct/extant biosphere in close proximity to ancient water environment. Of special attention should be search for life signatures beneath the surface shielded from harmful UV solar radiation and based on metabolic processes similar to those of deep-seated biosphere on Earth.

New missions to Mars are on the agenda of the different national space agencies, which stimulates continuing activity in planetary exploration. An important step forward is the most recently launched NASA Mars Atmosphere and Volatile EvolutioN (MAVEN) spacecraft focused on the Martian aeronomy-upper atmosphere study with the goal to reveal atmospheric formation and loss history. ESA announced its planning of INSPIRE—a future Mars network science mission with deployment of several coherently operated landers on the Mars surface—which was earlier preliminarily designed at the NASA Mars Mission Research Center. ESA and ROSCOSMOS joined efforts around the two challenging ExoMars missions (Fig. 2.56) including a

Fig. 2.56 Joint ROSCOSMOS-ESA project of Mars study ExoMars (Courtesy of ROSCOSMOS and ESA)

Fig. 2.57 Panorama of the Martian surface at the Pathfinder landing site in Ares Vallis. Note the *gray* rocks covered by *red-brown* dust (Courtesy of NASA)

Mars rover to be followed by the above-mentioned new Phobos-Grunt (preliminarily called Boomerang) and Mars Sample Return (MSR) missions.

In the contemporary epoch, Mars with its hostile natural conditions is an inhospitable desert world that would not be easy for humans to adapt to (Fig. 2.57). It is, however, the most suitable place among other solar system planets to visit and possibly to habitat. Robotic missions have returned to Earth invaluable information on Mars. New ambitious missions of progressively growing complexity and capabilities will further improve our knowledge on the present and past of this intriguing world. They are also regarded as precursors of future human flights and the long-term goal of Mars exploration and utilization. Different mission profiles and scenarios have been suggested for flying to Mars with humans before the middle of this century. Are they realistic? The bottom line is that such a mission can be undertaken when new technologies become available, first of all, new propulsion systems involving nuclear engines and/or electric plasma jets. Needless to say, we should ensure minimum risk for astronauts in terms of safe return flights and protection from a hazardous radiation environment. Also, a human flight to Mars will require the investment of enormous financial resources, which are currently estimated as at least 600 billion dollars. Can we afford this ourselves now when humanity faces so many serious problems on the home planet? Would it be justified in terms of promoting future technology development and potential trade-offs? The first goal is, therefore, to examine how to pool financial and technological resources most efficiently to undertake such a mission internationally.

Nonetheless, the purpose of the coming decades is to develop foundations at a program level architecture for robotic exploration of Mars that is consistent with the challenge of human missions. It would incrementally develop an infrastructure on the Mars surface that would be needed for the future human flight. This would involve not only detailed studies of the selected sites and their natural conditions throughout several years to ensure that future astronauts could avoid unexpected situations, but also the organization of a reliable navigation/beacon system, preparation for some necessary logistics deployment, and preliminary study of the potential for production and storage of breathable oxygen and propellant from atmospheric carbon dioxide and soil to facilitate safety and reduce the cost of the human expedition. A well-developed robotic infrastructure on Mars would be extremely supportive and valuable for humans and it would advance human-machine symbiosis in general.

Undoubtedly, sooner or later the human flight to Mars will be implemented. It should be undertaken as an international endeavor and the common target of our civilization manifesting the next step in its evolution (Fig. 2.58). Projects of Mars terraforming are also discussed for the very far future.

Some Problems of Our Neighbor Planets' Evolution

Let us now return to the key questions intrinsically related with the formation of natural conditions on our neighbor planets. Why did Venus and Mars take different evolutionary paths than Earth? What formed the peculiar thermal regime of Venus, and what catastrophic event caused the dramatic change of an assumed originally clement climate on Mars?

Earth is located in a comparatively narrow zone of circumsolar space where the development of favorable (for life's existence) climatic conditions is possible. This is called the habitable zone in the solar system. Its inner boundary is located 10–15 million km closer to the Sun than the Earth's orbit, while its outer boundary extends

Fig. 2.58 First astronauts on Mars. Artists concept (Credit: NASA)

to approximately the orbit of Mars. The orbit of Venus is outside this zone, at a distance that is almost triple the critical value. Obviously, if the Earth were moved to the location of Venus (actually, even not as far), then it would probably evolve according to the Venusian scenario. Let us also recall that the whole solar system is fortunate to be located in a quite limited zone of our Galaxy where conditions suitable for biota appearance could be principally developed (see Chap. 10).

Indeed, let us assume that the initial albedo of early Earth was mainly determined by surface and corresponded to the lunar one ($A = 0.07$). Then based on the average annual balance between the solar radiation absorbed by the terrestrial surface and the thermal (bolometric) radiation emitted by Earth (at the reduced luminosity of the young Sun) the effective (equilibrium) temperature would be ~255 K, which is even below the freezing point of salt water. We therefore need to assume that a greenhouse effect developed in the original atmosphere and substantially contributed to the early stages of our planet's evolution (currently, the greenhouse effect raises the mean temperature near the ground to 288 K). This means that even under reasonably low atmospheric pressure (~0.01 atm) Earth could retain its water that was concentrated in the primordial water reservoirs and atmosphere. In turn, carbon dioxide would accumulate in the terrestrial hydrosphere and carbonates of sedimentary rocks through its binding with metal oxides incorporated into the ocean crust and upper mantle (with the formation of aqueous silicates) and, biogenically, through the deposits of lime skeletons of sea organisms.

Venus. Now, addressing Venus, one must be aware that at the same postulated initial albedo, the equilibrium temperature turns out to be at least 325 K, which is above the water boiling temperature down to a pressure of 0.2 atm. To retain water on the surface, Venus should have an initial atmosphere denser than that of Earth by about an order of magnitude. At equal rates of mantle material degassing and atmospheric dissipation into outer space, this is unlikely. More likely, carbon dioxide gradually accumulated in the atmosphere together with water vapor. This, in turn, contributed to a further rise in surface temperature and eventually the runaway greenhouse effect developed, resulting in the transport of increasingly large amounts of CO_2 and H_2O into the atmosphere, up to some equilibrium state. The latter is characterized by the relationships between mineral phases and volatiles on the surface, carbonate-silicate interaction in the upper layer of the planetary crust being the most important.

Let us emphasize that the above scenario is characteristic of a system with positive feedback and internal instability when initial perturbation is not suppressed but, on the contrary, fairly rapidly enhanced. Therefore, it is not coincidental that such an effect on Venus is called a runaway greenhouse effect leading ultimately to the high surface temperature. At this temperature, carbon dioxide turned out to be not bound in carbonates of sedimentary rocks, as on Earth (and, probably, on Mars), but was released into the atmosphere, giving rise to very high pressure. According to the model estimates, the amount of carbon dioxide locked in the Earth's sedimentary cover is comparable to the content of CO_2 in Venus's atmosphere.

Whereas the abundance of carbon dioxide on Venus can be explained in terms of a fairly simple equilibrium model (though the real geochemical processes involving sulfur-bearing and other compounds are much more complex), the situation with water is even more difficult to understand. Assuming "geochemical similarity" in evolution scenarios with the involvement of both endogenous and exogenous processes, the volumes of volatiles and, accordingly, the reserves of water on Venus, must have corresponded to the volume of the terrestrial hydrosphere, which is approximately 1,370 million km^3 or more than 1.37×10^{24} g. Meanwhile, water could not be preserved on the surface of Venus at a temperature above the critical one (647 K). The same is valid for aqueous solutions (brines) with a slightly higher critical temperature (675–700 K). Regarding the atmosphere, the amount of water it contains at an average relative water vapor content of 5×10^{-4} does not exceed 3.5×10^{20} g. This is considerably greater than the total water content in the terrestrial atmosphere (because of the much higher gas volume) but is less than the reserves of water in the hydrosphere by almost four orders of magnitude.

Thus, the key problem of evolution we address for Venus is whether the early Venus had water, if so, how and where it was stored, and when and how it was lost. We may reasonably assume that Earth and Venus received approximately the same reserves of volatiles, including water, through degassing from the mantle that accompanied the interior differentiation process. Heterogeneous accretion due to migration of planetesimals from the formation zone of the giant planets and their fall to inner planets as comets and asteroids of the carbonaceous chondrite class could contribute significantly to this process as well. Then one might expect Venus to have possessed an ocean at the initial phase of its evolution comparable in volume to the terrestrial one—an ocean that was somehow lost. The deuterium-to-hydrogen

D/H ratio measured in the Venusian atmosphere, which turned out to be larger than that in the Earth's atmosphere by two orders of magnitude, confirms this concept. This high deuterium enrichment can be explained by an efficient thermal escape of the lighter hydrogen isotope during the evaporation of the primordial ocean followed by dissociation of water molecules by solar UV radiation and its dissipation from the atmosphere. However, adoption of this mechanism requires an assumption that enormous masses of liberated hydrogen and oxygen were evacuated from the atmosphere and bound by surface rocks, which seems hardly realistic though, in order to accommodate a huge amount of hydrogen evacuation, the very efficient hydrodynamic blow off assist mechanism was suggested.

There is also a different viewpoint that Venus was initially formed as a "dry" planet, though the introduction of volatiles through the mechanism of heterogeneous accretion is assumed to play an important role as well. In this scenario, the water content was essentially constant throughout the geological history of Venus, remaining approximately at the current level while the efficiencies of heterogeneous accretion and hydrogen dissipation from the atmosphere were considerably lower. In such a case, the observed enrichment of the Venusian atmosphere with deuterium during the separation of isotopes in the process of their dissipation from the atmosphere can also be explained.

As we see, at present, it is difficult to answer the question of whether Venus had a primordial ocean and, thus, to choose between the above scenarios for its evolution due to limited experimental data to back the models. Both scenarios basically lead to the development of a runaway greenhouse effect and the present-day climatic conditions. At the same time, the possibility that Venus possessed a more clement climate at the earliest stages of its evolution must not be ruled out. The model postulates an existence of negative feedback stabilizing some equilibrium state rather than positive feedback enhancing temperature growth and being responsible for the runaway greenhouse effect. The negative feedback may be attributable to the atmosphere-lithosphere interaction pattern controlled by the carbonate-silicate cycle and, in principle, could exist on Venus in a certain period, before the luminosity of the Sun rose by about 30 % (when it passed to the main sequence on the Hertzsprung-Russell diagram, see Chap. 6) and the atmospheric humidity threshold had been overcome. The mechanism underlying this scenario was called the "wet" greenhouse effect. As calculations showed, the transition from the wet greenhouse effect to the runaway one might not happen if the increase in the planet's albedo to its current value occurred earlier than the rise in solar luminosity and, thus, compensated for the increase in the influx of solar energy. Venus would then possibly possess a wet carbon dioxide atmosphere with a pressure near the surface of only several atmospheres and a temperature of less than 100 °C. Such a planet could be suitable for the emergence of at least primitive life forms.

This consideration leads us to the idea that marginal possibilities containing both deterministic and stochastic components existed in the evolution of Venus, which eventually led to the runaway greenhouse scenario. The processes responsible for this scenario included the planet's formation at a certain distance from the Sun, orientation of its rotation axis in space, geological evolution and lithosphere-atmosphere interaction involving decomposition of carbonates and the loss of water, the retention of major and minor components in the atmosphere maintained by volcanic activity which provided high opacity, the circulation of sulfur-bearing compounds and halogens between the surface, atmosphere, and clouds, formation of the peculiar features of the planetary circulation, etc. Altogether, these processes led ultimately to the existing unusual nature of Venus. We cannot exclude, however, that a different evolutionary scenario of the planet as a nonlinear dissipative system could happen, provided that the essence and sequence of the above processes would somehow break and then the planet would pass to a stable state with different natural conditions.

Unfortunately, the current state of Venus seems more stable than the climatic state of Earth. Our main concerns are focused on mankind's invasion of the environment which eventually becomes comparable with the natural processes. Climatic changes could lead to uncontrolled heating involving carbonate decomposition, water evaporation, and, accordingly, a sharp rise in the atmospheric temperature and pressure. The case for Venus thus serves as a warning for Earth to avoid such unfavorable processes and dangerous scenarios. In this connection, it is pertinent to recall the statement quoted by Nobel Prize winner Ilya Prigogine in the book *The End of Certainty (2001)* that human existence consists of the continuous creation of unpredictable innovations. We hope that the innovations of our civilization will not dramatically impact the nature of our home planet.

Mars. Let us now address the problem of Mars evolution. As we saw, Mars took a completely different path of evolution from that of Venus. First of all, it is necessary to answer the question about mechanisms that exerted a critical impact on the pleasant natural conditions that presumably were on Mars at an early epoch, which is backed by the morphological features preserved on the planet, in particular, the ancient cratered terrain of the southern highlands which are thought to hold clues to the planet's early differentiation. Important geochemical clues to the early geological history of Mars have been provided by the SNC meteorite quite reliably associated with the Mars origin. Some clasts (e.g., NWA7533 found in northwest Africa) were identified with a Martian regolith breccia of unique composition (different from that of the Moon) containing zircon crystals[5] for which an age of 4,428,625 million years was measured. This confirms early crustal differentiation with contribution from impact melts and chondritic (CI) input. In addition, it implies that the Martian crust and its volatile inventory, formed during the first 100 million years of Martian history, coeval with the earliest crust formation on the Moon and the Earth. This allows us to unravel the geologic history of early Mars.

Assuming the integral albedo of early Mars was similar to what we postulated for early Earth and Venus, its equilibrium temperature would not exceed 220 K. At this temperature Mars would be able to retain only frozen rather than liquid water on its surface. However, there is widespread evidence of the presence of liquid water flows, meaning a considerably denser atmosphere and much more favorable climate

---

[5]Zircon crystals typically form during magma crystallization, and they are impervious to the impact melting processes influencing the composition of their host rocks.

than the one today. The hypothesis about the favorable climate of ancient Mars immediately poses the following fundamental question: Were its present-day natural conditions formed as a result of long and complex evolution, or did these changes occur suddenly on the geological time scale? The composition of SNC meteorites (in particular, the measured D/H ratio) imposes certain constraints on the standard model of the planet's thermal evolution. The model suggests the separation of an iron sulfide core shortly after the end of accretion, the differentiation of constituent material into shells (though less complete than that on Earth), convective heat transport in the mantle that provided early volcanism, generation of the magnetic field through the dynamo mechanism in the core, as long as it remained liquid, and collapse of the dense atmosphere. However, it is not clear whether this model is in accord with the formation of Mars from the most ancient primordial matter, similar to other terrestrial planets. The primordial matter is assumed to constitute chondrite meteorites for which the iron and silicon contents equal 1.7 which, in particular, corresponds to the terrestrial value. There is some evidence that this is not the case for Mars, which probably experienced disturbance 1,712,685 million years after formation, an evidence brought by SNC study. The original difference of evolution of Mars from that of Venus and the Earth was probably predetermined by its formation in a ring clump of a turbulent gas-dust cloud closer to Jupiter, which could influence the composition of its original matter. Answering this question would have important implications for resolving the problem of the solar system's origin.

Let us note that the nature of the present-day dry Mars cannot be explained by the fact that it is 0.5 AU further away from the Sun than Earth. More likely it should be associated with Mars's size, which is approximately half the Earth's. Accordingly, the mass of Mars is almost an order of magnitude lower than that of the Earth. As we said, this should have led to early depletion of radiogenic isotopes, which served as energy sources, and this would have determined its thermal history and geology, and eventually the collapse of the atmosphere. Mars began to cool, volcanism ceased, and the atmosphere reduced. In other words, a limited reserve of radiogenic isotopes rather than distance from the Sun dramatically impacted the Mars evolution.

This model seems fairly justified. At the same time, it is also hypothesized that Mars experienced a great catastrophe in its history: collision with a large asteroid that produced the Hellas basin of about 2,100 km across—the region in the Southern Hemisphere antipodal to the Tharsis upland, on which the largest shield volcanoes in the solar system originated. The hypothesis suggests that the release of magmatic rocks through the young Martian crust and volcanism on Mars were triggered by such a catastrophic event. In turn, this led to a radical change of the relief in the Northern Hemisphere and may have contributed to the formation of an atmosphere. The follow-up loss of the atmosphere and the transition of Mars from an evolutionary path similar to the Earth's in the first billion years to a completely different scenario of the follow-up evolution are associated with another, more recent, global catastrophe that was also produced by the collision with a large asteroid. Both these hypotheses are largely speculative. Nonetheless, the possibility of catastrophic events, likewise in the case of the Moon, should not be ruled out, especially bearing in mind the higher efficiency of impact bombardment near Jupiter.

We conclude this section by addressing, once again, the importance of extending geophysics studies focusing on Earth to all terrestrial planets. Turning to our immediate cosmic environment, we naturally wish not just to understand how the solar system came into being, but what is responsible for its stable configuration, as we state in Chap. 1. We should also understand what sets the Earth, with its unique natural conditions, apart from other terrestrial planets, primarily its nearest neighbors Venus and Mars, and what the limits are for the existing regulation mechanisms for the feedback on Earth to prevent unfavorable scenarios for its subsequent evolution. In other words, we want to know whether, for example, the accumulation of negative anthropogenic impacts on the surrounding natural medium will lead to a radical change of conditions associated with the bifurcation of the state of an open nonlinear dissipative system. An integration of Earth and planetary sciences aimed at a better understanding of the present, past, and future of the Earth based on a comparative planetology approach is called for to answer this key question. Concurrently, this must help to solve the cardinal problems of planetary cosmogony and, in particular, to impose much more stringent constraints on the range of parameters used in developing models for the origin and evolution of Earth and solar system as a whole.

Chapter 3
The Giant Planets

General View

The gaseous giant planets (Fig. 3.1) with their numerous satellites and rings are worlds that are completely different from the terrestrial planets. These planets also underwent differentiation of their interiors and, as a result, rather large rocky cores and extending gaseous-icy mantles emerged, the outer shell being referred to as an atmosphere (Fig. 3.2). Their effective temperatures range from 135 K (Jupiter) to 38 K (Neptune). Because of the continuing gravitational contraction of these massive planets, they radiate energy into space, shrinking and cooling; however, unlike stars, there is no unique relationship between their luminosity and mass (see Chap. 2). The heat released from their interiors exceeds the incident solar flux by about a factor of 2. It is interesting to note that this energy loss results in the continuing contraction of Jupiter by about 2 cm per year. The internal heat (which is many orders of magnitude more than that of the terrestrial planets) is responsible for many specific features of atmospheric circulation on these planets, including the system of zones and belts along latitudes, creating very strong shear turbulence and eddy formation. Brilliant examples of cyclonic mode eddies are the long-lived Great Red Spot (GRS) on Jupiter and the relatively short-lived Great Dark Spot (GDS) on Neptune (see Fig. 3.1). The basic atmospheric composition of the giants is hydrogen, helium, and hydrogen-bearing compounds, particularly water, ammonia, and methane.

In contrast to the terrestrial planets, which were formed near the Sun from the heavy high-temperature fraction of the original protoplanetary cloud material, the giant planets accumulated much lighter species corresponding to the cosmic abundance of elements. Accordingly, Jupiter and Saturn, composed mainly of the most abundant elements hydrogen and helium, have a nearly solar composition. Only the very massive planets with a threshold mass $M > 10\ M_E$ (M_E is the mass of Earth) were capable of accreting these light gases from the protoplanetary cloud and retaining them. Although direct measurements of the chemical composition of the interiors are not possible, we may reasonably assume that it is close to the measured

© Springer Science+Business Media New York 2015   93
M.Ya. Marov, *The Fundamentals of Modern Astrophysics*,
DOI 10.1007/978-1-4614-8730-2_3

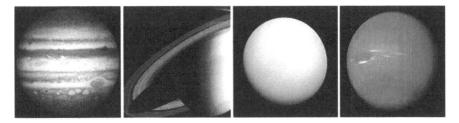

Fig. 3.1 Giant planets Jupiter, Saturn, Uranus, Neptune (not in scale) (Courtesy of NASA)

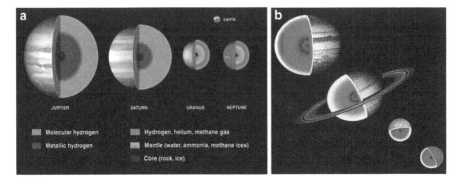

Fig. 3.2 Internal structure of the giant planets. (**a**) The constituent gases and ices ratio mostly defines the giant planets interiors; (**b**) Gaseous Jupiter and Saturn and icy Uranus and Neptune (from *upper left* to *bottom right*) (Adapted from Wikipedia)

atmospheric composition and the protoplanetary disk matter in general. However, the atmospheres of the giant planets are enriched with heavier elements, in particular carbon, whose abundance exceeds the solar one by a factor of 3 on Jupiter and even by a factor of 30 on Uranus and Neptune. This may be explained by their being bombarded by planetesimals for a long time after completion of the main accretion phase.

The giant planets are gas–liquid and icy bodies with a low mean density (despite the enormous pressure in their interiors), without any solid surface, and with an outer gas layer called an atmosphere. Detailed information about the atmospheric composition of these planets is summarized in Table 3.1. Clearly, the abundances of the "parent" molecules are controlled by thermochemical equilibrium and atmospheric transport. Under low temperatures at the visible levels only the most volatile molecules shown in the table may survive there and interact with solar ultraviolet radiation; most equilibrium constituents condense deep in the atmosphere. Therefore, photochemistry on the outer planets is centered on small amounts of volatile molecules that contain elements such as carbon, nitrogen, phosphorus, and sulfur, as well as different hydrocarbons.

Table 3.1 Composition of the atmospheres of the giant planets

Gas	Jupiter[a]	Saturn	Uranus	Neptune
H_2	86.4 ± 0.3 %	88 ± 2 %	$\sim 82.5 \pm 3.3$ %	$\sim 80 \pm 3.2$ %
^4He	13.6 ± 0.3 %	12 ± 2 %	15.2 ± 3.3 %	19.0 ± 3.2 %
CH_4	$(1.81 \pm 0.34) \times 10^{-3}$	$(4.7 \pm 0.2) \times 10^{-3}$	~ 2.3 %	$\sim 1-2$ %
NH_3	$(6.1 \pm 2.8) \times 10^{-4}$	$(1.6 \pm 1.1) \times 10^{-4}$	< 100 ppb	<600 ppb
H_2O	520^{+340}_{-240} ppm	2–20 ppb		
H_2S	67 ± 4 ppm	<0.4 ppm	<0.8 ppm	<3 ppm
HD	45 ± 12 ppm	110 ± 58 ppm	~ 148 ppm	~ 192 ppm
$^{13}CH_4$	19 ± 1 ppm	51 ± 2 ppm		
C_2H_6	5.8 ± 1.5 ppm	7.0 ± 1.5 ppm		
PH_3	1.1 ± 0.4 ppm	4.5 ± 1.4 ppm		
CH_3D	0.20 ± 0.04 ppm	0.30 ± 0.02 ppm	~ 8.3 ppm	~ 12 ppm
C_2H_2	0.11 ± 0.03 ppm	0.30 ± 0.10 ppm	~ 10 ppb	60^{+140}_{-40} ppb
HCN	60 ± 10 ppb	<4 ppb	<15 ppb	0.3 ± 0.15 ppb
HC_3N			<0.8 ppb	<0.4 ppb
C_2H_4	7 ± 3 ppb	~ 0.2 ppb[b]		
CO_2	5–35 ppb	0.3 ppb	40 ± 5 ppt	
C_2H_6			10 ± 1 ppb	$1.5^{+2.5}_{-0.5}$ ppm
CH_3C_2H	2.5^{+2}_{-1} ppb	0.6 ppb	0.25 ± 0.03 ppb	
CO	1.6 ± 0.3 ppb	1.4 ± 0.7 ppb	<40 ppb	0.65 ± 0.35 ppm
CH_3CN				<5 ppb
GeH_4	$0.7^{+0.4}_{-0.2}$ ppb	0.4 ± 0.4 ppb		
C_4H_2	0.3 ± 0.2 ppb	0.09 ppb	0.16 ± 0.02 ppb	
AsH_3	0.22 ± 0.11 ppb	2.1 ± 1.3 ppb		

[a] ^3He 22.6 ± 0.7 ppm, Ne 21 ± 3 ppm, Ar 16 ± 3 ppm, Kr 8 ± 1 ppb, Xe 0.8 ± 0.1 ppb
[b] Assuming a total stratospheric column density of 1.54×10^{25} cm^{-2}
Credit: K. Lodders and B. Fegley; P. Mahaffy; S. Atreya; K. Lodders; M.Wong

Generally, hydrogen-bearing compounds such as water, ammonia, and methane ("ices"), which entered mainly into the composition of Uranus and Neptune in various ratios, condensed most efficiently at very low temperatures in the solar system regions between 10 and 30 AU. Even heavier elements and compounds concentrated only in the cores of the giant planets onto which lighter elements and compounds accreted. In Jupiter and Saturn, they account for a relatively small fraction of the planetary mass, only from 3 % to 15 %, and reach 80–90 % in Uranus and Neptune. This is apparently explained by a lower content of primordial gases in this disk region at the later stage of these planets' evolution.

The observed differences in chemical composition generally satisfy the models of giant planet formation including the original accumulation of a massive ($\sim 10–30 M_E$) rocky core. This is followed by the accretion of gases on the core from the protoplanetary cloud which is composed of the most abundant hydrogen and helium. The earlier formed massive Jupiter and Saturn accreted the lion's share of the primordial gases; Uranus and Neptune reached their critical masses to accrete the remaining gases much later. That later date together with the lower temperatures in the zones of their formation influenced their chemical composition, in particular the enrichment of heavier elements. The process of core accumulation and accretion was estimated to occupy ~ 100 million years, the accretion phase being much shorter. However, we must mention a caveat of this rather simple model in the case of icy planets Uranus and Neptune, whose genesis is still not fully understood but which has been hard to fit into existing timelines. Their genesis is presumed in the prevailing Nice model to have begun closer to the Sun in the Jupiter-Saturn feeding zone and then migrated outward. The currently existing resonances in the Kuiper Belt, particularly 2:5 and some other resonances (see Chap. 4), which could set up in due course of Neptune's migration towards its present position, as well as chaotization of the Kuiper Belt objects and possibly late heavy bombardment, support this idea.

We will first briefly characterize these planets, then discuss the unique dynamics of their atmospheres, and finally, address the fascinating world of their numerous satellites and rings with their peculiar properties.

Jupiter

Jupiter (the name for the Roman supreme god, known as Zeus in Greek mythology) is the fifth planet from the Sun, orbiting the Sun at about 5 AU. Because of its great size and mass it is often called the king of planets. The semimajor axis of Jupiter's elliptical orbit is 778.57 million km (5.2 AU), and the eccentricity is 0.05. Jupiter lies 740.52 million km (4.95 AU) from the Sun at perihelion and 816.62 million km (5.46 AU) at aphelion. Its sidereal period of revolution (relative to the stars) is 11.86 years (Jupiter's year), while the synodic period of revolution (relative to the Earth) is only 398.88 days. Jupiter's orbit is inclined to the plane of the ecliptic by 1.03°, and the inclination of its axis of intrinsic rotation is 3.13°, which means that there are essentially no seasonal changes on the planet. In the night sky Jupiter looks like a very bright star, yielding (at the periods of great opposition, repeating every 13 months) only to the Moon and Venus. Its visual stellar magnitude changes from -2.94^m at the opposition to -1.61^m at the aphelion in its orbit. The integral spherical albedo of Jupiter is 0.34, very close to that of Earth.

Jupiter's equatorial radius R_J is 71,492 km and its polar radius is 66,854 km, making it more than ten times larger than Earth (Fig. 3.3). The large difference between the equatorial and polar radii (the oblateness is 0.065) is caused by the very fast rotation of Jupiter around its axis—less than 10 hours, faster than all the other planets and more than twice as fast as Earth. Thus, Jupiter has an approximately 10-hours day. Let us note that the equatorial regions rotate faster than the polar ones by 5.5 minutes (9.83 hours as compared to 9.92 hours). This is called differential rotation and is related to the atmospheric processes. Because of its mostly gaseous composition, Jupiter's density is very low, only 1.33 g/cm^3 (very similar to that of the Sun). Nonetheless, its mass (1.90×10^{27} kg) is 318 times more than the Earth's mass, 2.5 times more than the mass of all the other planets together, and nearly

Fig. 3.3 Comparison of sizes
of Jupiter and Earth
(montage) (Adapted from
Wikipedia)

0.1 % the mass of the Sun. Jupiter is primarily responsible for the common gravity center of the solar system (the barycenter) not coinciding with the center of the Sun, but being shifted by about a solar radius towards Jupiter. The mass of Jupiter is only about two orders of magnitude short of the mass needed for the thermonuclear reactions in the hydrogen-deuterium ($M = 0.11\ M_O$) and hydrogen-helium ($M = 0.8\ M_O$) cycles, respectively, to begin. Therefore, it is believed that Jupiter occupies a position near the lower boundary of stellar evolution, being close to a brown dwarf (see Chap. 6). The enormous acceleration due to gravity on Jupiter of 24.79 m/s^2 (2.535 g) gives rise to a large value for the escape velocity of 59.5 km/s. The scale of the natural phenomena on Jupiter such as atmospheric eddies, storms, lightning, and polar auroras exceeds those on Earth, with a magnetosphere an order of magnitude higher.

The chemical composition of Jupiter is known mainly from measurements relating to the upper levels of the gaseous shell, its atmosphere. Besides molecular hydrogen (~90 %) and helium (~10 %), which make up the bulk composition of the planet, different admixtures have been found: carbon, nitrogen, sulfur, oxygen, and phosphorus in such compounds as water, methane, ammonia, hydrogen sulfide, and phosphine. There are also the noble gases neon, argon, krypton, and xenon, their abundance exceeding those on the Sun (except neon), as well as some hydrocarbons and possibly organics. Minor constituents like sulfur and phosphorus are mainly responsible for the rich coloration of Jupiter's visible disk attributed to the upper clouds. Bands of different colors on the disk of Jupiter are the most prominent feature of the planet. They reflect the very complicated dynamics of its atmosphere, including strong winds of variable directions, convective cells, well-developed turbulence, and enormous eddies like the Great Red Spot (GRS) and white ovals of different sizes, which we will discuss below. Lightning of enormous size and power is observed in the eddy cores: thunderstorms occupy areas of thousands of kilometers, and the power of the lightning exceeds terrestrial standards by three orders of magnitude.

The model of Jupiter's structure and, specifically, its interior (see Fig. 3.2) is based on the assumption that the planet is in both hydrodynamic and thermodynamic equilibrium. Three main regions are distinguished: upper shell including atmosphere, mantle, and core, although with no distinct boundaries between them. In the atmosphere, the tropopause is conditionally chosen at 50 km where the pressure $P = 100$ mbar and the temperature $T = 120$ K. Temperature and pressure progressively grow downward with the involvement of phase transitions. The structure of the clouds includes three layers of condensates located in the atmosphere between the pressure levels 1–20 atm and the temperature range 200–450 K. The upper layer is composed of ammonia, the middle layer is composed of ammonium hydrosulfide (NH_4HS), and the lower layer is composed of water crystals and droplets. Below the clouds, the first phase transition occurs at ~ 20,000 km at a temperature of ~ 6,000 K when hydrogen transfers from a gaseous to a liquid state. Then, under a pressure exceeding one million atmospheres (1 Mbar), hydrogen transfers to a metallic state, and this region extends for about 45,000 km. The central part of Jupiter is occupied by a rocky (iron-stone) core of about 1.5 Earth's radius in size and nearly 10 times the Earth's mass. The temperature here is ~ 25,000 K and the pressure is ~ 80 Mbar.

Jupiter, like the other giant planets except Uranus, possesses an internal heat source caused by the continued shrinking of the planet since the time of its accumulation and resulting in a gravitational energy release. This internal heat source exceeds the energy Jupiter receives from the Sun nearly twofold, and this is why its equilibrium (effective) temperature ($T = 135$ K) is higher compared to what it would be if the Sun were the sole energy source. The internal heat release results in Jupiter's contracting by about 2 cm per year.

The region of metallic hydrogen has a high conductivity and is mainly responsible for the generation of Jupiter's powerful magnetic field, comprising a mainly dipole structure though with significant contributions from multipole (quadrupole, octupole) components. The axis of the magnetic dipole is tilted to the axis of the planet's intrinsic rotation by $10.2°$, and the field strength is about 4 oersted (Oe) at the equator and 11–14 Oe at the poles, more than an order of magnitude higher than the Earth's. The magnetic field with the trapped particles (electrons and protons) of the solar wind plasma forms the great magnetosphere of Jupiter, which extends about 20 R_J of the planet on the day side and on the night side reaches well beyond Saturn (Fig. 3.4). Radiation belts within the magnetosphere possess enormous energy exceeding that of the Earth's. Their emission in radio wavelengths (5–40 MHz) is produced by the synchrotron mechanism of electron interaction with the magnetic field. Other interesting phenomena are the interaction of Jupiter's magnetosphere and the ionosphere, resulting in nearly permanent auroras in the polar regions, and especially the magnetosphere's interaction with the torus formed along the satellite Io's orbit at 5.9 R_J and composed of ions of sulfur and oxygen (S^+, O^+, S^{2+}, O^{2+}) that are supplied by Io's volcanoes. Such interactions significantly influence the magnetosphere dynamics of Jupiter.

Jupiter has numerous satellites and a ring system. By now 67 satellites are known, the most spectacular being the four largest moons discovered by Galileo and thus called the Galilean satellites. Their unique properties are discussed later. Seven

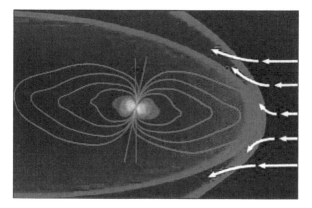

Fig. 3.4 Magnetosphere of Jupiter. Its size is about 20 R_J of the planet from the day side and it extends at the night side well beyond Saturn orbit at 10 AU. Trapped charge particles form enormous radiation belts within the magnetosphere having great radiation hazard (Adapted from Wikipedia)

spacecraft have visited Jupiter en route to other planets, the most remarkable being Pioneers 10 and 11 in 1973–1974 and Voyager 1 and 2 in 1979 flybys. In 1989, Galileo was placed into orbit around Jupiter and also entered its atmosphere with the probe jettisoned from the spacecraft, which descended to about 22 km and made in situ measurements of the atmospheric parameters. Besides the planet, the orbiter also studied Jupiter's satellites, specifically the Galilean satellites, and the planet's ring system. In August 2011, NASA launched a new spacecraft, Juno, which will continue the study of Jupiter from a polar orbit. It is specially targeted to investigate the huge magnetosphere of the planet and related phenomena, such as the polar auroras. NASA, ESA, and RSA (ROSCOSMOS) are planning to launch spacecraft to study in detail the Galilean satellites Europa and Ganymede. The preliminary name of the study is the Europa Jupiter System Mission (EJSM)-Laplace project.

Saturn

Saturn (in Roman mythology the god of agriculture) is the sixth planet from the Sun, located twice as far from the Sun as Jupiter at a mean distance of 9.58 AU (Figs. 3.1 and 3.5). It is smaller than Jupiter; its equatorial radius ($R_s = 60,268$ km) is about 11,000 km less. The orbital characteristics of Saturn are as follows: semi-major axis of the orbit is 1.433 billion km, eccentricity is 0.06, inclination of the orbit to the plane of the ecliptic is 2.46°. The sidereal period of revolution (relative to the stars) is 29.46 years (Saturn's year), and the synodic period of revolution (relative to the Earth) is 378.09 days. Because of its elliptical orbit, Saturn approaches the Sun as close as 1.353 billion km (9.05 AU) at perihelion and moves out to 1.513 billion km (10.12 AU) at aphelion. The distance between Saturn and Earth changes

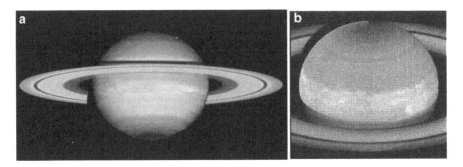

Fig. 3.5 (**a**) Saturn in contrast-enhanced artificial colors. The *blue, red-* and- *orange*, and *green* colors correspond to the main clouds, upper-level clouds, and the overcloud haze, respectively. Three main rings are distinguished in the rings surrounding Saturn: C, B, and A. There is the Cassini division between the A and B rings. A Hubble Space Telescope image. (**b**) Saturn atmosphere. The band structure of the cloud layer attributable to planetary circulation and the turbulized structure of the clouds near the equator are clearly distinguishable (Image from the Cassini spacecraft. Credit: NASA and ESA)

from 8.0 AU to 11.1 AU; the mean distance in the opposition is 1.28 billion km. In contrast to Jupiter, the inclination of the axis of intrinsic rotation of Saturn is quite high (26.73°); therefore, there are significant seasonal changes on the planet. The orbital motions of Saturn and Jupiter are locked in a 2:5 resonance.

The visual stellar magnitude of Saturn changes from -0.24^m to $+1.47^m$, so its brightness is weaker than that of Jupiter, though its integral spherical albedo is just the same (0.34). Saturn rotates around its axis nearly as fast as Jupiter—one turn in 10.57 hours (the Saturnian day)—which results in a difference in the equatorial and polar radii of the planet R_S (60,268 km vs. 54,364 km) and a large polar oblateness (0.098), exceeding that of Jupiter. This means that Saturn is the most compressed planet in the solar system. The density of the gaseous Saturn is less than Jupiter's and even less than the density of water; it is only 0.69 g/cm^3. No other planet has such a low density. If we found a tub of enormous size, filled it up with water, and put Saturn in it, the planet would float! The total mass of the planet yields to that of Jupiter; it is 5.68×10^{26} kg (95 times Earth's mass). The acceleration due to gravity is 10.44 m/s^2, and the escape velocity is 35.5 km/s.

Our knowledge about Saturn and the system of its satellites and rings was greatly advanced by space exploration. Four spacecraft have visited the planet: Pioneer 11 (in 1979), Voyagers 1 and 2 (in 1980–1981), and Cassini-Huygens (since 2004), the latter returning back to Earth the most detailed information. The extremely successful Cassini-Huygens mission allowed us to land on the surface of Saturn's largest satellite Titan for the first time, and to make the first in situ measurements and study of its exotic nature. The Cassini mission is scheduled to run into 2017. NASA and ESA also have a proposal for a joint mission to Saturn to make an in-depth study of the planet and especially its most interesting satellites Titan and Enceladus. This expedition is tentatively called the Titan Saturn System Mission.

Like Jupiter, Saturn consists mostly of hydrogen and helium, although its helium mixing ratio is about three times smaller. The minor components are primarily water, methane, and ammonia, which add to the composition of the clouds of ammonia, ammonium hydrosulfide (NH_4HS), and frozen water. On Saturn's visible disk a system of zone and belts is also distinguished, though it is less pronounced than Jupiter's and dimmer in color (see Fig. 3.5). Various features observed in the atmosphere move with different velocities. Similar to Jupiter, strong lightning activity is present in Saturn's atmosphere. At high latitudes peculiar hexagonal-shaped structures form. Powerful storms thousands of kilometers across and giant eddies like the Large White Oval gather, though they appear irregularly and have a limited lifetime. A more detailed description of Saturn's atmospheric dynamics as well as a discussion of the nature of its peculiar satellites and rings will be given below.

Saturn's interior is generally similar to that of Jupiter (see Fig. 3.2), hydrogen and helium being mainly responsible for its structure. Only the upper layer of the planet associated with the atmosphere is accessible to observation. The temperature and pressure progressively grow from $T = 134$ K at pressure level $P = 1$ bar downward, meeting the first hydrogen phase transition from gaseous to liquid state and then to metallic state when the pressure reaches $P = 3$ Mbar. According to the model, there is a massive rocky (iron-stone) core in Saturn's center having radius ~12,000 km and temperature ~12,000 K. Its mass may reach ~20 Earth masses. Continuing gravitational compression of the planet produces a large inner thermal flux radiated into outer space, which exceeds the solar radiation flux Saturn receives from the Sun by 2.5 times.

Saturn possesses a significant magnetic field, its strength being very similar to that of Earth (0.21 oersted vs. 0.35 oersted) at the equator. The field is generated by the dynamo effect in the metallic hydrogen shell and outer metallic core. It has a generally dipolar structure, with the axis of the magnetic dipole and the axis of intrinsic rotation nearly coinciding. This results in a remarkable symmetry of the structure of Saturn's magnetosphere. The magnetosphere fills up with both solar wind particles and ions produced by the planet's satellites, Enceladus being the most dominant contributor. Numerous features of magnetospheric activity and its interactions with the solar wind appear to be caused by the magnetic field line reconnections. A powerful aurora is observed nearly permanently at both poles, exhibiting asymmetry of the plasma processes involved. The boundary of the magnetosphere interaction with the solar wind (magnetopause) is located at about 22 R_S, while the magnetospheric tail extends well beyond hundreds of equatorial radii.

Uranus

Uranus (the god of the sky in Greek mythology) is the seventh planet from the Sun (see Fig. 3.1). Unknown to ancient civilizations, it was discovered only in 1781 with the telescope by the great astronomer William Herschel, although he first mistook it for a comet.

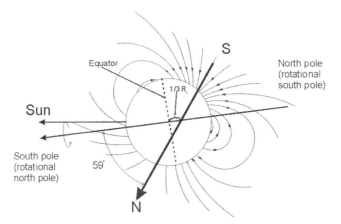

Fig. 3.6 Uranian magnetic field dipole position relative to the axis of rotation. Besides their different orientation there is also magnetic dipole off set from the geometric center of the planet (Adapted from Wikipedia)

Uranus orbits the Sun at a distance of about 20 AU. It has a moderately elliptical orbit (the eccentricity is 0.04) with a semimajor axis of 2.88 billion km (19.23 AU) and an inclination to the plane of the ecliptic of 0.77°. Uranus lies 2.75 billion km (18.38 AU) from the Sun at perihelion and 3.00 billion km (20.08 AU) at aphelion. Its sidereal period of revolution (relative to the stars) is 84.32 years (Uranus's year), while the synodic period of revolution (relative to the Earth) is only 369.66 days. The inclination of the axis of intrinsic rotation is 97.77°, which means that the axis lies nearly in the plane of Uranus's orbit. This means that the planet faces the Sun successively with both poles (each pole is illuminated and screened for 42 years each, peaking at the solstices) and middle latitudes. The equator is illuminated for short times during equinoxes when "normal" diurnal variations occur. All this results in many peculiar phenomena on the planet and, specifically, solar energy redistribution over the globe and interaction of the magnetic field with solar plasma (Fig. 3.6). The reason for Uranus's unusual position in space remains enigmatic; to explain this phenomenon, the hypothesis of a catastrophic collision with a large body in the early epoch dramatically tilting the axis of rotation was proposed.

In the night sky Uranus looks like a dim star. Its angular size is about 4″, and its visual stellar magnitude changes from 5.3^m at opposition to 5.9^m at the aphelion of its orbit. Its integral spherical albedo is 0.3. In size Uranus yields only to Jupiter and Saturn, in terms of mass it is similar to Neptune. Its equatorial radius R_U is 25,559 km, its polar radius is 24,973 km, and its polar oblateness is 0.02. Uranus rotates around its axis more slowly than Jupiter and Saturn: the period of intrinsic rotation (the Uranian day) is 17 hours 14 minutes (0.72 of Earth's day). The bulk composition of Uranus is a mixture of gases and ices, giving a total mass of 8.68×10^{25} kg (14.5 times the Earth's mass), the mean density is 1.27 g/cm³, the acceleration due to gravity (at the equator) is 8.2 m/s², and the escape velocity is 21.3 km/s. Uranus receives very little energy from the Sun: ~1/400 of the solar constant at Earth's orbit.

Voyager 2 was the only spacecraft to fly by Uranus, in 1986. The transmitted images of the planet have since been complemented by observations by the Hubble Space Telescope. Unlike other giants, Uranus looks featureless, shrouded with clouds and exhibiting only limited traces of atmospheric motion and storms. However, strong winds measured in the clouds brought some evidence about planetary circulation involving atmospheric super-rotation. In the Southern Hemisphere two prominent features were distinguished: a narrow ring in the middle latitudes and the dark polar hood. Distinct differences in the cloud structures of both hemispheres were found, the clouds having a limited lifetime. A big dark spot was noticed on the disk, though its origin and nature are not clear. As we have said, seasonal changes are well pronounced, changing four times during the 84-Earth-year-long Uranian year and recurrently illuminating the Northern and Southern Hemispheres and equatorial regions. The atmosphere of Uranus is less dynamic than that of Neptune, though they are very similar in size and mass.

Hydrogen and helium compose most of the upper shell of Uranus including its cloudy atmosphere, helium abundance amounting to a high 15 % (by volume). There is significant methane abundance—more than 2 %. Interestingly, the fraction of helium by mass is 0.26 ± 0.05, which is close to the protostar helium mass fraction (0.28 ± 0.01) and favors the idea that planetary composition is closer to the original one in the outer parts of the solar system. The clouds on Uranus appear to have a multilayered structure and consist of methane, ammonia, ammonium hydrosulfide (NH_4HS), water, and hydrogen sulfide (H_2S). Methane is responsible for the green-blue (aquamarine) color of the planet. There are also traces of hydrocarbons more complicated than methane (probably products of photolysis in the upper atmosphere), CO_2, and CO. At a pressure level of 1 bar the temperature is 76 K; this level is conditionally referred to as the surface of the planet (there is actually no solid surface at all). The temperature at the tropopause at a 0.1 mbar pressure level drops down to 49 K (-224 °C). This is the lowest temperature of all the solar system planets, even lower than that of more distant Neptune, and it is caused by the very specific thermal balance on Uranus. Another interesting peculiarity of Uranus is its unusual upper atmosphere, specifically, a hot thermosphere that exhibits steady temperature growth with height. It is composed of molecular and atomic hydrogen and is extended by more than two radii of the planet. The ionosphere of Uranus lying within the thermosphere is even denser than those of Jupiter and Saturn.

The composition of the planet's interior (see Fig. 3.2) is represented by the high-temperature modifications of hydrogen-bearing ices (water, methane, and ammonia), that is, a hot and dense liquid made of these compounds (although they are called "ices" they do not fit the usual definition of ice, being a fluid mix). These ices (sometimes also called an "ammonia-methane ocean" but containing water as well) comprise up to 90 % of the total mass of Uranus (9–13 Earth masses). This is why Uranus and Neptune are classified as "icy giants," in contrast to the "gaseous giants" Jupiter and Saturn. Modeling suggests a rocky core in the center of the planet of less than 3–4 % of its total mass. Thus, the three-layer model of Uranus is accepted: a small rocky core whose radius is about 5,000 km with a density of 9 g/cm^3, an extended icy mantle, and a thin gaseous hydrogen-helium envelope.

The pressure and temperature at the boundary between core and mantle are estimated to be 8 Mbar and 5,000 K, respectively.

The magnetosphere of Uranus is defined by a magnetic field dipole that is not aligned with the axis of the intrinsic rotation of the planet (Fig. 3.6) and represents a very interesting phenomenon. The axis of the magnetic dipole is inclined relative to the axis of rotation by 59° and offset from geometrical center towards the South Pole by about one-third of the planet's radius. This results in an asymmetry of the magnetic field, its strength varying from 0.1 gauss (G) in the Southern Hemisphere to 1.1 G in the Northern Hemisphere, unlike on other planets. This could be explained by different mechanisms of the magnetic field generation in Uranus (and also Neptune): in these planets it could occur in a liquid conductive layer (for example, in the ammonia-methane ocean) in the mantle rather than in the core. Nonetheless, the Uranus magnetosphere generally has a quite regular structure despite a rather peculiar interaction with the solar wind plasma. Its magnetopause is located at 18 R_U, and there are well-developed radiation belts and a long magnetic tail. The radiation belts contain energetic protons and electrons and also some fraction of molecular hydrogen. Uranus's satellites exert a strong influence on the magnetosphere, forming cavities in the magnetic field structure. Polar auroras are clearly seen as bright arcs around the magnetic poles.

Neptune

Neptune (see Figs. 3.1 and 3.7) is the outermost planet of our solar system (the eighth planet from the Sun). Its name honors the god of the sea in Roman mythology (the Greek equivalent is Poseidon). The planet's existence was predicted by mathematical calculations made independently by Urbain Le Verrier and John

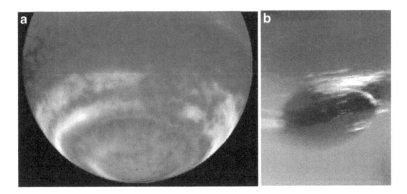

Fig. 3.7 (a) Neptune with the system of bands attributable to zonal circulation at the cloud level. The aquamarine color of the planetary disk is explained by the methane absorption of the red spectral region; (b). The Great Black Spot (*GBS*) on Neptune (Courtesy of NASA)

Couch Adams; the event is often referred to as "the discovery at the tip of a pen." It was soon confirmed by telescopic observations when Johann Gottfried Galle discovered the planet very close (1°) to the predicted position in September 1846.

Neptune orbits the Sun at a mean distance of about 30 AU. The semimajor axis of its orbit is 4.50 billion km (30.10 AU), its eccentricity is 0.01, and its inclination to the plane of the ecliptic is 1.77°. With a nearly circular orbit, Neptune lies 4.45 billion km (29.77 AU) from the Sun at perihelion and 4.55 billion km (30.44 AU) at aphelion. Its sidereal period of revolution (relative to the stars) is 164.79 years (Neptune's year), while its synodic period of revolution (relative to the Earth) is 367.49 days. The inclination of the axis of intrinsic rotation is 28.32°; this means that the planet experiences substantial seasonal changes, similar to those on Earth and Mars. However, each season lasts 40 years!

Neptune is inaccessible to visual observations: its angular diameter is only $2.2''$–$2.4''$, its visual stellar magnitude is between $+8.0^m$ and $+7.7^m$, and its albedo is 0.29. Neptune is a bit smaller than Uranus in size, while it slightly exceeds Uranus in mass. Its equatorial radius $R_N = 24,764$ km, its polar radius is 24,341 km, and its polar oblateness is 0.02. Neptune rotates around its axis faster than Uranus: the period of its intrinsic rotation (Neptune's day) is 15 hours 58 minutes (0.67 of Earth's day). Similarly to Uranus, the bulk composition of Neptune is a mixture of gases and ices. With a larger fraction of heavier (icy) components, its mean density is 1.64 g/cm^3. This explains why Neptune is more massive than Uranus, which is larger in size. Neptune's total mass is 1.02×10^{26} kg (17.1 times the Earth's mass), its acceleration due to gravity is 11.15 m/s^2, and its escape velocity is 23.5 km/s.

As we will see below, the powerful gravitational field of giant Neptune exerts a strong influence on the orbital behavior of icy dwarf planets and other small bodies that populate the nearby (located at about 40 AU) Kuiper Belt. In particular, it is responsible for the belt's structure, stability, and secular resonances. Secular resonance means that commensurability of the periods of revolution of a given body and Neptune have a simple integer relation with a shared period, which is expressed as the ratio of natural numbers (e.g., 1:2, 2:3, 3:5). Note that this mechanism is comparable with Jupiter's influence on the main asteroid belt located between the orbits of Jupiter and Mars.

Neptune is very similar to Uranus, both being icy giants that share a certain composition and structure of the atmosphere and interior. However, the meteorology of Neptune is much more complicated than that of Uranus. Neptune's atmosphere exhibits more dynamical properties and various patterns in the clouds. This was clearly seen on the images transmitted by Voyager 2, which flew by Neptune in 1989, and also by Hubble Space Telescope observations. Strong variable winds in the clouds (even more severe than those on the gaseous giants Jupiter and Saturn) and great storms were measured on Neptune, which determine its weather. The most spectacular was the giant eddy imaged by Voyager 2 which resembled the Great Red Spot on Jupiter and thus was called the Great Dark Spot (Fig. 3.7b). It was originally assumed that this anticyclone would also have a very long lifetime, but it disappeared within five years, as the Hubble observations showed. Instead, different large and small dark spots appeared, providing new evidence about the instability of Neptune's atmosphere, the variability of the cloud formations, and the high cyclonic activity.

Neptune receives much less energy from the Sun than Uranus: ~1/900 of the solar constant (incident energy) at the Earth's orbit. More efficient in the planet's thermal energy balance is its internal heat. Whereas Uranus radiates in space only about 1.1 times more heat than it receives from the Sun, Neptune's radiation exceeds the solar incident flux by a factor of 2.6. This means that the internal heat flux from Neptune's interior supplies an additional 160 % of the total energy available on the planet. It hardly seems possible that Neptune's inner heat is caused by the continuing compression of the planet since the time of accumulation as in the case of Jupiter and Saturn. Rather it could be related to radiogenic decay in the interior or chemical processes with involvement of hydrocarbons. Peculiarities of the energy exchange determine a wealth of properties of the Neptunian atmosphere and its dynamics, which is much stronger and more pronounced than in the case of Uranus. Neptune's atmosphere exhibits a strong differential rotation that is expressed more explicitly than in the atmospheres of all other planets that show a similar effect. Its period of intrinsic rotation (about 16 hours) was deduced from the measured rotation of the magnetic field fixed to the planetary body. However, the equatorial regions of the atmosphere rotate more slowly (with a period of ~ 18 hours), whereas the polar regions rotate much faster (period ~ 12 hours). This strongly affects the patterns of planetary circulation, specifically latitudinal variations of the wind velocities.

Like Uranus, hydrogen and helium compose the upper shell of Neptune and its cloudy atmosphere, the helium mean abundance being even higher: 19 % by volume. There is also significant methane abundance, about 1.5 %. Intense absorption by the methane bands of the red-yellow part of solar rays in the visible spectrum explains the bright azure color of the planet. Neptune's clouds have a multilayered structure and are composed of methane, ammonia, ammonium hydrosulfide (NH_4HS), hydrogen sulfide (H_2S), and water. There are also traces of hydrocarbons more complicated than methane (ethane, acetylene) that appear to be produced by the process of photolysis in the upper atmosphere. Their abundance in the equatorial regions is hundreds of times more than that over the poles. Obviously, photochemical processes producing complex hydrocarbons in the atmosphere and clouds are triggered in the methane atmosphere of Neptune even at a comparatively low flux of solar ultraviolet photons at the distance of 30 AU. While there is no solid surface on the planet, the pressure level of 1 bar and temperature 72 K is conditionally referred to as the "surface" or rather as the reference level. The tropopause is at the 0.1 mbar pressure level where the temperature drops down to 55 K. It quickly rises upward reaching an unusually high value (~750 K) in the thermosphere. The dissipation of inner gravitational waves from the interior may possibly contribute to the upper atmosphere heating.

Similar to Uranus, the composition of Neptune's interior (see Fig. 3.2) appears to be represented by the high-temperature modifications of hydrogen-bearing ices (water, methane, and ammonia), which are only conditionally called "ices," as we said above. Ices comprise more than 90 % of the total mass of Neptune. A model of the Neptunian interior starting from the outer gaseous shell (atmosphere) includes a deep icy mantle, where the temperature increases to ~ 2,000–5,000 K, and a rocky (iron-silicate) core in the center, where the pressure amounts to ~ 7 Mbar and the temperature is ~ 6,000–7,000 K. The core is estimated to be 1.2 times as massive as Earth.

The magnetosphere of Neptune generally resembles that of Uranus. It looks like a "tilted rotator," the axis of the magnetic dipole being deflected from the axis of intrinsic rotation by 47° and offset from the geometric center of the planet. Therefore, a cone-shaped configuration of the rotating magnetic field of the planet forms that dramatically influences the magnetospheric structure and the patterns of its interaction with the solar wind plasma. Its magnetic field of 1.4 G could originate from convective flows in the electroconductive liquid ammonia-methane mantle that result in the mechanism of a magnetohydrodynamic dynamo. The field is quite complicated, and its generally dipolar structure is strongly distorted by a quadrupole component. The boundary of solar wind interaction with the magnetosphere (the magnetopause) is located at a distance of ~ 25 R_n, and the magnetic tail extends by more than 100 R_N in the antisolar direction.

Atmospheric Dynamics

We shall now address in more detail some general properties of atmospheric dynamics responsible for the climate and meteorology on the giant planets. Their common features and their differences, as revealed by observations and modeling, are of great importance in order to better understand the key physical mechanisms and hydrodynamic processes of different scales operating in nature. The main property of atmospheric circulation on the giant planets is alternating prograde (eastward) strong equatorial jets on Jupiter and Saturn and retrograde (westward) jets on Uranus and Neptune. These jets have different speeds and widths and are responsible for the presence of the above-mentioned ordered system of light zones and dark belts at low and middle latitudes most clearly exhibited on Jupiter and Saturn (Fig. 3.8).

The most detailed investigation and comprehensive analysis of the mechanism of planetary dynamics on these planets was made by the Galileo and Cassini missions. They revealed most reliably the patterns of motion in the Jovian and Saturnian atmospheres. Three main dynamical regions—near the equator, mid-latitudes, and poles—with distinct features were distinguished. In the equatorial region a broad eastward jet dominates the dynamics (vortices are not found), whereas narrower zonal jets with alternate eastward and westward motion coexist with vortices in the mid-latitudes. Here the cloud bands correspond to cyclonic and anticyclonic shear zones between the jets. The morphology in the polar regions is dominated by a fully turbulent flow with numerous interacting and merging vortices. Saturn exhibits a more regular zonal organization extending up to the poles and culminating with a large cyclonic vortex at each of the poles. The jet speeds on Saturn are also twice as fast as those on Jupiter.

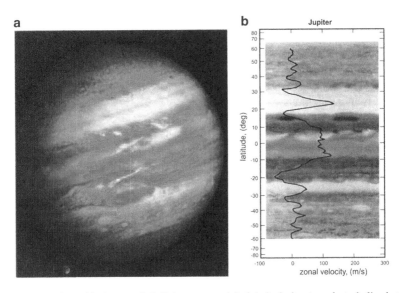

Fig. 3.8 (**a**) Jupiter with characteristic light zones and dark belts in its strongly turbulized atmosphere at the upper cloud deck; (**b**) Structure of zones and belts on Jupiter. The lighter regions of upflows are zones and the darker regions of downflows are belts. The *solid curve* indicates winds of a variable direction (with the velocity both in the direction of the planet's rotation in zones and in the opposite direction in belts); strong shear flows emerge in the transition regions. Disordered motions of various scales observed in the cloud structure are superimposed on this image (Courtesy of NASA)

In contrast to Earth, where circulation is caused by the difference of solar heating at the low and high latitudes, circulation on Jupiter and Saturn is mainly driven by ascending heat fluxes from the interior. Since the Rossby number $Ro \ll 1$ for these rapidly rotating planets, strong zonal flows emerge between zones and belts due to the Coriolis interaction of meridional flows. Gas moves upward in zones and downward in belts. The gas flows in zones being affected by the Coriolis forces are extended along meridians and deviate either eastward or westward depending on the meridional flow direction (which is different in the Northern and Southern Hemispheres). Opposite edges of these zones moving in different directions produce the strongest winds of variable direction in the transition regions, creating vortices and shear turbulence. Here the largest temperature gradients are also observed against a background absence of a noticeable temperature difference between equator and poles. The mean zonal wind velocity on Jupiter is ~150 m/s, while on Saturn it reaches ~500 m/s (see Fig. 3.8). Even stronger winds were measured in Neptune's atmosphere. On Jupiter, a growth in zonal wind velocity with depth from 70 to 175 m/s between the levels with a pressure of 0.4 and 5 bar was measured by the Galileo probe. Let us note that the Earth's atmosphere has basically the same structure of air flows but with much lower velocities.

Another important mechanism of planetary dynamics is natural convection, which is also attributable to the existence of a heat source in the interiors and resulting in disordered motions and numerous coherent vortex structures (convective cells) observed at high latitudes. This mechanism is similar in many respects to the classical problem of the Rayleigh-Taylor hydrodynamic instability of horizontal fluid flow heated from below (Rayleigh-Bénard convection). However, in the case of the giant planets, the convective interiors are in close dynamical interaction with the overlaying gas layer where the solar energy is absorbed. In other words, convection generated by the deep intrinsic heat flux penetrates the upper troposphere where equatorial Rossby waves arise. This leads to a very complex flow pattern with numerous vortex structures and a high degree of turbulization, as is observed most clearly on the disk of Jupiter. In turn, the formation of zonal flow and the maintenance of planetary circulation on the rotating planet can be explained by the nonlinear mechanism of vortex energy transformation into the kinetic energy of mean motion. Indeed, both three-dimensional general circulation modeling and differential rotation reproduced in laboratory experiments gave evidence that there are mechanisms universal across the giant planets that produce various flow patterns, provided differences of radiative heating and intrinsic heat flux are taken into account. Nonetheless, we cannot yet explain the very complicated dynamics and flow configuration on the giant planets in an energetically consistent manner.

As we have seen, clouds are an important indicator of various motions in the strongly turbulized atmospheres of Jupiter and Saturn (Fig. 3.9). The clouds in the ascending flows of zones on Jupiter are located about 20 km higher than in the downstream belts. Ammonia crystals in the cold zones explain their bright white color. However, the composition of the main condensates does not explain the observed palette of colors on the disks of these planes, specifically in the lower warmer belts, which represent a truly unique natural phenomenon. Therefore, the presence of more complex compounds should be assumed, including phosphine (PH_3), hydrocarbons (C_nH_n), and organic polymers presumably formed under solar ultraviolet radiation and lightning discharges in the atmosphere and clouds. On Uranus and Neptune, the upper cloud layers at lower effective temperatures are

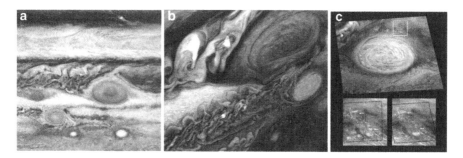

Fig. 3.9 (**a, b**) The Great Red Spot (*GRS*) in Jupiter's atmosphere (~25,000 × 12,000 km in size). Strongly developed turbulence near the Great Red Spot (*GRS*) exhibits a rich gamma of cloud colors. There are also strongly turbulized flow regions westward and southward from it. Smaller vortices ("*white ovals*") are on the *right*. The maximum image resolution is 95 km. (**c**) Some details of motion inside the GRS (Courtesy of NASA)

composed of methane, with ammonia and sulfur-bearing compounds located in the clouds below. Dynamical processes involving chemical transformations are typical of the multicomponent turbulent media of these planets as well. Obviously, the chaotic structure of strongly developed turbulence in the clouds of all the giant planets, accompanied by energy exchange at the microscopic level, emerges in the formation of ordered vortex structures attributable to self-organization processes.

Just as on Jupiter and Saturn, convection acting as a thermal mechanism when energy is transferred from depth must also play an important role in Neptune's atmospheric dynamics where, in contrast to Uranus, the internal heat source is also nearly twice the energy influx from the Sun. In contrast, the thermal regime and atmospheric dynamics of Uranus are determined by its orientation in space. However, despite the large difference in inclination and energy exchange, qualitatively identical meridional temperature and zonal wind profiles are observed on both planets at the level of clouds, although winds on Uranus are weaker. On both these planets winds blow in a retrograde direction in the equatorial and middle latitudes and are prograde near the poles. Of special interest is the fact that the wind velocities in Neptune's atmosphere are nearly 2.5 times higher than those in the atmospheres of Jupiter and Saturn (up to 600 m/s at the equator, i.e., the fastest of all the planets!), while the power energy source per unit area on Neptune is approximately a factor of 20 lower than that on Jupiter. This probably occurs because the Neptunian atmosphere has very low turbulent viscosity and, accordingly, a low level of energy dissipation of wind motion and shear flow turbulization. Interestingly, in contrast to Neptune, our terrestrial atmosphere has the greatest level of dissipation, in which processes related to the hydrological cycle, along with small-scale convection and surface friction, play a major role. Therefore, although the Earth derives incomparably more solar energy, its wind velocities are almost an order of magnitude lower than those on Neptune. Another interesting feature is that the direction of winds on Neptune and Uranus is opposite to the direction of their rotation. This distinguishes them from Jupiter, Saturn, and Venus (as well as the Sun and Titan), all of which are characterized by equatorial super-rotation.

Numerous large-scale vortex structures whose nature is associated with variable directions of zonal winds under the Coriolis force stand out in particular among the remarkable features of the cloud structures on the giant planets. The best known is the Great Red Spot (GRS) on Jupiter (Fig. 3.9), whose size is currently 15,000 km in the latitudinal and 30,000 km in the longitudinal directions, although it was nearly 1.5 times larger in earlier times. There are also numerous oval-shaped structures in the strongly turbulized atmospheres of Jupiter and Saturn, and these structures also originate from the strong zonal and meridional gas transport.

The GRS on Jupiter is a giant anticyclone whose lifetime, as estimated from similarity parameters, is several thousand years, as opposed to terrestrial cyclonic structures with a typical lifetime of the order of 1 or 2 weeks. It is located in the atmosphere above the surrounding cloud layer due to ascending motions and the release of latent vaporization heat that may serve as an additional energy source. The GRS has a very bright range of colors and a complex morphology of its internal vortex flows. It is assumed that red phosphorus enters its composition, though a higher abundance of other species cannot be ruled out. Particularly strong flow turbulization and gas and cloud particle exchange between the vortex and neighboring zones are observed on the periphery where the velocities exceed 100 m/s. The vortex makes a complete turn in 7 days, while the GRS itself periodically drifts in latitude forward and backward, having turned three times around the planet in the twentieth century. There are no such large structures on Saturn, but as we mentioned above, numerous vortices (ovals) with smaller sizes and shorter lifetimes exist, just as on Jupiter and Neptune. Besides, powerful storms occur more frequently in the Saturnian atmosphere. The motions in ovals are clockwise, while their longitude-latitude oscillations resemble the motion of the upper part of a vortex in a stably stratified shear flow. Just like ordered zonal flows, it is natural to consider them from the standpoint of the formation of a hydrological cycle in a stratified gas–liquid medium by taking into account its chemical composition, energy, and the fulfillment of the stability criterion.

Another giant eddy structure similar to the GRS was found in the dynamical turbulized atmosphere of Neptune during the Voyager 2 flyby (Fig. 3.7). It was called the Great Dark Spot (GDS), and originally it was thought to be a long-lived anticyclone similar to the GRS. However, it disappeared in a few years, as observations with the Hubble Space Telescope showed. However, different large and small dark spots were discovered, bringing convincing evidence about the instability of Neptune's atmosphere, the variability of the cloud formations, and high cyclonic activity.

We have seen that more or less stable structures periodically emerge in the atmospheres of planets against the background of chaotic (turbulent) gas motions within an open nonlinear system exchanging energy with the environment. Cyclones and anticyclones on Earth, super-rotation of the atmosphere on Venus and Titan produced by various energy sources, and stable structures like the GRS and other cyclonic ovals in the atmospheres of Jupiter and Saturn are the most characteristic stable structures. There are also less stable phenomena, such as the powerful dust storms on Mars or the GDS on Neptune, and numerous examples of instability and/or sporadic activity in the form of relatively small clouds appearing and disappearing for several hours in a turbulent medium against the background of small-scale chaotic motions. Based on the concept of nonlinear dynamics of complex dissipative systems, these phenomena of atmospheric dynamics can be considered from the standpoint of self-organization in the structures of chaotic turbulent media.

Satellites and Rings

All the giant planets have satellites (moons) and rings. They form very complicated structures involving satellite-ring tidal influence and gravity oscillations.

Satellites

The satellite-ring systems around Jupiter and Saturn are shown in Figs. 3.10 and 3.11. Due to enormous progress in the observing methods and technologies of

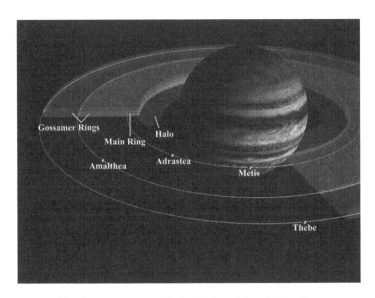

Fig. 3.10 The satellite-ring system around Jupiter (Adapted from Wikipedia)

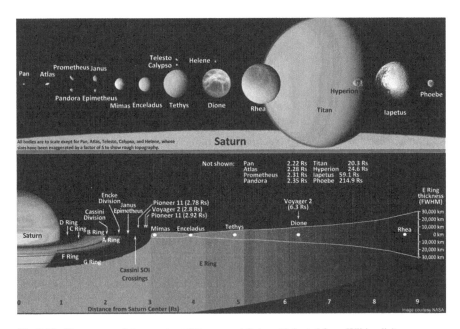

Fig. 3.11 The system of rings and satellites around Saturn (Adapted from Wikipedia)

ground-based astronomy, the number of satellites discovered in recent decades has increased several fold. 169 satellites were known at the beginning of 2012: 67 for Jupiter (the highest number of all the solar system planets), 62 for Saturn, 27 for Uranus, and 13 for Neptune. Thin structures of the rings were also found.

The names of the moons of these giants come either from the mythology related to the planet's name or from William Shakespeare characters. The size of most of the satellites does not even exceed tens or hundreds of kilometers, but some are comparable to the Moon and even Mercury. These are the four Galilean satellites of Jupiter (Io, Europa, Ganymede, and Callisto), Saturn's satellite Titan, and Neptune's satellite Triton. Nevertheless, for a single planet the ratio of the total mass of its satellites to the mass of the planet itself is no more than 0.01 %. Our Moon is the exception: its mass is more than 0.1 % that of Earth. All large satellites of the giants experience synchronous rotation because of tidal interactions; hence, they face their planet with one side. Unlike the gaseous or icy planetary body, the satellites are solid bodies having a rigid surface, although their relatively low mean densities suggest a significant fraction of ices, primarily water ice, in their bulk composition. Obviously, the large satellites were formed from the material of the same disk as the planet itself and much closer to it, subsequently increasing their radial distances through tidal interactions with the formed planet. The satellites considerably farther from the planet are mostly captured asteroids and comet nuclei. The system of satellites around each of the planets lying near the plane of its equator resembles the solar system in miniature.

Satellites of Jupiter. We first address the Galilean satellites of Jupiter: Io, Europa, Ganymede, and Callisto (Fig. 3.12). They have different bulk compositions manifested by the mean density dropping down from 3.53 g/cm^3 for Io to 2.99 g/cm^3 for Europe, 1.94 g/cm^3 for Ganymede, and 1.83 g/cm^3 for Callisto. This indicates a change from rocky to rocky-icy composition with a progressively larger proportion of water ice. The flights of the Voyager and Galileo spacecraft made a decisive contribution to the understanding of their fascinating nature. The Galilean satellites were probably formed simultaneously with Jupiter in its close vicinity and subsequently moved outward by the influence of its gravitational field. The theory of their

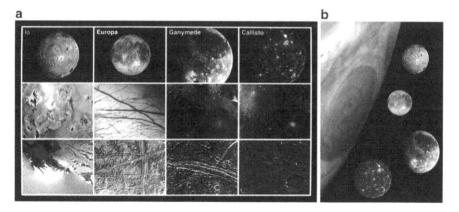

Fig. 3.12 (**a**) Galilean satellites Io, Europa, Ganymede, and Callisto. Most prominent unique features of the satellites are shown in the images of the surface areas under each of them; (**b**) Galilean satellites against the background of Jupiter's Great Red Spot (montage) (Images from the Voyager and Galileo spacecraft. Courtesy of NASA)

motion has a characteristic feature found long ago by the outstanding French mathematician Pierre-Simon Laplace: there is a triple (1:2:4) resonance in the system of these satellites; i.e., the revolution periods of Io, Europa, and Ganymede around Jupiter are in this multiple ratio. In addition, there exists a relationship between the mean motions causing libration of the satellites, while they themselves experience perturbations due to their strong gravitational interaction with each other. As a result, certain unique features are inherent in each of the Galilean satellites, but Io and Europa, which are nearest to Jupiter with revolution periods of 1.77 and 3.55 Earth days, respectively, are especially distinguished. Clearly, the unique features in both these and other satellites of the giants resulted from self-organization processes that initially gave rise to resonances and, in due course of subsequent evolution, to peculiar natural complex formation.

Global widespread volcanic activity continuing at the present epoch was unexpectedly detected on Io (Fig. 3.13), whose radius (1,821 km) is comparable to that of the Moon. As a rule, several active volcanoes, the largest of which were named Prometheus and Pele, are simultaneously observed on its surface, which is virtually devoid of any traces of impact craters because of its continuous refreshment at an estimated rate of about 1 cm/year. Regions of the vast valleys are covered with deposits of sulfur and its allotropes produced by phase transitions in the volcanic eruptions and imparting a characteristic yellow-orange-brown hue to the surface (see Figs. 3.13a, b). Apart from the lava itself, the condensation of volatiles and pyroclastics probably contributed to the formation of these properties. Mountains up to 15 km in height and numerous calderas from 10 to 200 km in width and up to 1 km in depth are distinguished on Io's surface, while the traces of lava flows extend to hundreds of kilometers. Interestingly, there are numerous thermal anomalies against the background of the surrounding cold surface, among which the Loki Patera region particularly stands out.

Another spectacular phenomenon observed on Io is the eruption of lava and gases into the outer space vacuum with a speed of about 1 km/s, which under the low-gravity conditions on the body forms plumes more than 200–400 km in height

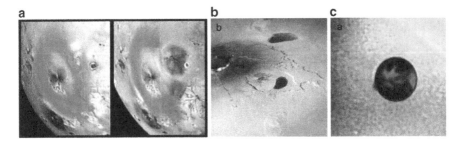

Fig. 3.13 Widespread volcanism on Io. (**a**) Portion of Io's surface with an active volcano in the images obtained by Voyager in 1981 (*left*) and Galileo in 1996 (*right*). The surface underwent small changes; an absence of impact craters is indicative of its young age attributable to active volcanism. (**b**) Lava outflow on the Io surface (An image from the Galileo spacecraft) (**c**) Volcanic eruption (plumes) on Io's limb (An image from the Voyager spacecraft. Courtesy of NASA)

(see Fig. 3.13c) The lava is composed mainly of silicates and sulfur compounds, which means that volcanism on Io differs little by its nature from the silicate explosive volcanism on the terrestrial planets, except that it is generated by SO_2 rather than H_2O and CO_2 gases. Both the recorded high lava temperature (more than 1,500 K) and the model of the Io interior are consistent with silicate volcanism. The interior structure includes an iron or iron sulfide core (~20 % by mass), a partially melted convective silicate mantle (possibly enriched with magnesium), and a lithosphere ~ 30 km in thickness. However, in contrast to the Earth, Venus, and Mars, where the volcanism is attributable to radiogenic heat, the radioactive isotopes on Io, just as on the Moon, have long been exhausted because of its small size. The cause of the volcanic eruptions on Io and their maintenance is of a completely different nature: it is the dissipation of tidal energy due to the above-mentioned gravitational interaction of Io with other Galilean satellites during the orbital motion in Jupiter's gravitational field, which causes periodic strong deformations of Io's shape. According to the existing estimates, this mechanism exceeds in efficiency other possible energy sources by approximately two orders of magnitude. The fairly high mean density of Io (3.53 g/cm^3) suggests that it consists almost entirely of rock; its ice-water envelope (if it ever existed) was apparently lost through the heating of its interiors at an early evolutionary stage. Interestingly, the energy emitted by Io into outer space (2.5–5 W/m^2) exceeds the internal heat fluxes on Earth (0.08 W/m^2) and the Moon (0.02 W/m^2) by approximately two orders of magnitude. In addition, because this energy at the present epoch is almost an order of magnitude greater than the equilibrium value (0.8 W/m^2), one may assume that the tidal energy dissipation rate underwent noticeable changes throughout the time of evolution.

Because of its volcanic activity, Io possesses a sulfur dioxide atmosphere, though a very tenuous one, and numerous clouds. Because of strong sunlight scattering, the sodium emission is most characteristic of these clouds in the sulfur and oxygen background. The same components in the ionized state form Io's ionosphere and a plasma torus along its orbit, which actively interacts with Jupiter's massive magnetosphere. In particular, this interaction leads to the well-known modulation of Jupiter's decametric emission, which was recorded back in the middle of the last century. Being located at a radial distance of 5.91 R_j, Io is literally immersed in Jupiter's magnetosphere. As a result, neutral particles are ejected from Io's surface by energetic magnetospheric ions (an effect called *sputtering*). This serves as the main mechanism for the loss of material from the satellite and replenishment of the atmosphere and the torus. Interaction of the Io plasma envelope with Jupiter's magnetosphere generates currents as strong as a million amperes projected from the torus into the planetary ionosphere.

The second of the Galilean satellites, Europa (radius 1,561 km, density 3 g/cm^3), is regarded as one of the most intriguing bodies in the solar system (Fig. 3.14). While it is located slightly farther from Jupiter and is smallest in size among all the Galilean satellites, it experiences similar but weaker tidal interaction. However, whereas intense heating and related volcanism on Io led to the loss of its original icy envelope (as on other Galilean satellites), on Europa it presumably turned into a water ocean ~ 50–100 km in depth beneath an ice shell ~ 10–20 km in thickness

Fig. 3.14 (**a**) Jupiter's Galilean satellite Europa. The surface is crisscrossed by ridges, troughs, and faults whose relief does not exceed several hundred meters in height. The absence of craters is indicative of a young surface. The section on the *left* corresponds to the present-day model of Europa's internal structure: there is a water ocean ~50–100 km in depth under a relatively thin ice crust ~10–15 km in thickness and a silicate mantle and a core composed of rocks lie below it (An image of the surface from the Galileo spacecraft taken on December 17, 1996). (**b**) A 70 km × 30 km area of Europa's surface (the Conamara region). The colors are enhanced to emphasize the relief features; the Sun is on the *right*. The *white* and *blue* regions correspond to a fresh surface partially covered with dust, while the *brown* ones probably owe their origin to mineral deposits. The areas ~ 10 km in size bear the traces of displacements of the upper ice crust layer, which can be associated with the presence of water or soft ice at a comparatively small depth (An image from the Galileo spacecraft. Courtesy of NASA)

(see Fig. 3.14a). An estimated heat flux of about 5 K/km caused by dissipation of the tidal energy makes the appearance of liquid water at such a depth possible. This is also evidenced by the tidally driven asynchronous (with respect to the interior) rotation of Europa's outer envelope as well as its unusually smooth figure and surface morphology: the presence of numerous cracks in the ice, the formation of regions with a chaotic pile-up of blocks that can be likened to icebergs, and the unusual shape of impact craters. Moreover, the ocean can warm up through periodic energy release in the satellite's interior and contain such gases as carbon dioxide and oxygen. Therefore, it could be a suitable biogenic medium and even harbor some primitive life forms. Interestingly, if the idea of Europa's ocean and estimates of the volume of water are correct, then it should considerably exceed the Earth's reserves of water.

Several close encounters of the Galileo spacecraft with Europa provided quite convincing support of the idea. The images transmitted back to Earth showed that its icy surface is literally crisscrossed by countless low ridges and faults up to 3,000 km in length, up to 70 km in width, and several hundred meters in depth—all existing against the background of the unique smoothness of the Europa form (see Fig. 3.14b). These geological features of the relief devoid of any order were apparently formed comparatively recently, as suggested by the absence of ancient impact craters on the surface. The origin of faults can be attributed to convective and other dynamical processes in the ocean triggering ice motions and, possibly, a partial outflow of water outward where the forming cracks are filled with fresh brash ice. This idea is concurrent with the most recent discovery by the Hubble Space Telescope of plumes spouting what appears to be vaporized ocean water into space.

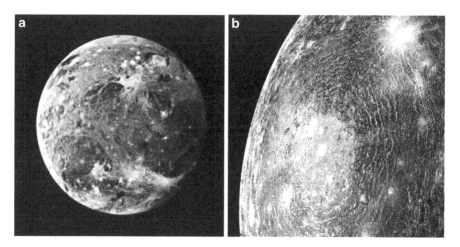

Fig. 3.15 Jupiter's Galilean satellites Ganymede and Calisto. (**a**) *Bright white* regions on Ganimede attributable to relatively fresh ice crust denudations under impact bombardment are distinguished on the dark heavily cratered icy surface covered with mineral dust with numerous ridges, young troughs, and large depressions, some being associated with cryovolcanism. Likewise Europa, Ganymede is assumed to contain water ocean beneath its icy-rocky surface (**b**) A heavily cratered terrain though a partially liquid water ice mantle is also assumed to exist. The most prominent feature on the surface of Callisto—huge impact basin Valhalla with the system of concentric ridges of frozen ice-rock melts (An image from the Galileo and Voyager spacecraft. Credit: ESA and NASA)

Obviously, the fairly fast erosion (smoothing) of the traces of Europa's meteorite bombardment is associated with the same processes. Another important argument is the existence of a comparatively weak magnetic field that is most likely due to the presence of an electrically conducting salty ocean where electric currents are induced, causing Jupiter's magnetosphere to experience noticeable perturbations.

Magnetic fields were also detected on the two other Galilean satellites, Ganymede and Callisto (Fig. 3.15). Ganymede is the largest satellite among the Galilean group and in the solar system as a whole. Its size is larger than that of the terrestrial planet Mercury. Ganymede is locked in a Laplace orbital dynamical resonance, but the experienced tidal interactions are weaker compared to those of Io and Europa. The morphology of the Ganymede surface reflecting its geological history is very complicated. Light and dark regions that probably originate from the denudation of water ice or, vice versa, the deposits of rocks caused by meteorite bombardment of the surface including large asteroids and comets, are clearly distinguished. There are also systems of cliffs, troughs, and fractures probably associated with tectonic activity (Fig. 3.15). Apart from the influx of tidal energy during the resonant interaction with Europa and Io, it may also be attributable to the preserved source of radiogenic heat in the silicate mantle of this rather large body controlled by convective transport. In any case, these energy sources led to differentiation of Ganymede's interiors, which is obviously responsible for its magnetic field. The latter can be generated both by a dynamo in a partially melted iron or iron sulfide core whose radius is estimated to be ~ 1,000 km and by induction in a presumable water-ice outer envelope ~ 800 km in extent. In other words, similar to Europa, we may assume an existence of an even more vast ocean on Ganymede. Sending submersible probes to these satellites promises to return invaluable scientific information.

Callisto is only slightly smaller than Ganymede in size and very close to it in bulk density. It is the least geologically modified body among all of the Galilean satellites. The degree of its surface cratering is close to saturation, suggesting high efficiency of impact bombardment in the vicinity of Jupiter. Large old craters degraded much less than those on Ganymede were preserved on Callisto. Of particular interest is the prominent Valhalla region with a system of concentric rims and crests several hundred meters in height and several thousand kilometers in extent, which were probably produced by the fall of a large asteroid that formed a vast basin similar to the lunar ones (Figs. 3.15b and 3.16). However, in contrast to the Moon, the icy surface of this satellite retained the sequence of wave propagation from the explosion epicenter owing to the ice crust's plasticity. Callisto's surface is covered with layers of darker material (probably of an exogenous origin) containing carbon-bearing compounds such as CH, CO_2, and CN, as well as SO_2 and possibly SH radicals.

Since, in contrast to the other three Galilean satellites, Callisto is not subjected to tidal heating due to the Laplace resonance, there are no traces of endogenous activity on its surface. Obviously, for this reason the degree of differentiation of rocks constituting Callisto is considerably lower than that for Ganymede. This idea is supported by the dimensionless moment of inertia for the body $I = 0.359 \pm 0.005$; for Ganymede it is $I = 0.3105 \pm 0.0028$.

In the absence of an external energy source, it is hard to expect liquid water to be preserved at depth, because the interior should have been cooled over geological time by subsolidus convection controlled by ice viscosity, though via a less efficient

Fig. 3.16 One of cliffs surrounding impact basin Valhalla on Calisto (Voyager image. Credit: NASA)

process than the one on Ganymede. However, there is also a different viewpoint backed by the absence of noticeable seismic traces of the impact from the fall of a large body on the side opposite to the Valhalla basin, which can be explained by the absorption of impact energy by a liquid interior layer (a salt ocean with a high electrical conductivity analogous to the oceans on Europa and Ganymede). The existence of a magnetic field on Callisto (also comparable in strength to those on Europa and Ganymede) and its variation depending on Callisto's orientation relative to Jupiter's magnetosphere provide evidence in support of this hypothesis.

Jupiter's other moons, located inside and outside the Galilean satellite orbits, are much smaller objects, of predominantly rocky composition (see Fig. 3.10). Of special interest is Amalthea of about 200 km in size, whose density argues for an inhomogeneous interior, which is probably caused by former disruption of the body and then its recompilation. Some moons at the outer edge of the satellite system (which extends roughly 24 million km) revolve in a retrograde direction and are assumed to be either asteroids or comets captured by Jupiter's gravitational field. These satellites are distinguished in name by a few groups of irregular satellites to which they belong, such as Ananke, Carme, and Pasiphae.

Satellites of Saturn. As in the case of Jupiter, the majority of the 62 Saturnian satellites are relatively small bodies ranging from hundreds to tens of kilometers in size (see Fig. 3.11). Larger satellites of several hundred kilometers across are Mimas, Enceladus, Tethys, Dione, Rhea, Hyperion, Iapetus, and Phoebe. However, Titan, with a diameter of 5,150 km, is one of the largest satellites in the solar system, yielding in size only to Ganymede but exceeding even Mercury (although it is less than half as large by mass). Similar to Jupiter's satellites, all Saturn's large satellites are in synchronous orbital-rotational motion, facing the planet with one side. Among the medium-sized satellites of special interest is Iapetus, which exhibits an albedo difference of its leading (bright) and trailing (dark) sides by nearly an order of magnitude, a phenomenon awaiting explanation. Another attraction is Hyperion, a body of very irregular shape locked in orbital resonance with Titan. Unexpectedly, relatively young surfaces were found on Uranus's satellite Miranda, which is less than 500 km in diameter, and even on Jupiter's Thebe, which is slightly larger than 200 km in diameter - one of the four known small moons that orbit closer to the planet than the four vastly larger Galilean moons (Fig. 3.17). They presumably were modified by endogenous and/or exogenous processes, although the real source remains unclear. Many small satellites at the outer edge of the system are in retrograde rotation and as we said above, are probably captured asteroids and/or comets.

The great attractions from the viewpoint of natural conditions and their evolution involving pre-biogenic organics are Titan and Enceladus. Enceladus is quite small, an almost entirely icy satellite only about 500 km across with a density as low as 1.12 g/cm^3, but it turns out to be an extremely active body. Numerous geyser fountains from the interior to high altitudes were observed by the Cassini orbiter (Fig. 3.18). They are assumed to feed the innermost ring E of Saturn and hence maintain this quite tenuous formation. Much of the surface is strongly modified by the continuing geological processes which left behind ridges and troughs of different sizes, while an absence of craters and the satellite's albedo value (close to unity)

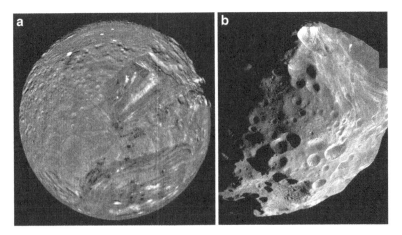

Fig. 3.17 Satellites of Saturn Miranda of 480 km in diameter 9 (*left*) and Thebe (220×230 km) in size (*right*) having relatively young surface presumably modified by both endogenic and exogenic processes (Credit: NASA and ESA)

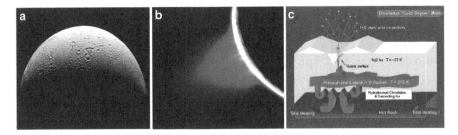

Fig. 3.18 (**a**) Saturn's moon Enceladus of 500 km in diameter; (**b**) Plumes above the limb of Enceladus feeding the E ring of Saturn. The jets of water emanate from "tiger stripes" near the South Pole. These jets provide our first glimpse into the interior of an icy satellite (Image from Cassini. Credit: ESA); (**c**) Model representation of Enceladus's cryovolcanism (Adapted from Wikipedia)

suggest that its surface is very young. The preserved unique geologic activity of this relatively small cold body is caused by the slightly elliptical orbit of Enceladus locked in a 2:1 resonance with another Saturnian satellite, Dione. The tidal effect from Dione is sufficient to heat the interior of Enceladus to a temperature of 176 K, corresponding to the melting temperature of a water-ammonia eutectic, thought to be the component in the erupting geysers. This mechanism also serves as the source of particles filling up Saturn's E ring within which Enceladus is located.

Certainly, the most spectacular and intriguing body among the Saturnian satellites is Titan (Fig. 3.19). Titan's bulk density (2.04 g/cm^3) is close to those of Ganymede and Callisto and, similarly, it consists of half rocks and half ices, probably mainly water ice with different types of crystallization. Titan has an absolutely unique nature that distinguishes it in the outer solar system. Unlike other satellites

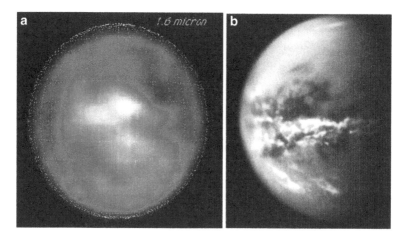

Fig. 3.19 (**a**) Saturnian satellite Titan as it is seen from outside (Cassini image at Saturn approach). The surface is shrouded by thick clouds composed of hydrocarbons giving Titan orange color. (**b**) Titan has an active meteorological cycle, producing methane clouds, rain, streams, and lakes (Credit: ESA)

of the giant planets, it has a thick nitrogen-argon atmosphere with methane and ethane admixtures extending for more than 400 km, with several inversion layers and a thick cloud deck. The atmospheric pressure at the surface is 1.6 atm and the temperature is 94 K. This temperature is close to the triple point of methane at which phase transitions may occur on the surface. Since the acceleration due to gravity on Titan is approximately one-seventh of that on Earth, the mass of Titan's atmosphere must be an order of magnitude larger than the terrestrial one to produce 1.6 atm pressure. The presence of ^{40}Ar in the atmosphere implies an existence of volcanic activity. Moreover, the morphological features of the surface do not rule out the possibility that it was affected by tectonic processes.

Another interesting feature of Titan is the circulation of methane, including the formation of methane clouds in the atmosphere with other hydrocarbon admixtures, and precipitation in the form of methane rains on the surface. The existence of such a methane cycle was hypothesized previously based on computer modeling and radar measurements of surface properties as well as observations with the Hubble Space Telescope. The concept of the methane cycle was confirmed by surface images transmitted from the Huygens lander that was jettisoned from the Cassini spacecraft and landed on Titan in January 2005 (Fig. 3.20). The images showed many peculiar features including rounded boulders composed, probably, of water and methane ices and possibly other organic compounds; valleys resembling the beds of rivers flowing down uplands; rows of dunes consisting of "hydrocarbon dust" particles apparently formed by strong winds; and separate filled or dried lakes hundreds of kilometers in size whose formation can actually be associated with the precipitation of liquid methane from the atmosphere (Fig. 3.21).

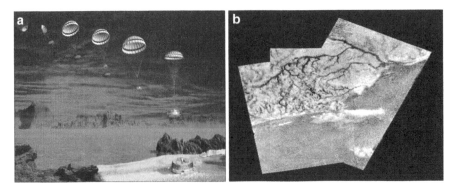

Fig. 3.20 (a) Huygens lander descent on Titan (artist's view). (b) Image of Titan's surface at Huygens descent from 10 km. Methane channels (river beds) on the rough terrain are clearly seen (Credit: ESA)

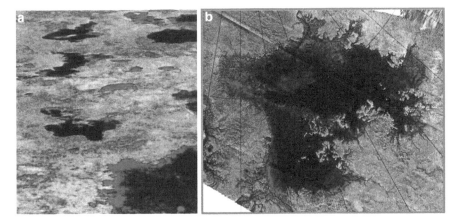

Fig. 3.21 (a) The surface of Titan. The blue patches on the *yellow* surface are methane (with possibly other hydrocarbons) lakes, which is in accord with the concept about methane cycle between the surface and atmosphere. It is the only world other than Earth with open bodies of liquid on its surface. The colors in the images are distorted to emphasize the contrast of features. This is composite of images obtained by Huygens lander. (b) Methane lake Ligeia Mare on the Titan's surface (Credit: ESA)

Radar images from the Cassini orbiter revealed structures which look like volcanoes on Titan's surface. They are thought to spew icy "lavas" and pump methane into the atmosphere. However, this idea is not compatible with the view of Titan's crust as a shell of rock-hard ice tens of kilometers thick that would prevent methane-laden lavas from rising from the interior and flowing across the surface. The question of how to distinguish between these two options has important implications in terms of determining whether Titan is geologically young or old and the nature of the store of atmospheric methane as the source of precipitation which fills its lakes and cut canyons. It is also relevant to the problem of the lack of correlation of the surface topography with the gravity field: the lower gravity anomalies correspond to higher topographic areas, and vice versa. One explanation assumes lighter ice shells tens of kilometers thick floating above a very deep ocean beneath Titan's surface at a depth of ~50 km, a possible abode for life. Basically, there is a consensus among planetary physicists that at least one cryovolcano Sotra Facula is evident on Titan's surface, whereas others possibly had no chance to appear on the vast smooth plains because lavas from frozen outpourings of eruptions were too fluid to form volcanic domes.

Methane condenses into clouds at altitudes of several tens of kilometers, and weak frost continuously precipitates from these clouds to the surface. As we said above, Titan's clouds also contain hydrocarbons which impart a characteristic red-orange color to the atmosphere (see Fig. 3.19a). Larger "rain" drops appear to precipitate from particularly dense clouds, as observed near the South Pole, being compensated by evaporation from the surface. Such cyclic exchange with methane resembles the hydrologic cycle on Earth. Obviously, no liquid water can exist on the Titan surface at very low temperatures, but it may be present at depths in the subsurface layer, just as on the Galilean satellites.

We also must mention the unique dynamics of the Titan atmosphere. The insolation on Titan is too low to provide the development of intense dynamical processes, but, nevertheless, they exist. The tidal effects from Saturn, which are stronger than the lunar tides on Earth by a factor of 400, can be assumed to serve as the main energy source. The assumption about the tidal mechanism of wind motions is backed by the orientation of the rows of dunes ubiquitously encountered on Titan. As measurements on the Cassini spacecraft showed, the preferential direction of winds is indicative of super-rotation in the atmosphere of Titan similar to that on Venus. The altitude profile of the wind velocity in the Titan atmosphere increases from several to ~ 30 m/s in the range of altitudes 10–60 km and is generally similar to the Venus atmosphere pattern. Interestingly, the motions weaken greatly above this region, and strong turbulization of the atmosphere is observed starting from an altitude of ~120 km. What causes such unusual dynamical properties of the atmospheric medium cannot yet be explained. They may be associated with the formation of a layered structure of hydrocarbon clouds in this region of Titan's atmosphere.

Titan draws special attention as an environment where complex organic compounds could evolve towards more complicated pre-biogenic structures and might even be a cradle of primitive life forms. Indeed, Titan has wonderful organic storage: apart from methane much more complex unsaturated hydrocarbons (ethane, ethylene, acetylene, diacetylene, methyl acetylene, and cyan acetylene) were discovered on Titan's surface. Also, propane, hydrocyanic acid, and other organic compounds have been found to form in the upper atmosphere and in the hydrocarbon clouds under ultraviolet radiation through methane photolysis processes. It is thus believed that favorable conditions for the initial stages of biogenic synthesis on Titan might be similar to those that existed on the early Earth. But we address Titan as a world that "never grows up."

Satelites of Uranus and Neptune. All satellites orbiting Uranus are of moderate and small size. The largest are Miranda, Ariel, Umbriel, Titania, and Oberon, though

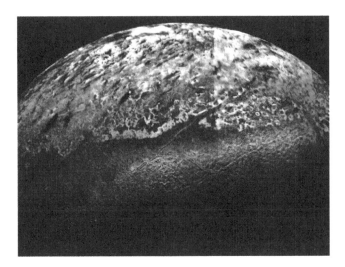

Fig. 3.22 Neptune's satellite Triton. Active nitrogen geysers were detected on the methane surface (An image from the Voyager spacecraft. Courtesy of NASA)

their size is less than half of the Moon's size. They have low albedo values (0.20–0.25) and densities corresponding to about a half-and-half ice and rock composition, perhaps with ammonia and carbon dioxide ices mixed with the water ice. No special features of geological activity were distinguished on the cratered surface of the satellites except Miranda, whose surface exhibits chaotic terrain, terraces, and deep canyons possibly caused by the collision of Miranda with another large body. The set of inner satellites identified with the beautiful names of Shakespearean characters (Cordelia, Ophelia, Bianca, Cressida, Desdemona, Juliet) are in close gravitational interaction with the whole system of Uranian satellites and control their structure. The outer small satellites are mostly of irregular shape, and some of them could be captured small bodies.

In the system of Neptune's satellites the most spectacular is Triton (Fig. 3.22), which is one of the largest moons in the solar system. Though not as large as Titan and only slightly smaller than the Moon, it is similar to Titan in its mean density (2.07 g/cm^3) and bulk icy-rocky composition. It comprises nearly the whole mass of the satellite system, since the mass of all other Neptune satellites is only 0.5 %. Another moon of interest is Nereid, a body of irregular shape with unusually high eccentricity (0.75) that results in an apocenter distance seven times more than the pericenter one and probably speaks in favor of its capture by Neptune's gravitational field. Four inner small satellites (Naiad, Thalassa, Despina, and Galatea) are located within Neptune's rings and are closely related with the rings' dynamics.

Triton has virtually no atmosphere (the pressure at the surface does not exceed 15 μbar) and the temperature of its nitrogen-methane surface is only 38 K. A number of signatures make Triton akin to Pluto which, as we said earlier, lost the status of ninth solar system planet and has been transferred to the new category of plutoids: large Kuiper Belt bodies. Meanwhile, there are also great differences between these two bodies that became obvious after the Voyager 2 flyby of Neptune. First of all, dozens of dark bands, some of which were identified with geyser-like ejections of liquid nitrogen to an altitude of several kilometers (*cryovolcanism*), were detected on the surface of Triton in the southern polar cap region (see Fig. 3.22). Deposits of dust particles on the frozen methane surface and their transport by predominant winds even in the highly rarefied atmosphere are probably also associated with geyser activity. In addition, structures resembling frozen lakes with nitrogen-methane coastal terraces of ~1 km in height were detected on the surface. Their formation may be associated with successive melting-freezing epochs caused by the changing insolation conditions throughout the long Neptunian year or (more likely) may be due to the tidal interactions of Triton with Neptune. Just as on the Galilean satellites, the interior heating through the dissipation of tidal energy apparently serves as the main source of cryovolcanism on this very cold body. The small number of impact craters on the relatively young surface of Triton provides evidence for its preserved geological activity.

Triton has a very unusual orbit, which acutely poses the question about its origin. It is highly inclined to the ecliptic plane and has a nearly zero eccentricity, while the motion of Triton itself, in contrast to all other large planetary satellites, is retrograde (clockwise). The peculiarities of Triton's orbital motion suggest that it was initially formed in the Kuiper Belt, just like Pluto, and was subsequently captured by Neptune. However, as calculations show, an ordinary gravitational capture is unlikely. Therefore, it is additionally hypothesized that Triton was a member of a binary system or gradually decelerated in the upper atmosphere of Neptune. This hypothesis is corroborated by the fact that when passing to the orbit around Neptune, Triton should have experienced a strong tidal effect from the planet and the system of its satellites (in particular, Nereid), which led to the melting of its predominantly water-ice interior. It is quite probable that the ongoing tidal effect from Neptune and Nereid at the present epoch continues to heat up the planet; as a result, the thermal flux from Triton's interior exceeds the insolation value almost by a factor of 3. Another consequence of this scenario is that Triton is gradually approaching Neptune, and in the distant future (~10–100 million years) it will enter inside the Roche limit and be torn apart.

Rings

When addressing giant planets we usually speak about the systems of their satellites and rings jointly because of their close coupling and interactions. The rings of giant planets are fascinating phenomena and a remarkable example of self-organization in nature. The rings of all the giants except Jupiter are clearly highly structured, and each one has some peculiarities.

A ring can be roughly described as a system of particles (an infinite number of tiny satellites) that are in orbital motion around a planet and experience simultaneously chaotic interactions. Collective processes and inelastic collisions between the macroparticles in the disk system are mainly responsible for ordering in the ring configurations. In other words, self-organization is part of the system itself, although some satellites embedded in the ring structure exert an additional "stimulating" influence through the tidal influence. These satellites are called "shepherds."

Besides their influence, particles of the rings are locked in resonances with larger planetary satellites outside the ring system. This disrupts the homogeneous structure of the rings and in particular, gives rise to the formation of the gaps inside them.

Rings of Jupiter and Saturn. Jupiter's rings (see Fig. 3.10) represent a very tenuous formation (optical depth $\sim 10^{-6}$). They are populated by micrometer-size dark particles and are poorly visible when observed from the outside. Nonetheless, three main rings (main, web, and halo) are distinguished in structure. The main (most pronounced) ring occupies the region between 122,000 and 130,000 km from the center of Jupiter. Two satellites, Metis and Adrastea, are located just at the halo's outer edge. Meteorite bombardment of their surfaces and those of two other inner satellites, Amalthea and Thebe, is assumed to be mainly responsible for producing a beehive of microparticles which eventually enter the rings. The lifetime of particle replacement is estimated to be very short, several ten thousands of years.

The most spectacular is the system of rings of Saturn, with their spacious ($\sim 250,000$ km in diameter) but very thin (less than ~ 1 km) formation within the Roche limit (see Figs. 3.11 and 3.23). Their thickness-to-width ratio is similar to that of a large sheet of cigarette paper. The rings are tilted by $28°$ to the plane of the ecliptic, which gives us an opportunity to see them either nearly face on or edge on in due course of Saturn's annual motion around the Sun (the effect of "rings opening"). Three main rings form the ring structure: outer A, middle B, and inner C. The B ring is the widest, densest, and brightest of all, while the C ring is nearly transparent. A fourth very thin ring, F, of irregular shape is located beyond the A ring. Also the very tenuous E ring is distinguishable close to the planet, within

Fig. 3.23 Saturn's rings. Their fine structure, the set of isolated ringlets that result from the process of self-organization in the ring structure, is clearly distinguished. The radial dark bands (dark spokes) in the B ring are attributable to the formation of dusty plasma located above the plane of rings and controlled by the planet's magnetic field (An image from the Voyager-2 spacecraft. Courtesy of NASA)

which Enceladus is located. Rings A and B are separated by the famous Cassini Division of about 4,000 km in width. Another region from which particles are wiped out by tidal forces is the Encke gap within the A ring. As we will see when discussing small bodies in Chap. 4, these gaps are similar in nature to the Kirkwood gaps in the main asteroid belt. The same mechanism of tidal interactions is responsible for generation of the density waves, as well as for the formation of the hierarchical structure of rings and their stratification into thousands of thin spiral ringlets inside the ring's main structure due to the development of gravitational-dissipative instability. The ringlets often are compared with tracks on a phonograph disk.

The bodies populating Saturn's disks range from micrometer and centimeter size particles to boulders of 1–10 m across. They are composed mostly of water ice (~93 %), carbon (~7 %), and some silicates. The total mass of the rings' material is small. A single body accumulated of all particles would be less than 100 km in size. Some Saturnian satellites control the ring structures. For example, Mimas and the Cassini Division are in 2:1 resonance, which prevents particles from entering the division, while Pan is mainly responsible for the Encke gap formation. The shepherd satellites Prometheus and Pandora tightly interact with the F ring, which causes its irregular structure.

Rings of Uranus and Neptune. Another interesting phenomenon and brilliant example of self-organization processes in nature is the formation of the system of rings around Uranus and the mode of their tight interaction with the nearby planetary satellites (Fig. 3.24). Here the ring particles are forced to concentrate in very

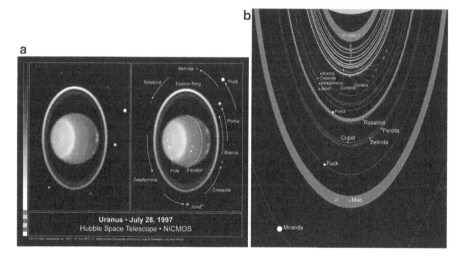

Fig. 3.24 The rings of Uranus from Hubble Space Telescope observations. The *left* picture (**a**) was taken in 1997; the *right* one (**b**) was taken later on. The motions of satellites and atmospheric structures are shown in it. The formation of Uranus's rings is even more pronouncedly than in the case of Saturn affected by the satellites (shown in the *right* image) located near the rings, along with the collective particle interaction processes. They are responsible for the emergence of additional order in the nine main rings structure. (**b**) Nomenclature of the Uranus rings structure shown in the *right* image. (Credit: NASA)

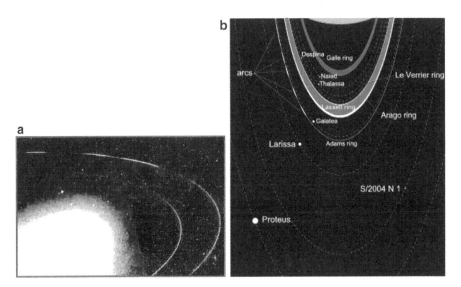

Fig. 3.25 The rings of Neptune consisting of five tenuous main rings. (**a**) Image obtained by Voyager 2 spacecraft during Neptune's fly by. (**b**) Schematic view of the position and irregular structure of Adams ring with "arcs". The arcs occupying a narrow range of orbital longitudes are remarkably stable. The stability is probably caused by the resonant interaction between the Adams ring and its inner shepherd satellite Galatea (Credit: NASA and Wikipedia)

narrow rings by gravitational focusing. The system of closely located satellites serving to ensure this order was first predicted theoretically and then confirmed. There are 13 thin rings of Uranus, named with letters of the Greek alphabet, the brightest being the ring ε. As in the case of Saturn, the rings of Uranus are populated by micrometer-millimeter size particles and meter-size bodies of varying composition, as is suggested by the different color of the inner and outer rings.

An even more fascinating structure is exhibited in the system of rings around Neptune (Fig. 3.25). Similarly, gravitational forcing exerted by the nearby satellites seems to be responsible for the nonuniform distribution of particles in the form of separate "arcs" along the orbit, arcs slowly drifting in the azimuthal direction. The two most characteristics rings are called Adams and Le Verrier in honor of Neptune's discoverers ("on the tip of a pen"), whom we mentioned earlier. The arc formation mechanism is not completely understood. Resonance (42:43) of ring particles with the eccentricity and inclination of the Neptunian inner shepherd moon Galatea was invoked to explain the nonuniform distribution of particles along the orbit of the Adams ring, though such a mechanism needs to be confirmed.

The problem of the planetary rings' origin is not yet fully resolved. They could be either a swarm of primordial particles whose formation into a satellite was hindered by gravitational forces inside the Roche limit or, in contrast, remnants of an asteroid or a comet that broke up when captured by the gravity field of the planet and eventually entered the Roche limit. Jupiter's ring filled with very small particles and surrounded from outside and inside by diffuse nebulae serves as a characteristic example and confirmation of such an event. The breakup hypothesis was put forward based on the estimated limited lifetime of the rings, ~ 0.5 Gyr, which is much less than the age of the solar system. In this case, the rings should be considered not as relics of the accretion stage of the planet itself, but as periodically appearing and disappearing structures due to the accidental catastrophic event of the gravitational capture of a small body by the planet and its subsequent destruction within the Roche limit, which, in general, seems quite likely and justifiable. This hypothesis is also supported by the fact that, for example, the predominantly icy particles of Saturn's rings have a high albedo, i.e., they were not covered with dark micrometeoric matter, as it would happen to the relic rings over the lifetime of the solar system.

Chapter 4
Small Bodies

Nature and Dynamics

"Small bodies" is the general term for the most numerous celestial bodies populating the solar system. These are comets, asteroids, meteoroids, and meteor (interplanetary) dust. The main reservoirs of asteroids and comets in the solar system are shown in Fig. 1.3. They are of primary interest from the viewpoint of the solar system's origin and evolution, because many of them contain preserved primordial matter. As the carriers of this primordial matter from which the solar system was formed, they are of key importance in cosmochemistry and cosmogony. There is a huge variety of these objects, every one reflecting certain characteristic features of cosmic media and the diversity of the complex physicochemical processes involved. They also allow us to reveal some specific properties which concurrently contributed to the formation of planets and systems of their satellites. Undoubtedly, the relationship between order and chaos played an important role in these processes, and their study is aimed at answering the key questions of planetary astronomy.

Small bodies, whose size extends from several hundred kilometers (large asteroids) to meters-centimeters (meteoroids) and (dust grains), are the most dynamic objects in the solar system. They played and continue to play an important role in the transport of matter due to their migration processes and are also responsible for numerous collisions with planets and satellites. Comets are of particular interest because they presumably preserve chemical composition essentially unmodified since their origin. Thus, comets attract progressively increasing interest as possible carriers of primary life forms. By now, quite a few asteroids and comet nuclei have been visited by spacecraft (Fig. 4.1).

Meteorites are collisional fragments of asteroids that survive atmospheric entry and fall to the Earth's surface. The Earth and other planets, their satellites, and asteroids themselves experience permanent bombardment by bodies of different size—the lower the size, the higher the impact probability. The cratered surfaces of the

© Springer Science+Business Media New York 2015
M.Ya. Marov, *The Fundamentals of Modern Astrophysics*,
DOI 10.1007/978-1-4614-8730-2_4

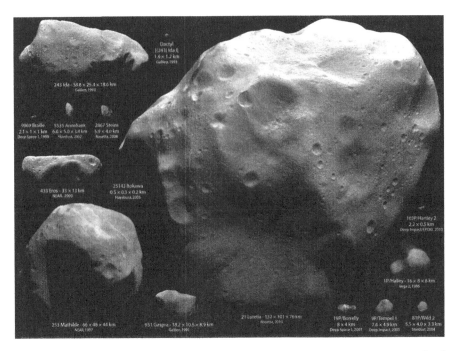

Fig. 4.1 Asteroids and comet's nuclei visited by spacecraft. The names and sizes are properly of bodies are shown below every body (Courtesy of NASA/J. Green)

Moon and planets bear clear evidence of such impact events throughout their geologic history. Of special interest are meteorites identified as being ejected from the Moon and Mars due to great impact events with large asteroid-like bodies.

Asteroids

The main bulk of asteroids in the inner solar system are located in the main asteroid belt (MAB) between the orbits of Mars and Jupiter from ~2.8 to 3.2 AU (see Fig. 1.4b). This region is close to the so-called "snow line" in the protosolar nebula (between 3 and 3.5 AU), the boundary where water's freezing point is located, with important implications for mineralogy and primary bodies' formation. Bodies in the main asteroid belt experience resonances due to tidal interaction with Jupiter (commensurabilities of Jupiter and asteroid orbital periods) resulting in the formation of Kirkwood gaps (Fig. 4.2). These are the regions where asteroids are "prohibited" to enter. The number of bodies of size larger ~1 km in the MAB is estimated at ~10^5, with an exponential increase of smaller size bodies. Bodies less than a few meters to centimeters in size are called meteoroids. There are also large bodies of a few hundred kilometers across, the largest being the dwarf planet Ceres (900 km) and close to it in size the asteroid Vesta (500 km). In Fig. 4.3 asteroids are compared

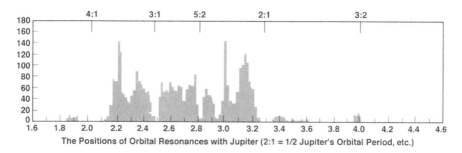

Fig. 4.2 Kirkwood gaps in the main asteroid belt. The regions of orbital resonances with Jupiter (voids in the diagram) are the regions from where asteroids are expelled (Adapted from Wikipedia)

Fig. 4.3 Middle and large size asteroids in comparison with Pluto and the Earth's Moon . They are scaled relative to two USA states (*bottom left*) (Courtesy of NASA)

with the Moon and Pluto, and in Fig. 4.4 Vesta is compared with other asteroids visited by spacecraft. Even the middle-size asteroids such as Ida, Eros, and Gaspra essentially dwarf the majority of the family (Fig. 4.5). Let us emphasize that despite their large number, the total mass of bodies in the MAB is less than $10^{-3}\,M_E$ (M_E is an Earth's mass). Asteroids in the MAB were sometimes regarded as a relic of an unformed planet (called Phaeton) or a planet that was destroyed by great tidal perturbations from nearby Jupiter. However, the physical dichotomy and different chemical compositions of the asteroid belt bodies as well as the negligible mass left behind do not support a former planet hypothesis.

The substantial diversity in asteroid abundances is comparable to solar system meteorite classes, as the latter are their fragments. There are water-rich asteroids

Fig. 4.4 Synopsis of asteroids as compared to Vesta by size (Courtesy of NASA)

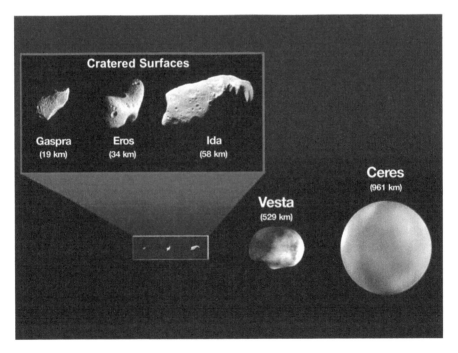

Fig. 4.5 Largest asteroid Ceres (~900 km) and Vesta (~500 km), Ceres being classified according to IAU nomenclature as a dwarf planet. For the comparison some asteroids visited by spacecraft (fly by) are shown. Sizes of these largest asteroids are compared with representatives of the main population (Courtesy of NASA/J. Green)

Fig. 4.6 Vesta close up view
from the Dawn spacecraft.
The asteroid exhibits heavily
cratered terrain and
inhomogeneities of its gravity
field, anomalies showing in
color and scaling in relative
units at the bottom (Courtesy
of NASA/J. Green)

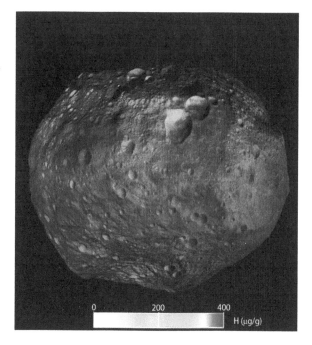

like Ceres that is assumed to have an interior's water content similar to the Earth's oceans. The close-up view of asteroid Vesta from the Dawn spacecraft, which entered an orbit around the body, exhibited heavily cratered terrain, different physical properties of surface regolith, and inhomogeneities of the gravity field (Fig. 4.6). Surprisingly, a central peak 2.5 times taller than Mount Everest was revealed inside Vesta's Rheasilvia basin (Fig. 4.7). This peak is comparable in height with the Mars shield volcanoes and is unusual for this much smaller body. Vesta's interior appears to have an iron-nickel core and a magnesium silicate mantle.

Some asteroids are probably former comets covered with a rather thick refractory crust after a multitude of perihelion passages, and they therefore contain near-surface water ice. Dynamical simulations showed that such asteroids in the MAB can preserve stability > 100 Myr. Because of the permanent supply to MAB bodies from the outer solar system, they are possibly composed of pristine material indicative of a dumbbell shape (resembling that of a comet nucleus) and with a rough surface structure (see Fig. 4.8). Moreover, an asteroid can be an example of a "rubble-pile" body, which earlier experienced partial destruction due to collision with other asteroids and then coalesced; therefore, it may contain numerous pieces or even be composed of former fragments.

Three special groups of asteroids at eccentric orbits that approach or even cross the Earth's orbit (Apollo, Amor, and Aten) are specifically distinguished and referred to as near-Earth objects (NEOs), see Fig. 4.9. These asteroids reach a few kilometers in size and may impact the Earth, representing a potential threat to the

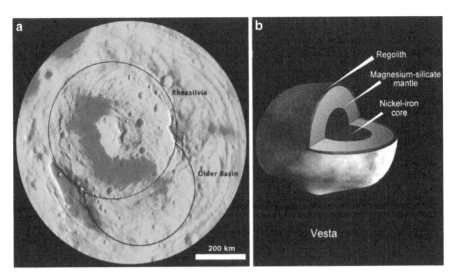

Fig. 4.7 (**a**) Vesta's Rheasilvia basin with the central peak inside of 2.5× taller than Mount Everest; (**b**) Vesta interior; it appears to have an iron core and a silicate mantle (Courtesy of NASA/J. Green)

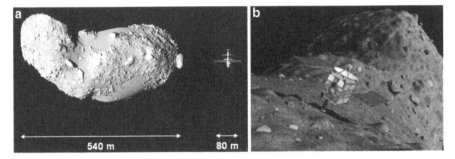

Fig. 4.8 (**a**) Asteroid Itokawa close-up view. Itokawa is of dumbbell shape (resembling that of Halley comet) with a sort of straight arc of smooth surface which is contrasted to otherwise rough surface of the whole body pitted up with boulders of different sizes. It is probably a rubble pile and may contain pieces trying to coalesce; (**b**) Image returned by Habayusa spacecraft shown flying over the asteroid in artists concept (Courtesy of JAXA)

home planet. The probability of impact strongly depends on the size of the body: for large ones it is rare. Meteorites are collisional fragments of asteroids that have fallen and been found on the Earth's surface. Asteroids and meteorites are commonly classified by their composition as stony, iron, and iron-stony. The most primitive stony meteorites (carbonaceous chondrites) come from the outer part of the MAB; they have encapsulated primordial material and, hence, appear to bear the most important information from the time of the solar system's origin. These meteorites have never been heated to melting temperatures (except for small refractory inclusions or

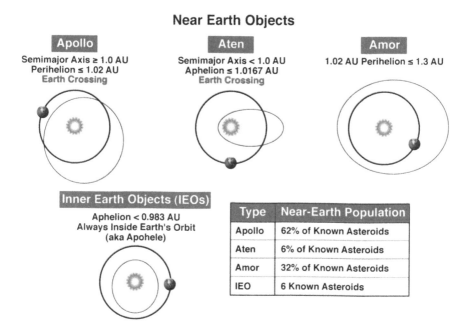

Fig. 4.9 Near Earth Objects (NEO)—three groups of asteroids at eccentric orbits approaching or even crossing the Earth' orbit: Apollo, Amor, and Aten (Courtesy of NASA)

chondrules imbedded in their matrix in the course of formation) and they are thought to retain the composition of the proto-Sun and not be affected by settling of heavy elements.

Another reservoir of asteroids (and also comets) is the Kuiper Belt (see Figs. 1.4 and 4.10). It is located near the ecliptic plane beyond the orbit of Neptune and extends from 40 AU to approximately 100 AU. The most abundant region of the "classic" trans-Neptunian objects (TNOs) is within 50 AU, the regions at 30, 40, and 45 AU experiencing some scattering. The overall number of bodies in the Kuiper Belt (which can also be regarded as the region where a planet much larger than Pluto failed to be formed) is estimated to be ~10^8. More than 10^4 of that number could be larger than 200 km in size. About a thousand TNOs have been discovered to date. Among them bodies comparable to Pluto in size (Eris, Makemake, Haumea, Sedna, Quaoar, etc.), which are characteristic representatives of this family of icy miniplanets, have been detected. They are called "plutinos" because Pluto is one of the group (Fig. 4.11). New Horizons, the first space mission to the dwarf planet Pluto and the Kuiper Belt, is now approaching its final destination. It will explore Pluto and its four moons during a well-choreographed flyby (Fig. 4.12).

Bodies in the Kuiper Belt are stabilized in their orbits and retain their positions for a long time (in some cases comparable with the age of the solar system) because of mean motion resonances with Neptune, which are revealed as integer ratios of their periods of revolution around the Sun (1:2, 3:4, 2:5, 3:5, 4:7, etc.); see Figs. 4.13 and 4.14. For example, for 1:2 resonance an object in the Kuiper Belt will move

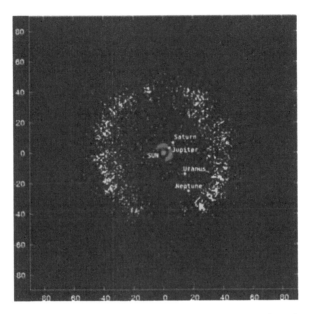

Fig. 4.10 The Kuiper belt around the Sun (*red spot*). The scales along horizontal and vertical axis are in AU. The four outer planets are shown (*blue dots*). The known objects in the Kuiper belt are colored *green* and scattered objects outside are colored *orange*. Jupiter's Trojans are colored *pink*, whereas Neptune's few known Trojans are *yellow*. The scattered objects between Jupiter's orbit and the Kuiper belt are known as Centaurus. The pronounced gap at the bottom is due to difficulties in detection against the background of the plane of the Milky Way (Courtesy of Minor Planet Center and Wikipedia)

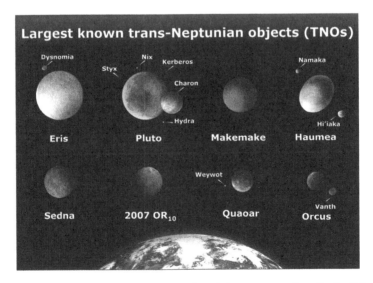

Fig. 4.11 Mini-planets ("plutinos") in the Kuiper belt as compared to Earth: Eris, Pluto, Makemake, Haumea, Sedna, 2007 OR$_{10}$, Quaoar, Orcus. Artistic montage (Adapted from Wikipedia)

Fig. 4.12 New Horizons—
first mission to the dwarf
planet Pluto and the Kuiper
Belt—will explore Pluto and
its (now) four moons during a
well choreographed flyby
(Courtesy of NASA, ESA,
and M. Showalter)

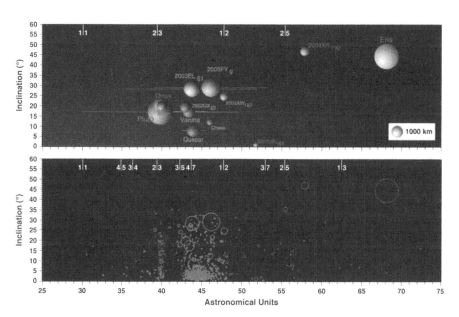

Fig. 4.13 Resonances of TNO objects (Plutoids including) in the Kuiper Belt. The bodies are plotted with regard inclinations of their orbits (NASA Courtesy)

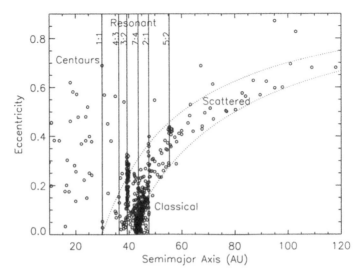

Fig. 4.14 Resonance structure of the Kuiper Belt. Classical resonant zone, Centaurs, and scattered objects are distinguished in semimajor axis—eccentricity plot (Credit: S. Sheppard)

twice as slowly as the planet. More than 200 quite large bodies located in the densest part of the Kuiper Belt populated with plutinos are locked in 2:3 resonance with Neptune. In particular, while the orbits of Pluto and Neptune intersect, they avoid collisions because of this resonance.

TNOs are characterized by a great variety of physical processes that manifest themselves in differences of their albedo A (from 2.5 to >60 %) and spectral features, which allow the properties of their surfaces to be diagnosed. In particular, H_2O, CH_4, and N_2 ice deposits were detected on some of the bodies, making them akin to Pluto and Triton, although the mean density of these objects is appreciably lower and closer to that of the nuclei of comets (~0.5 g/cm^3). Interestingly, about 11 % of the investigated objects are binary systems; Pluto even has four satellites.

There are special groups of asteroids called Trojans in the Jupiter and Neptune orbits (see Fig. 4.10). As we mentioned in Chap. 1, they are located at the libration (Lagrange) points L4 and L5 along a planet's orbit. Trojan objects share a planet's orbit, straying not far from 60° ahead of L4 or behind L5. They are in 1:1 resonance with the planet and move with it around the Sun. We know of nearly 1,600 Trojans of Jupiter and even more bodies in the Lagrange L4-L5 points of Neptune. They may well be bodies from the Kuiper Belt captured by the forming planet at an early chaotic phase of the solar system evolution (in the first 400–500 Myr), contrary to the traditional viewpoint that they were formed as the planet itself was growing and were then captured into 1:1 resonance. This might have occurred when Jupiter and Saturn were initially in 1:2 resonance and then reconstructed their orbits resulting in Kuiper Belt disturbance and some scattering (Fig. 4.14). The similarity between the color albedo of Trojans and comet nuclei (the growth in reflectivity with wavelength) and the absence of distinct spectral signatures, as in most trans-Neptunian bodies, provide evidence for the hypothesis of their connection with TNOs. Similar properties are characteristic of the groups of bodies located between the orbits of Jupiter and Neptune called the Centaurs, which are apparently also genetically related to the Kuiper Belt and partially evolve to the orbits of Jupiter-family comets (see Fig. 4.10). According to numerical modeling, a co-orbital fraction of ~2.4 % among the Centaurs is expected under trans-Neptunian supply. It was also shown that about 0.4 % and 2.8 % of captured bodies in the Centaurs population remain in temporary Uranian and Neptunian co-orbitals, respectively, for about one million years before becoming Centaurs.

Comets

Comets are distinguished as the most pristine bodies—remnants of the gaseous-icy outer planets formation (planetesimals)—which were thrown out by gravity perturbations to the periphery of the solar system beyond and which (some of them) irregularly come back approaching the Sun. It is therefore natural to consider the comets as "probes" of the Galactic regions nearest to the solar system. Because they contain primitive matter, comets are studied to unlock the very secrets of solar system formation. Historically, they were associated with the threat of the heavens to humans and with different catastrophes (see Fig. 4.15). Comets are classified as short-period (orbiting the Sun for less than 200 years) and long-period (having a period of more than 200 years). Short-period comets, such as Halley's Comet (Fig. 4.16a), belong to

Fig. 4.15 Picture from popular magazine before Comet Haley encounter with Earth in 1910 (Public domain)

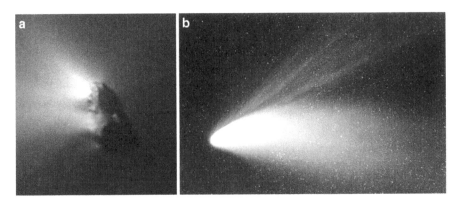

Fig. 4.16 (**a**) The nucleus of Halley comet from the distance of 4,000 km. The mean size of the nucleus is ~15 km. The streams of gas and dust (jets) from the icy surface of the nucleus as the comet approached the Sun are seen on the left. An image from the Giotto spacecraft. ESA courtesy. (**b**) Image of Haley Bopp comet during its encounter with the Sun. A small nucleus (~10 km) is hidden deep inside a bright region—coma (cometary atmosphere) tens of thousands of kilometers across produced by the sublimation of gas and dust from the icy surface of the nucleus. Extended type I and II tails are clearly seen (Courtesy of D. di Cicco)

giant planet families, specifically to the well-known Jupiter and Neptune families, and to the Kuiper Belt. The main family of comets is located in the Oort Cloud at 10^4 to 10^5 AU (see Fig. 1.4). Comets in the Oort Cloud occupy a very vast region well beyond the Kuiper Belt: the outer boundary extends to approximately quater the distance to the nearest stars (~60,000 AU). They periodically experience gravitational perturbations from giant interstellar gas-dust clouds and the Galactic disk, as well as from random encounters with stars. Some of the bodies from the cloud then pass to highly elliptical orbits and can be observed as long-period comets as they approach the Sun (Fig. 4.16b). Subsequently, under the influence of gravitational perturbations from planets, they either replenish the well-known families of short-period comets regularly returning to the Sun or leave the solar system forever by passing to parabolic or hyperbolic orbits. However, the main source of short-period comets is the Kuiper Belt. It experiences gravitational perturbations from Neptune, causing a relatively small fraction of icy bodies to migrate, as will be seen from the subsequent analysis, into the inner solar system, replenishing bodies in the MAB.

Comets are icy bodies consisting of a nucleus, atmosphere (coma), and tail. The nuclei of comets have small sizes, from several to several tens of kilometers, and a very low mean density, which usually does not exceed fractions of grams per cubic centimeter. This suggests a porous structure of these bodies composed predominantly of water ice and some other low-temperature condensates with an admixture of silicates, graphite, metals, and organic compounds. The water-ice composition of comets is explained by the fact that the water molecule is very abundant in the solar system and, not coincidentally, H_2O ice also accounts for a significant fraction of the mass of the satellites around the giant planets and other small bodies. The processes involving liquid water at the initial evolutionary stages of the solar system are evidenced, in particular, by the mineral composition of meteorites and, consequently, their parent bodies located in the MAB.

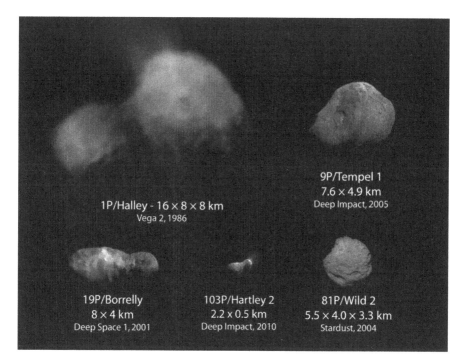

Fig. 4.17 Comet's nuclei visited by spacecraft (Courtesy of NASA)

Based on the measurements made during the encounter of spacecraft with several comets (Fig. 4.17), we can say that, on the whole, the nucleus corresponds in its physical properties to the model of a dirty snowball (or a ball of frozen dirt) proposed back in the middle of the past century by American planetary scientist Fred Whipple (Fig. 4.18). An atmosphere (coma), halo, and tail are developed due to ice sublimation when a comet traveling along its usually very eccentric orbit approaches the Sun and the insolation grows. Sublimed particles are partially ionized. The coma is tens of thousands of kilometers in extent, the hydrogen halo is hundreds of thousands of kilometers in size, and the tail stretches in the antisolar direction for millions of kilometers. There are two types of tails: one is mainly composed of fine dust pushed away from the coma by the solar light pressure, and the second consists of mainly ionized particles influenced by the solar wind plasma and interplanetary magnetic field. Most molecules pertinent to comets have been detected in the coma (Fig. 4.19). It is composed mainly of hydrogen, water, hydroxyl, carbon-, nitrogen-, and sulfur-bearing molecules, and hydrocarbons and forms the comet head which shines due to luminescence processes and is partially ionized by short-wavelength solar radiation. Dust is streamed away into the atmosphere during ice sublimation. Following the terminology introduced by the Russian astronomers F. Bredikhin and S. Orlov, sublimed submicrometer particles affected by the solar light pressure produce a comet's tail of type I, while the plasma forming in the outer coma when interacting with the solar wind produces a tail of type II. Both tails are clearly seen in Fig. 4.16b.

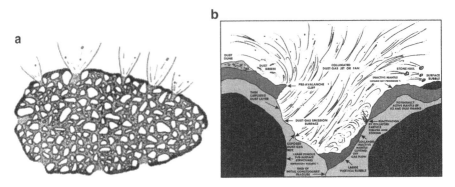

Fig. 4.18 (**a**) Model of comet nucleus as a dirty snowball; (**b**) Gas and dust sublimation from irregular comet icy surface when comet travelling along very eccentric orbit approaches the Sun, insolation grows and atmosphere (coma), halo and tail are developed (Courtesy of F. Whipple)

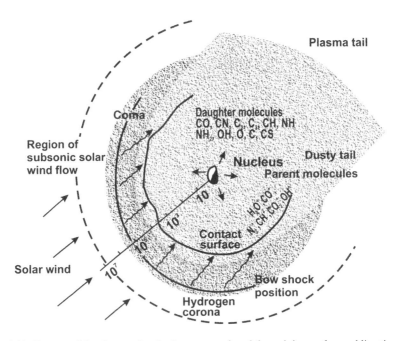

Fig. 4.19 Parent and daughter molecules in coma produced through icy nucleus sublimation and model of interaction of comet with solar wind and plasma tail formation (Courtesy of L. Marochnik)

Studying the nonstationary heat and mass transport processes in a porous nucleus and the formation of an inhomogeneous structure of the surface from which an icy conglomerate sublimates and a gas-dust coma is formed is of key importance in understanding the nature and evolution of comets at various heliocentric distances.

As the kinetic modeling results showed, the flow near the nuclei of active comets in the entire daytime hemisphere is close to an equilibrium one, the gas density decreases rapidly as one recedes from the surface of the nucleus, with the temperature dropping to a few kelvins at a distance of ~ 10 km from the nucleus due to adiabatic expansion into a vacuum followed by the fast temperature increase, and the well-defined jet is formed near the symmetry axis. Several jets attributable to intense removal of gas and dust are seen in the image for the nucleus of Comet Halley obtained by the Vega and Giotto spacecraft during their flyby excursions (see Fig. 4.16a). Such inhomogeneous sublimation from the surface of the nucleus can be explained by thermal deformations producing faults and fractures in the dust-ice crust formed during successive approaches of the comet to the Sun.

It is certainly a challenge to probe a comet nucleus and/or measure physical properties and chemical composition of fragments of pristine composition. An attempt has been undertaken with NASA's Deep Impact spacecraft launched in 2005. At its close-up encounter with Comet Tempel 1, it delivered a 372-kg copper projectile which impacted the nucleus with a speed of 10.3 km/s and investigated the aftermath of the collision. The image of the event is shown in Fig. 4.20. Later in 2010 the spacecraft flew by Comet Hartley 2 and then in 2013 imaged long-periodic Comet ISON coming from the outermost reaches of the solar system before its perihelion approach. Astronomers expected to observe all phases of this bright comet's (Fig. 4.21) encounter with the Sun. However Comet ISON was destroyed when approaching the Sun in November 2013, emerging from behind the Sun as a diffuse cloud of dust.

We may conclude that comets are challenging objects in terms of both the physicochemical processes involved and especially as carriers of the pristine matter they encapsulate in their composition, which would allow us to reconstruct the

Fig. 4.20 Comet Tempel 1 up close with Deep Impact spacecraft at the moment of projectile impacted comet on July 4, 2005 (Courtesy of NASA/M. A'Hern)

Fig. 4.21 Astronomical observations the bright long-period comet ISON that was destroyed when approaching the Sun in November, 2013 emerging from behind the Sun as a diffuse cloud of dust (Courtesy of NASA/HST)

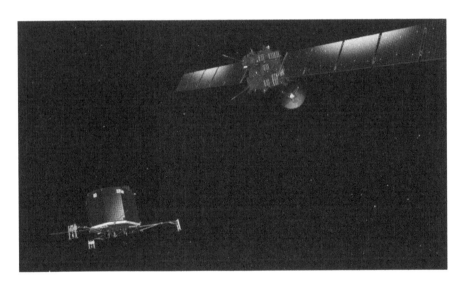

Fig. 4.22 ESA Rosetta mission to Churyumov-Gerasimenko comet. Spacecraft (bus) and probe to land on the comet surface are shown. Artist's concept (Courtesy of ESA)

solar system formation. ESA's Rosetta mission encountered Comet Churyumov-Gerasimenko in August, 2014 and it is intended to land a capsule on the nucleus for direct in situ measurements in November during the comet's approach of the Sun (Fig. 4.22). It is hoped that this will manifest the next breakthrough in our knowledge about comets. See the summary information sheet in Fig. 4.23.

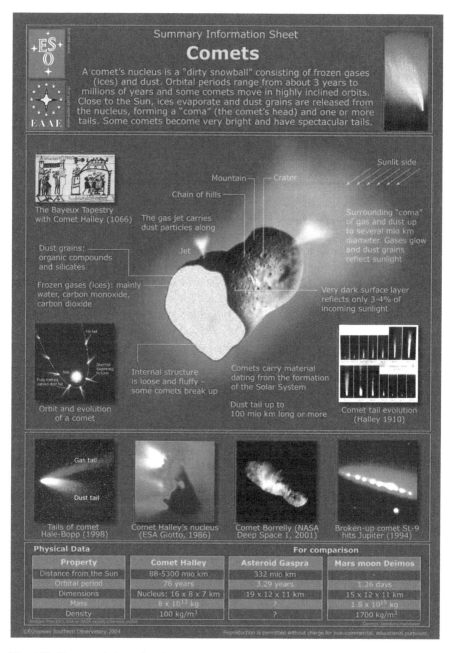

Fig. 4.23 Summary information sheet on comets. Some distinguished comets are shown; comet Halley nucleus is compared with asteroid Gaspra and Mars' satellite Deimos (Courtesy of ESA/ESO/EAAE)

Interplanetary Dust

Asteroid collisions and dust particles ejected by comets are the main source of interplanetary dust particles—another category of small bodies. They range in size from
nanometer- to millimeter-sized grains, though the lower threshold is as small as
molecular clusters. The millimeter-centimeter boundary is used as a conditional
boundary between dust particles and meteoroids. The surfaces of tiny interplanetary
dust particles are space-weathered by the solar wind, causing amorphous rims to
form on their surfaces. Implantation of protons may react with oxygen in the silicate
minerals of particles forming trace amounts of hydroxyl ions and/or water. Similarly,
amorphous rims up to ~ 150 nm thick were found on lunar and asteroid regolith
grains, whereas tiny pockets of water have been discovered in dust grains returned
by the Stardust space probe.

Dust particles accompanying the process of sublimation from a comet nucleus
are known as meteor showers. They are periodically observed in the night sky as
"shooting stars" when Earth intersects the dust torus left behind by a comet moving
along its circumsolar orbit. The radiants of the showers are named for the constellations they point to, such as Aquarids, Perseids, Quadrantids, etc. Earth periodically
meets some of these dust fluxes when orbiting the Sun and intersecting the respective torus. Some dust can be lifted from planets and their satellites or adjoin to them.
The most familiar is the zodiacal light seen from Earth as a diffuse glow in the west
after twilight and in the east before dawn. This wedge-shaped cloud is interplanetary dust that lies along the ecliptic and reflects sunlight (see Fig. 10.3c).

There is also interstellar dust sweeping through the solar system. Basically, dust
particles are randomly distributed in interplanetary space, where the smallest particles spiral towards the Sun under the influence of the Poynting-Robertson effect (an
inequality between absorbed and radiated solar energy for relatively small grains
that forces them to depart from Keplerian orbits), while the larger particles form
torus-like configurations along the comet orbits.

Meteorites

Like dust particles (but much larger), meteorites are collisional fragments of asteroids surviving atmospheric entry that have fallen and are found on the Earth's surface. They should be distinguished from meteors, which are particles that are
destroyed on atmospheric entry and observed like a flash ("shooting stars").
Different types of iron, stony, and iron-stony meteorites are distinguished; they are
generally classified as chondrites and achondrites (Fig. 4.24). The most prevalent
chondrites (of at least 14 different types) contain small, partly glassy spheres, called
chondrules, composed of refractory elements embedded in their matrix and preserved since the complex origin and evolution of the protoplanetary disk history.
Achondrites are meteorites lacking chondrules (commonly referred to as planetary
meteorites) which derive from current or former bodies of sufficient size which

Fig. 4.24 (**a–d**) Examples of several meteorites: (**a**) the largest known intact iron meteorite Hoba of mass ~60 ton found in Namibia; (**b**) an Esquel meteorite of the pallasite type; (**c**) ordinary chondrite with abundant chondrules and brecciation found in Algeria; (**d**) whole stone, showing wrinkled fusion crust and gray interior (Adapted from Wikipedia and M. Farmer). (**e–g**) Examples of Martian SNC (Shergottite—Nakhla—Chassignite) meteorites: (**e**) Shergottite; (**f**) Nakhla; (**g**) Chassignite (Courtesy of B. Fectay, N. Classen, and M. Horejsi)

contained enough internal heat to become differentiated with the emergence of a dense core, mantle, and crust. The disaggregated cores of such bodies of protoplanetary/asteroid size represent the relatively rare (~5 % of the whole population) iron meteorites and also the pallasites.

Of special interest are the most primitive carbonaceous chondrites composed of pristine matter. It is believed they retained this matter from the protoplanetary cloud and the gas-dust accretion disk in their composition, because they have undergone the smallest changes in the course of evolution, and their study is essential for understanding the early stages of the solar system. It is important to emphasize that various hydrocarbons, carbides, nitriles, polymers, and organics, including numerous isomers of amino acids, were found in the carbonaceous chondrites of several C classes (CIs, CMs, COs, CVs, and CBs) coming from different parent bodies. Their isotopic composition is unlike that of Earth, which means that they were not contaminated by being exposed to the Earth's environment after they were found, and this confirms their extraterrestrial origin. By their composition meteorites fit into many common classes of asteroids.

Another groups of particular interest are meteorites associated with the Moon and Mars origin found in Earth's deserts and Antarctic. This rather small collisional fragments ejected from planetary bodies when impacted by a large projectile is a new type of samples available and accessible for laboratory study of chemical/mineralogical composition and petrology of the Moon and Mars surface. There was found 135 lunar samples of total mass 55 kg which are well fitted to the main three types of lunar rocks: basaltic, ANT (anorthosite -norite-troelite) and KREEP (see Chap. 2). Martian SNC meteorites are genetically related shergottites, nakhlites, and chassignites (see Fig. 4.24) in 61 samples that were found, until recently, mostly in the Antarctic. A study of the physical process of fairly energetic collision by a

Fig. 4.25 Diagram of fitness of gas abundance (in log particles per cm³) trapped within shock glass veins and pockets of shergottite meteorite sample and in the Martian atmosphere measured by the Viking landers (Courtesy of T. Irving, University of Washington)

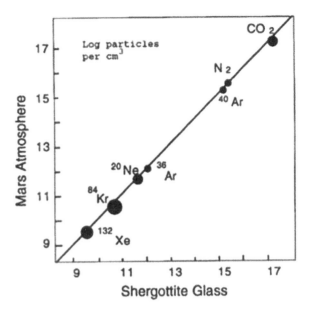

small asteroid with Mars (which has an acceleration due to gravity about 0.38 that of Earth) showed that it is possible to overcome the escape velocity and therefore, to excavate material consisting of rocks outcropping at the surface or even samples from a certain depth by a mechanism called spallation. This is confirmed by evidence that the material of Martian meteorites experienced moderate to high shock pressures. The most important evidence, however, is geochemical. Trapped within shock glass veins and pockets of several samples of ultramafic shergottite were found small amounts of different gases which turned out to have the exact same proportions as those in the thin Martian atmosphere measured by the Viking landers (see Fig. 4.25). Another line of evidence that the samples are Martian meteorites comes from other diagnostic criteria such as a narrow range of oxygen isotopic compositions different from those of any other achondritic meteorites; no metallic iron in iron-rich oxide minerals (magnetite, chromite, ilmenite); the distinctive ratios Fe/Mn within pyroxene and olivine minerals; and the presence of an iron sulfide mineral pyrrhotite instead of the troilite typical for iron metal-bearing meteorites. Special chemical analyses of mineral composition also confirm the unique properties of the samples and their probable relationship with Mars.

Migration and Implications

The orbital properties of small bodies reflect the dynamics of regular and chaotic motions in the solar system manifesting themselves through migration processes and permanent interplanetary matter transport. The preceding discussion of the

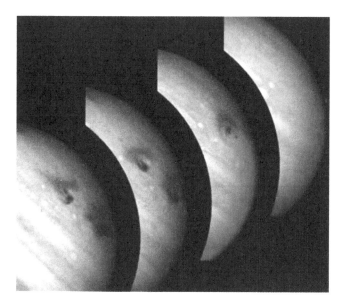

Fig. 4.26 The fall of fragments from Comet Shoemaker-Levy 9 torn asunder by tidal forces when approaching Jupiter in critical distance in 1994. This event clearly manifests the chaotic nature of migration and collision processes in the Solar system (Courtesy of NASA)

delivery of SNC meteorites from Mars to Earth is a convincing confirmation of these widespread phenomena. Migration is responsible for the collisions of comets and asteroids with the planets that leave behind numerous scars on the planetary surface. These processes are intrinsically related with dramatic events caused by impacts of the larger small bodies. There have been numerous disastrous events in the Earth's history, Chicxulub being the most familiar. The fall of a moderate-sized body in 1908 in Siberia over the Tunguska River is also well known. Elsewhere in the solar system, genuinely spectacular collisions of Comet Shoemaker-Levy 9 fragments with Jupiter occurred in 1994 (Fig. 4.26). This comet, which was earlier captured by the Jovian gravitational field and resided in Jupiter's orbit for nearly 50 years, was ultimately torn apart by tidal forces in the immediate vicinity of the planet. About 20 fragments fell down on the planet in 1994, entering the atmosphere with a velocity of 64 km/s. This was accompanied by a huge bright gas outburst illuminated in broad wavelengths, long-lived eddy formation, magnetospheric distortions, and polar auroras. This event, as well as bright spots periodically observed on the disk of Jupiter and presumably caused by asteroid impacts, once again persuasively confirms that collisional processes periodically occur in the solar system. It is a violent place containing objects that are real threats, particularly to Earth, so we should take warning and seek to determine what are real threats and how to mitigate them. In assessing a threat we first address the statistics, which is essentially the relationship between a body's size and an event's occurrence being probabilistic in nature, but with a good fit to the available data. An example is given in Fig. 4.27. A first step towards threat mitigation involves observations by the international

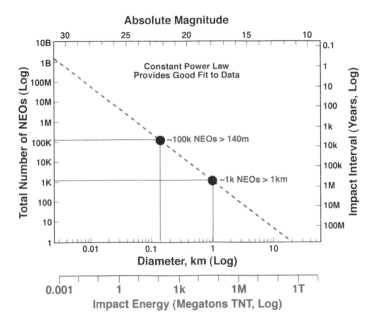

Fig. 4.27 Estimated number of different size NEOs and intervals of impacting Earth (Courtesy of NASA)

network of telescopes of potentially hazardous objects roughly more than 20–30 m in size, which are capable of bringing local devastation. The Chelyabinsk bolide[1] of 18 m across that exploded over a populated area in Russian Siberia in February 2013 was a real warning (Fig. 4.28). Unfortunately, objects sized below a few hundred meters are scarcely catalogued and, hence, poorly predictable. There is a concern now about NEO asteroid Apophis of a few hundred meters across which might collide with Earth in 2036. However, the recent most accurate calculations show that the threat seems to be exaggerated. However, the asteroid-comet threat problem and projects on how to mitigate it draw growing attention.[2]

We are confident that catastrophic events and mass transport were of great importance for the evolution of planets and their atmospheres. Indeed, numerical modeling has given convincing evidence that intense bombardment of early Earth and other planets by comets and hydrated asteroids about four billion years ago (period of late heavy bombardment, LHB) could contribute substantially to the volatile inventory, and therefore compensate for a volatile deficiency in the region of the

---

[1] A bolide is a very luminous meteorite passing through the Earth's atmosphere.

[2] Hunting for near-Earth asteroids is the goal of a Wide-field Infrared Survey Explorer (WISE) project called NEOWISE. In 2011 its all-sky survey was completed after the entire sky had been surveyed twice at infrared wavelengths. The WISE team has since combined all the data, allowing astronomers to study everything from nearby stars to distant galaxies. Significantly more sensitive than those previously released, these next generation all-sky images are part of a new project called AllWISE.

Fig. 4.28 Chelyabinsk bolide fall over populated area on Feb. 15, 2013. (**a**) Images of inverse trace before explosion at ~23 km height. (**b**) The largest fragment of 560 kg falling in the Chebarkule lake (Courtesy of S. Kolisnichenko)

terrestrial planets formation, invoking the mechanism of heterogeneous accretion. Actually, this implies accretion of primitive scattered planetesimals by a growing planet, whereas some planetesimals entered earlier into the planet-forming annulus. This allows us to explain sufficient volatile storage in the inner planets' atmospheres and hydrospheres. Finally, the migration of small bodies, including interplanetary particles, might be closely related to the intriguing problem of life origin, as comets are considered as the most probable original biotic seed carriers.

In the early solar system evolution it is generally believed that the original (primary) atmospheres on the terrestrial planets captured from the protoplanetary cloud at the stage of their accretion have been lost. The fact that the abundance and isotopic composition of inert gases in the existing secondary atmospheres differ sharply from the solar ones is convincing evidence for this loss. Undoubtedly, degassing from the interiors during the subsequent evolution of these planets made significant contributions to the formation of their hydrospheres and atmospheres of secondary origin. However, this source apparently was not unique and sufficient to compensate for the loss of low- and moderate-temperature volatiles (water, nitrogen, carbon, sulfur, etc.) through Jeans dissipation because temperatures in the inner part of the terrestrial planets formation zone (confined by the above-mentioned "snow line" at about 3 AU) reached ~1,000 K. In particular, Earth condensed at this high temperature. The low masses of these planets, along with this high temperature, also contributed to the escape of the lightest atmophile elements into outer space.

Thus, the existing, relatively high abundances of volatiles on the terrestrial planets can be explained by their intense bombardment by comets and asteroids from the outer solar system at an early evolutionary stage. This scenario is known as the above mentioned heterogeneous accretion mechanism. It involves orbital dynamics of small bodies, primarily the volatile-rich comets and the carbonaceous chondrites class asteroids, migrating from the Kuiper Belt inward. Modeling suggests that these bodies were initially captured on a Jupiter-crossing orbit and then migrated further into the inner solar system, many eventually becoming bodies of the MAB and NEO families. However, an even more efficient mechanism of direct migration could be from an original formation, which we now call the main asteroid belt (MAB), that had a much larger quantity of primordial bodies at the time of the planets' accumulation. A fairly high estimated collision probability of small bodies between themselves and with planets supports the idea of permanent fragmentation/ejection of material and its transport in the interplanetary medium. In particular, it turned out that one object could have collided with Earth every ~0.5 Myr, and similar estimates (by a factor of 3 lower) were derived for Mars and Venus. Interestingly, one of every 300 objects migrating inward was shown to fall on the Sun (the so-called *sun-grazers*).

The key corollary of the model is the confirmation of an important role of comets and carbonaceous chondrites as a source of volatiles in the evolution of the terrestrial planets. Based on the estimated lower limit for the collision probability of $\sim 4 \times 10^{-6}$ and the estimated total mass of the Jupiter-crossing planetesimals, ~100 M_E (M_E is the Earth's mass), we obtain the mass of bodies that collided with the Earth, $M \sim 0.0004\, M_E$. Assuming further that the content of water ice was about half of the total mass of an impact body, we derive the water mass delivered to Earth from the feeding zone of the giant planets as $\sim 2 \times 10^{24}$ g. This is almost a factor of 1.5 larger than the water reserves in the Earth's ocean. Modeling also argues that Venus and Mars obtained approximately the same volume of water per unit mass of the planet. This is in accord with the earlier discussed ideas about the existence of ancient oceans on the neighbor planets as well. However, an ocean was presumably lost due to the development of the runaway greenhouse effect on Venus and one was buried in the Martian cryosphere due to dramatic ancient climate change (see Chap. 2).

Naturally, the estimates of the amount of water supplied to the Earth by comets are valid only in order of magnitude, and we are far from asserting that this source was unique. Undoubtedly, a certain contribution was made through degassing from the interiors, although, as has been mentioned, so far it is hardly possible to determine the relative importance of the endogenic and exogenic sources. However, it is important to emphasize that an estimate with an error of 50 % and not by orders of magnitude is undeniable evidence that the comet model is realistic. Basically, it is constrained by the deuterium-to-hydrogen ratio D/H derived from the measurements of H_2O and HDO in comets. Earlier measurements of the ratio in six comets, in particular, Hyakutake and Hale-Bopp, turned out to be twice that in the terrestrial oceans, namely 2.96×10^{-4} compared to the standard oceanic value (Standard Mean Ocean Water, SMOW) of $D/H = 1.57 \times 10^{-4}$. However, these were long-period comets from the Oort Cloud, and one could reasonably assume that they differ significantly in genesis from those formed in the inner regions of the protoplanetary disk (say, Jupiter-family comets (JFC)) at different temperatures, which should have affected the D/H ratio. Indeed, most recent measurements of the D/H ratio in Comet Hartley 2 made with the Herschel Space Telescope gave a value of 1.57×10^{-4}, similar to that in the Earth's oceans. Let us emphasize again that in addition to comets another endogenic source, hydrated asteroids from the MAB could also supply a certain fraction of water and other volatiles to the terrestrial planets.

Apart from comets and asteroid-size bodies, interplanetary dust particles play an important role in matter transport and exchange. The relative fraction of the cometary and trans-Neptunian particles containing ~10 % of volatiles in their composition in the overall balance of the dust that fell on the terrestrial planets was estimated based on numerical modeling of dust migration as well. It turned out that, compared to small bodies, the contribution of dust to the supply of volatiles is smaller by three or four orders of magnitude. However, dust particles could be most efficient in delivering organic or even biogenic matter to the Earth because they are considerably less heated when entering a planetary atmosphere under a relatively small angle of attack.

An even more intriguing possibility is the mechanism of small bodies/dust exchange between different planetary systems. One may assume that "life seeds" were transported through space and sedimented on Earth by extrasolar meteoroids/planetesimals, and vice versa (lithopanspermia). Theoretical estimates involving low-energy (weak) transfer with quasi-parabolic orbits showed that transport of bodies originally embedded in a star cluster is feasible. Such a situation could exist for clusters of young stars with forming planetary systems and, in particular, for our Sun soon after its birth. This means that life-bearing (if any) extrasolar planetesimals could have been delivered to the solar system and could have served as seeds for the origin of an early microbial biosphere. In any case, small bodies are addressed as carriers of prebiotic matter and thus as a clue to answer the fundamental question of the origin of life, which will be addressed in more detail in Chap. 9.

Chapter 5
The Sun and Heliosphere

The Sun As a Star: General Properties

The Sun is an ordinary star of G2V class located in our Galaxy, the Milky Way, about two-thirds away from its center (26,000 ly, or ~ 10 Kpc and ~ 25 pc from the plane of Galaxy) and revolving around its center with a velocity of ~ 220 km/s (period 225–250 million years, a Galactic year) in a clockwise direction, as viewed from the north galactic pole (Fig. 5.1). The motion of the Sun against the background stars (and horizon) has been used to create calendars since ancient times, calendars which were used first of all for agricultural practices. The Gregorian calendar, currently used nearly everywhere in the world, is essentially a solar calendar based on the angle of the Earth's rotational axis relative to its local star, the Sun. The Sun's visual stellar magnitude is -26.74^{m}, and it is the brightest object in our sky. Its orbit around the Galaxy is expected to be roughly elliptical and perturbed due to the galactic spiral arms and nonuniform mass distributions. In addition, the Sun oscillates up and down relative to the galactic plane two to three times per orbit. This exerts a strong influence on stability at the outer edge of the solar system, and in particular, the invasion of comets from the Oort Cloud (see Chap. 4) inside the solar system, resulting in an increase of impact events.

The Sun is currently located close to the inner rim of the Orion Arm, traveling through the Local Bubble within the Local Interstellar Cloud of the Galaxy, which is filled with rarefied hot gas, possibly a supernova explosion remnant. As we will see in Chap. 10, this region is called the galactic habitable zone. The direction that the Sun travels through space in the Milky Way relative to other nearby stars (toward the star Vega in the constellation Lyra at an angle of roughly 60 degrees to the direction of the galactic center) is called the solar apex. Interestingly, since our Galaxy is also moving with respect to the cosmic microwave background radiation (CMB, see Chap. 10) with a speed of 550 km/s in the direction of the constellation Hydra, the Sun's resultant (residual) velocity with respect to the CMB is about 370 km/s in

© Springer Science+Business Media New York 2015
M.Ya. Marov, *The Fundamentals of Modern Astrophysics*,
DOI 10.1007/978-1-4614-8730-2_5

Fig. 5.1 Schematic view of the Sun position in the Galaxy. Note that because we may see our own Galaxy Milky Way only edge-on (see Fig. 10.3) similar to Milky Way Andromeda galaxy (Fig. 10.2a) is shown here to scale solar system position properly (Credit: SAO/G. Fazio)

the direction of the constellation Leo. Note also that the Sun's motion is complicated by perturbations from the planets, especially Jupiter, forming a barycenter as the center of mass of the solar system.

The Sun is classified as a yellow dwarf star, which appears yellow on earth because there is some excess yellow light in its spectrum caused by blue light scattering in the atmosphere. The letter V in the G2V designation means that the Sun resides in the main sequence (MS) of the Hertzsprung-Russell (HR) diagram (see Chap. 6) and fusion processes occur in the interior. Early in its life, the Sun appeared to resemble the well-known young T Tauri star evolution and before entering the MS followed the Hayashi track where it contracted and decreased its luminosity while remaining at roughly the same temperature.[1] After reaching the MS the Sun eventually increased its luminosity by about 25–30 %.

Basically, the Sun is a plasma ball of about 1.5 million km across: its equatorial radius is 6.955×10^5 km (diameter 1.392×10^6 km), i.e., it is more than two orders of magnitude greater than the size of the Earth and an order of magnitude more than the size of Jupiter. The angular size of the Sun when observed from Earth is 31.6′–32.7′, and the obliquity to the ecliptic is 7.25°. The Sun rotates around its axis counterclockwise (as viewed from the ecliptic north pole) with a velocity of 7.189×10^3 km/h (sidereal rotation period at the equator of 25.38 days), whereas the

---

[1] Less massive T Tauri stars also follow this track to the MS, while more massive stars turn onto the Henyey track.

period at the poles is much longer: 33.5 days. This difference is due to differential rotation caused by convection and nonuniform mass transfer from the core outwards and related with redistribution of the angular momentum. In turn, differential rotation influences the magnetic field structure and, in particular, produces twisting magnetic field lines over time. These magnetic field loops erupt from the Sun's surface, triggering sunspots and solar prominences. The twisting action creates the solar dynamo and an 11-year solar cycle of magnetic activity as the Sun's magnetic field reverses itself about every 11 years.

When viewed from the moving Earth, the apparent observed rotation period is about 28 days. The Sun's shape is nearly spherical and oblateness is negligible (about 9 millionths), which means that its polar radius is only ~ 10 km less than its equatorial one. The mass of the Sun is 1.99×10^{33} g (~330,000 times the mass of the Earth) and the mean density is 1.41 g/cm^3 (nearly 4 times less than the Earth's). The Sun comprises 99.86 % of the mass of the whole solar system. The acceleration due to gravity (at the equator) is 274.0 m/s^2 (27.94 g), and the escape velocity is 617.7 km/s (55 times that of Earth). The Sun's "surface" (effective) temperature (T_{eff}=5,778 K) is associated with the visible layer—the photosphere—while, as will be seen below, the temperature in the central core is ~1.57×10^7 K and the temperature of the outer atmosphere (corona) is ~5×10^6 K. Under such high temperatures, gases are in the plasma state. Basically, the photosphere is responsible for all of the emitted radiation: the gases above it are too cool or too thin to radiate a significant amount of light. The luminosity of the Sun is enormous: 3.85×10^{33} erg/s, and it nearly corresponds to black-body (Planck) radiation at a temperature of ~ 6,000 K.

Following the evolutionary scenario typical of low- to medium-mass stars, the Sun is about halfway through the MS stage of fusing hydrogen into helium in its nuclear fusion reactions, and it will continue in the MS stage for approximately another 5 billion years. The Sun loses 10^{-14} solar masses every year, or about 0.01 % of its total mass over its entire lifespan.[2] Following theoretical models, soon after entering the MS (between 3.8 and 2.5 million years ago), the Sun's luminosity increased by about 30 %. The much fainter and cooler Sun before this time would have posed a problem for the planets' evolution. In particular, for Earth, the necessary temperature for preserving liquid water on the surface and for the origin of life could be accomplished only if efficient heating were available due to the presence of a significant fraction of greenhouse gases in the Earth's atmosphere to compensate for the lower insolation. This assumption allows us to resolve the faint young Sun paradox. In the follow-up stage, the Sun's luminosity as well as its radius continued to grow slowly. It is estimated that the Sun becomes about 10 % more luminous every 1 billion years. Respectively, planetary surface temperatures (including those of Earth) are slowly rising. Presently the amount of power that the Sun deposits per Earth's unit area (the solar constant) is 1,368 W/m^2 at the top of its gas

---

[2] Very massive stars can lose 10^{-7}–10^{-5} solar masses each year, while stars less massive than 0.25 solar masses (red dwarfs) fuse nearly all of their mass as fuel. This significantly affects their evolution and lifetime, which is much shorter for the massive stars, and vice versa (see Chap. 6 for more details).

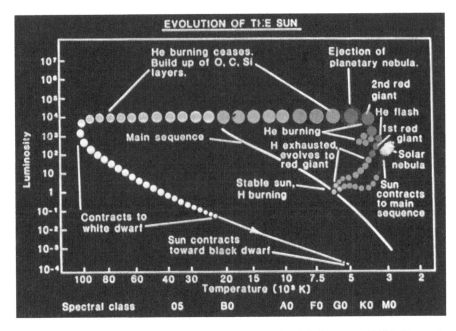

Fig. 5.2 Evolution of the Sun before, at and after residing the Main Sequence (NASA Courtesy)

envelope (atmosphere), which attenuates sunlight, and the power reaching the surface is nearly 30 % less (~1,000 W/m²) in clear conditions when the Sun is near the zenith.

By the end of its lifetime, the Sun will enter a red giant phase: its outer layers will expand as the hydrogen fuel at the core is depleted, replaced by helium ash, and the core will contract and heat up. Hydrogen fusion will continue along a shell surrounding a helium core, the shell steadily expanding as more helium is produced and the core temperature increases. Helium fusion in the core producing carbon will begin at ~ 100 million K. The mass of the Sun is not sufficient for later stages involving nitrogen and oxygen burning to begin, nor to explode as a supernova. Instead, intense thermal pulsations will cause the Sun to throw off its outer layers and form a planetary nebula, leaving behind a very hot core—a white dwarf—which will slowly cool and fade. The sequence of the processes of evolution for our Sun is shown in Fig. 5.2.

Structure and Energy

The Sun is divided into the following main regions: inner core, radiative zone, convective zone, photosphere, chromosphere, and corona, the latter two forming the solar atmosphere (Fig. 5.3). Historically, only the Sun's upper layers were accessible to direct observations; the interior was inaccessible to measurements. Great breakthroughs occurred during the last half century. Space exploration made it

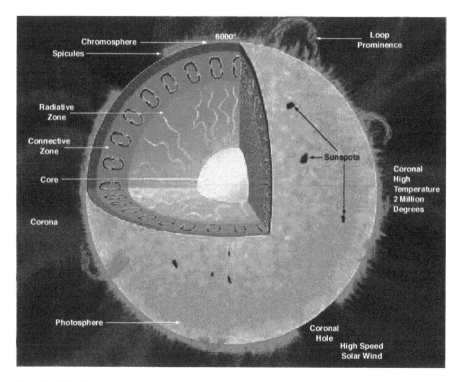

Fig. 5.3 The Sun's cross-section and the names of main regions. Some phenomena related to solar activity are shown at right (Credit: NASA)

possible to observe the Sun in the whole wavelength range, and the technique of helioseismology allows us to study the structure of the interior, which is opaque to electromagnetic radiation. Some 20 space missions have been undertaken (Pioneer, Skylab, Yohkoh, Ulysses, SOHO, CORONAS, Stereo, SDO, Hinode, to mention a few) which have allowed us to take a continuous survey of the Sun. They have provided detailed images of the Sun and its total spectral irradiance, magnetic activity, and dynamics, including those of the polar regions, which are especially informative when observed at different wavelengths. We have been able to monitor and quantify various phenomena of solar activity and how they affect Earth and other planets. Helioseismology has given us an opportunity to monitor the Sun's interior. Helioseismic measurements are made by ground-based and orbiting solar observatories and generally resemble the well-known seismology techniques used for the Earth. Helioseismology uses pressure (infrasound) acoustic waves traversing the Sun's interior to reveal and visualize its inner structure. The most informative are the waves having a period of 3–6 min generated inside the convective zone and experiencing reflection and refraction in the photosphere. These waves allow us to explore the temperature, pressure, and dynamics of different layers of the solar interior. This study is complemented by computer modeling. All these techniques have significantly advanced our understanding of the physical mechanisms underlying the Sun's behavior.

The core in the center of the Sun occupies about 20–25 % of the solar radius. Its temperature and density are above 15×10^6 K and 150 g/cm³, respectively. Under this high a temperature nuclear fusion with proton-proton (p-p) reactions converting hydrogen into helium (alpha particles) occurs. Hydrogen fuses to form helium in the proton-proton chain reaction:

$$4\,^1\mathrm{H} \rightarrow 2\,^2\mathrm{H} + 2\mathrm{e}^+ + 2\nu_e \left(4.0\,\mathrm{MeV} + 1.0\,\mathrm{MeV}\right)$$
$$2\,^1\mathrm{H} + 2\,^2\mathrm{H} \rightarrow 2\,^3\mathrm{He} + 2\gamma \left(5.5\,\mathrm{MeV}\right)$$
$$2\,^3\mathrm{He} \rightarrow\ ^4\mathrm{He} + 2\,^1\mathrm{H} \left(12.9\,\mathrm{MeV}\right)$$

These reactions result in the overall reaction:

$$4\,^1\mathrm{H} \rightarrow\ ^4\mathrm{He} + 2\mathrm{e}^+ + 2\gamma + 2\nu_e \left(26.7\,\mathrm{MeV}\right)$$

Here e^+ is a positron, γ is a gamma ray photon, ν_e is a neutrino, and H and He are isotopes of hydrogen and helium, respectively. One should be aware that the energy in tens of millions of electron volts (MeV) released by this reaction is actually only a tiny amount of energy. However, the overall energy produced constantly by the enormous numbers of these reactions is great, and it is sufficient to sustain the Sun's radiation output.

The process will be followed by a subsequent cycle involving helium-to-carbon burning. Currently, the conversion of protons into alpha particles occurs at the rate of 3.7×10^{38} protons/s. In order to maintain this reaction, the Sun fuses ~400 million tons of hydrogen each second, of which 0.7 % mass is converted into energy (~4.3 million tons, which is equivalent to 9.2×10^{10} megatons of TNT per second) to generate the aforementioned 3.85×10^{26} W. Interestingly, at this rate the Sun has already converted around 100 Earth masses of matter into energy. Nonetheless, that amount is negligible compared to the Sun's mass, and it will be able to run the fusion for another ~5 billion years. Let us not forget that the self-sustaining fusion reaction in the Sun's interior is slow—not like the reaction in a hydrogen bomb—and the tremendous power production of the Sun is explained by its large size rather than by a high power per unit volume release.

Energy generated with the p-p reaction inside the core transfers outward through the successive radiative and convective layers to the photosphere from which it is emitted as sunlight and the kinetic energy of plasma particles (Fig. 5.4). In contrast to the neutrinos that bear about 2 % of the total energy produced in the core with an escape time of ~2 s, a long time (about 10^7 years) is required for radiation to reach the photosphere because the solar interior is opaque to both gamma rays originally generated in the fusion reaction and the re-emitted photons of lower energy, in particular visible light escaping from the surface into outer space. It takes even more time (~30,000,000 years) for energy transport involving photons in thermodynamic equilibrium with matter. Each gamma photon is estimated to give birth eventually to millions of photons in the optical wavelength. Let us note that for years there was a neutrino paradox rooted in the discrepancy between the recorded number of neutrinos

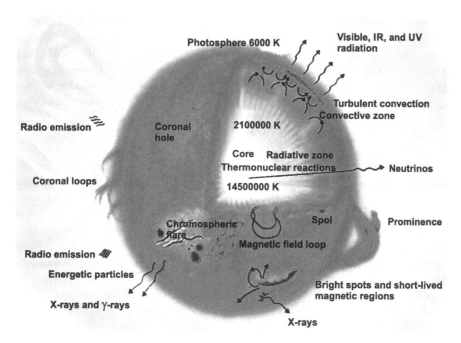

Fig. 5.4 Emissions from the different Sun's regions (Adapted from Wikipedia)

on Earth and the theoretical value (one-third to one-half of the number predicted by the standard solar model). The problem was successfully resolved recently when it was found that the neutrinos had changed "flavor" by the time they were detected. In other words, taking into account all three types of the existing neutrinos and their oscillation in matter allowed us to reconcile the model with the total neutrino emission rate.

The radiative zone extends from the core upper layer at about 0.25 to about 0.70 solar radii. It is still dense and very hot: the density drops a hundredfold (from 20 g/cm^3 to 0.2 g/cm^3), and the temperature drops down from 7 to 2 million K. However, the temperature lapse rate in the whole zone is subadiabatic, which prevents convection development. Heat transport upward is ensured by thermal radiation in the highly opaque medium when photons emitted by hydrogen and helium ions have a very limited free path before being reabsorbed by other ions, resulting in very slow heat transfer. A transition layer at the upper boundary of the radiative zone (the tachocline) separates the regions of nearly uniform rotation in the radiative zone from differential rotation in the convective zone above where shear motions and turbulence are created. This layer is assumed to be (at least partially) responsible for generation of the solar magnetic dynamo.

In the convective zone, heat exchange from a hot bottom layer to the relatively cool surface (photosphere) is performed by the hydrodynamic mechanism of convection rather than radiation. Convection is more efficient than radiation in thermal energy transfer because the plasma here is not sufficiently dense and hot: the temperature at the surface drops down to about 5,800 K and the density to 2×10^{-7} g/cm^3, i.e., only one ten-thousandth the density of air at sea level. The convection mechanism forming Bénard cells ensures both ascending fluxes of hot plasma upward to the surface and descending fluxes of cool plasma from the surface downward. Descending flux reaches the transition tachocline level where it heats, and then hot plasma lifts up again. Convective cells form thermal columns and imprint on the Sun's surface as numerous hexagonal prisms, a process called granulation. Granules have a higher temperature than their surrounding areas.

As we have mentioned, the photosphere is the visible surface of the Sun, tens to hundreds of kilometers thick (only 0.05 % of the solar radii), with nearly a black-body radiation spectrum at a temperature of 5,777 K (Fig. 5.5). It is slightly less opaque than the Earth's atmosphere due to sharply decreasing amounts of H$^-$ ions which are a strong absorber of light in the visible wavelengths. The effect of what is called limb darkening is pertinent to the Sun: the optical depth for the hotter radiation coming from below is larger at the solar disk edges than at the center, and so images of the Sun look brighter at the center compared to the limb. Also, dark spots having a temperature lower than the ambient temperature (sunspots) periodically appear on the solar surface. They are recorded using Wolf numbers and reflect the level of solar activity, though, as we will see, the decimetric solar flux $F_{10.7}$ is more widely used as a solar activity index. Besides sunspots, dramatic events like solar prominences are observed in the photosphere, which are caused by the Sun's differential rotation. This results in the twisting together of the magnetic field lines over time, and the resulting magnetic field loops formation is responsible for the eruption of energetic plasma from the solar surface.

Basically, the medium above the photosphere is transparent to visible sunlight, which easily leaves the Sun, propagating in space. However, an absorption by gaseous elements occurring in the tenuous overlaying plasma layers (in addition to protons and electrons) presents the signature of the solar composition (Fig. 5.6).

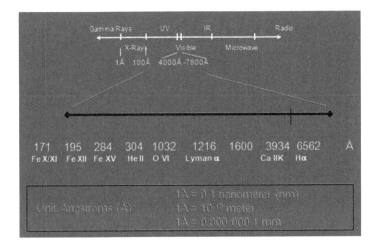

Fig. 5.5 Solar electromagnetic spectrum roughly corresponding to black body radiation under temperature 5,777 K

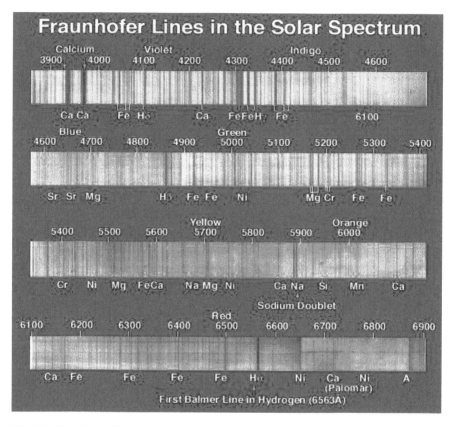

Fig. 5.6 Fraunhofer Lines in the solar spectrum indicating different elements of the Sun composition

This is a classical example of an absorption spectrum in which numerous lines of the chemical elements in different states of ionization are distinguished in the emitted background radiation. The spectral lines are called Fraunhofer lines, after the German physicist, Josef von Fraunhofer, who independently discovered this phenomenon as early as 1814.[3]

The layers located above the photosphere belong to the solar atmosphere. Here dramatic variations depending on solar activity are especially pronounced when the Sun is observed at different wavelengths (Fig. 5.7). The atmosphere involves the chromosphere and corona, separated by a thin transition region of about 200 km thick, and can be probed in the whole electromagnetic spectrum from gamma and X-rays through radio waves. The chromosphere is an irregular and very active region about 10,000 km thick (~1 % of the solar radii) located at a temperature minimum of about 4,100 K at 500 km above the photosphere, where simple molecules such as water and carbon monoxide have been detected. The temperature gradually increases to about 20,000 K at the top of the chromosphere. This is

---

[3] Note that British chemist William Hyde Wollaston first observed the dark lines in 1802.

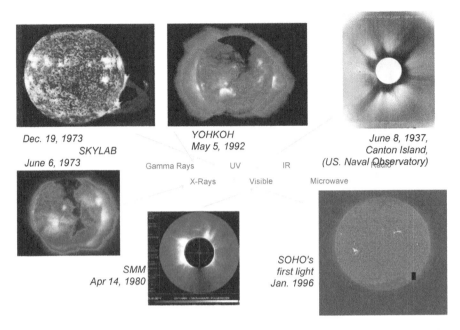

Fig. 5.7 Sun images in different wavelength. Granulation of the photosphere and different solar activities (especially in the corona) are clearly exhibited (Credit: NASA and ESA)

high enough for partially ionized helium to appear. Many other components are distinguished in both the absorption and emission spectra pertinent to this region. The chromosphere is observed in the emission lines of hydrogen and calcium and also as a colored flash at the beginning and end of the rather rare phenomenon of a total eclipse of the Sun (partial eclipses are more frequent, but they are not as informative). However, even more fascinating is the coronal dynamic structure clearly observed during total eclipses. The chromosphere is in permanent chaotic motion, exhibiting different features (spicules, filaments), the most spectacular being the chromospheric flares which originate here.

The corona is the outer part of the solar atmosphere extending in space well beyond the visible Sun and occupying a region exceeding its size by an order of magnitude. It is a source of X-ray emission. In the corona, particle density drops down to 10^9–10^{10} cm^{-3}, as compared to 10^{15}–10^{16} cm^{-3} in the chromosphere their free path is hundreds to millions of kilometers, and the kinetic temperature reaches several million degrees (up to 20×10^6 K in the hottest regions). Obviously, the corona is heated by something other than direct heat conduction from the photosphere. However, it is not yet clear what causes such a high temperature. Wave heating mechanisms were suggested involving sound, gravitational, or magnetohydrodynamic waves produced by turbulence in the convective zone, transferred by turbulent motions upward through the photosphere, and ultimately dissipating their wave energy in the corona. Another proposed mechanism involves the idea of magnetic heating caused by the continuous magnetic energy due to magnetized plasma motion in the photosphere and the energy release through the processes of magnetic reconnection in solar flares of different scales. The efficiency of both these mechanisms is still doubted. As a compromise, Alfvén waves—low-frequency oscillations of plasma ions perturbing the magnetic field along and transverse to the direction of propagation—were invoked because they are capable of reaching the corona more easily than other waves. However, Alfvén waves (which represent a kind of combined electromagnetic-hydrodynamic wave) do not easily dissipate in the corona because they are dispersionless. Therefore, it is doubtful that the energy carried by the Alfvén waves would be sufficient to heat the corona to its enormous temperatures. Thus, the longstanding question about the coronal heating mechanism remains one of the open-ended questions in astrophysics.

The corona is a very dynamic region: multiple plasma filaments of peculiar shape rise above the solar surface (Fig. 5.8). They either connect regions of different magnetic polarity or do not follow the regular magnetic field structure and are responsible for the most powerful solar activity events of coronal origin referred to as Coronal Mass Ejectios (CME). Under the influence of the magnetic field self-organization processes occur. A permanent plasma outflow from the corona, known as the solar wind, which consists mostly of ∼ 1-keV protons, electrons, and some heavier ionized atoms, carries the material with the embedded magnetic fields through the solar system. This outflow is filled with plasma and forms a cavity around the Sun, called the heliosphere, which extends to the outer boundaries of the solar system. The coronal plasma is retained by the closed magnetic field lines; however, coronal holes

Fig. 5.8 Image of the solar corona during powerful flare – Coronal Mass Ejections (CME). Numerous eruptive filaments connected with active regions on the solar photosphere and huge plasma bubbles extending to distances of more than two million km are observed. Self-organization processes under the influence of a magnetic field occur. Such frequently occurring events on the Sun give rise to magnetic storms on Earth. The image obtained in 2002 by SOHO spacecraft (Credit: ESA)

form in the regions of unclosed magnetic field lines, and here plasma is most easily ejected into outer space.

The energy of electrons is sufficient to dissipate from the corona, while proton emission is complemented by charge imbalance in the electrical field. The kinetic energy of solar wind particles is low (a few electron volts for electrons and a few kiloelectron volts for protons), but they dramatically increase their energy when they are accelerated in a planetary magnetic field (magnetosphere). The mean velocity of the solar wind plasma is ~400 km/s near the solar equator but ~700–800 km/s closer to the solar poles. When the solar activity increases, the typical velocity grows twofold to ~800 km/s at the equator, and the proton number density rises from a few to ~30 cm^{-3}. The magnetized solar wind plasma draws magnetic field lines from the Sun and, because of the Sun's intrinsic rotation, the carrying magnetic field lines become wrapped into an Archimedean spiral, the interplanetary magnetic field acquiring a sectorial structure. Solar wind particles travel along the magnetic field lines and interact with the planets when they meet them in regions of space corresponding to this field configuration; the Sun's magnetic dipole field strength reduces with the cube of the distance.

Composition

The Sun's composition corresponds to the cosmic abundance of elements in the universe. It consists mainly of hydrogen (73.46 %) and helium (24.85 %) by mass (~92 % and ~7 % by volume, respectively). Heavier elements (iron, magnesium, carbon, oxygen, nitrogen, silicon, sulfur, noble gases) compose only the remaining 1.69 % of the solar mass, which is still five times more than the mass of the Earth. Because the Sun is a heavy element-rich (Population I) star and thus contains a higher abundance of metals compared to the heavy element-poor (Population II) stars of earlier spectral classes, its metallicity is rather high ($Z=0.012$). Note that, in astronomy, heavy elements are called "metals," meaning all chemical elements heavier than hydrogen and helium. The most abundant are oxygen (~1 % by mass), carbon (0.3 %), neon (0.2 %), and iron (0.2 %). As will be noted in the next chapter, heavy elements are born in the processes of stellar nucleosynthesis—fusion in stellar interiors and/or supernovae explosions. In addition to metals in cosmic proportions that the Sun received from the original mostly hydrogen-helium protosolar nebula, there is evidence that its composition was enriched by the heavy elements implanted in the primordial matter due to a nearby supernova explosion. In the process of evolution a fraction of heavy elements sunk down from the solar surface to the interior, and therefore the photosphere is a bit depleted of metals. In turn, because of the nuclear fusion, the Sun's interior becomes progressively more abundant with helium, which is currently estimated as high as ~60 %, while metals are yet unchanged. Also, since the fusion process occurs inside the radiative zone rather than the convective zone, no fusion products transport upward to the photosphere. The solar atmosphere is believed to have the same composition as the whole Sun, with a possible small exception emerged from isotopic compositions of solar wind implanted noble gases.

Solar Activity

The Sun exhibits different kinds of activity and its appearance constantly changes, as revealed by numerous ground and space observations (Fig. 5.9). The most well-known activity is the 11-year solar activity cycle, which roughly follows the number of sunspots on the Sun's surface. Sunspots can be tens of thousands of kilometers across, usually exist as pairs with opposite magnetic polarity alternating every solar cycle, and peak at the solar maximum, appearing closer to the Sun's equator. As we mentioned, sunspots are darker and cooler than their surroundings because they are regions of the reducing energy convective transport from the hot interior, which is inhibited by strong magnetic fields. The polarity of the Sun's magnetic dipole changes every 11 years, such that the North magnetic pole becomes the South one, and vice versa. Because solar activity changes from one 11-year cycle to another, doubled cycles (22 years and longer) are also distinguished. Irregularity is specifically manifested by extended periods of a minimum number of sunspots and solar activity during several cycles, such as the one that occurred in the seventeenth century. This period was known as the Maunder Minimum, and it strongly impacted the Earth's climate. Some scientists believe that during the Maunder Minimum the Sun underwent a 70-year period with almost no sunspot activity. We also recall that during the last 11-year cycle an unusual solar minimum occurred in 2008 and lasted much longer and with a lower amount of sunspots than normal. Therefore, solar

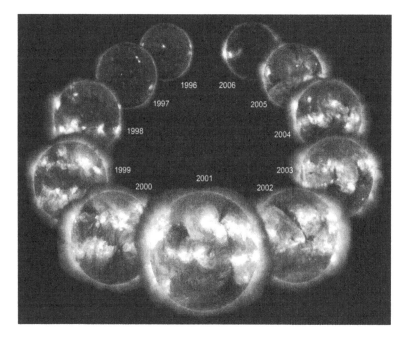

Fig. 5.9 The change of solar activity during the 1996–2006 11-year cycle (Credit: NASA)

activity recurrence is not stable. Moreover, theory claims that magnetic instabilities in the Sun's core could cause fluctuations with period of tens of thousands of years.

Solar flares, coronal mass ejections (CMEs), and solar proton events (SPEs) are the most characteristic phenomena of solar activity manifestation. Their activity rate is closely related with the 11-year solar cycle. These events are accompanied by huge ejected amounts of high-energy protons and electrons well exceeding the energy of quiet solar wind particles. They determine the state of the geomagnetic field and solar plasma interaction with the solar system bodies involving processes in the upper, middle, and lower atmospheres and on the planetary surfaces. Basically, solar activity events determine the space weather, which influences planetary environments and, in particular, life on Earth.

Solar flares are caused by the mechanism of tearing and reconnection of magnetic field lines (the B-field) in the chromosphere, accompanied by a rapid release of the stored energy (Fig. 5.8). A flare is a burst exhibited as an instantaneous and intense change of the brightness in an active area on the Sun's surface. The temperature inside a flare reaches 10^8 K and the energy release is equivalent to billions of megaton bombs. The flare duration may be as long as 200 min, accompanied by strong intensity variations in X-rays and powerful acceleration of electrons and protons, whose velocity approaches a fraction of the speed of light. Unlike the solar wind, particles generated in the flares reach Earth very quickly and strongly disturb its environment. This radiation is extremely harmful for astronauts orbiting the Earth.

CMEs are the most powerful phenomenon in the solar system (Fig. 5.10). They originate in the corona and represent outbursts of enormous volumes of the solar plasma and may be caused by the reconnection of magnetic field lines resulting in enormous energy release. Some of them are associated with solar flares or are related with the solar eruptive protuberances maintained above the solar surface by the magnetic fields. CMEs appear periodically and are composed of very energetic particles. Giant clots of plasma forming giant plasma bubbles expanding outward

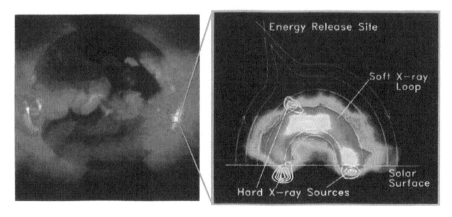

Fig. 5.10 Schematic mechanism of energetic solar plasma particles generation during powerful solar flare such as CME (Credit: NASA)

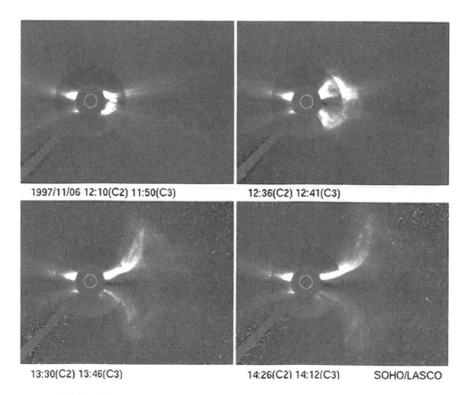

Fig. 5.11 Typical Coronal Mass Ejection (CME) development from SOHO satellite observations (Credit: ESA)

are thrown out in space (Fig. 5.11). Billions of tons of matter are ejected and travel through the interplanetary medium with a velocity exceeding ~ 1,000 km/s, forming a detached bow shock at the front. CMEs are responsible for powerful magnetic storms on Earth; the plasma inflow causes the magnetosphere size to decrease from ~12 R_E to ~6 R_E at the sunward direction. Like solar flares, CMEs carry very harmful radiation.

Another phenomenon is that of SPEs. They occur more often than solar flares and CMEs, and the energies of their generated protons are lower (energy E ~ 30 MeV, particle flux density ~10^{10} cm^{-3}), but their duration is longer, from a few hours to a few days. Whereas solar flares and CMEs are more characteristic for the maximum phase of the 11-year solar activity cycle, SPEs occur throughout the whole cycle; however, their influence on the space environment is much lower than that of CMEs.

The interaction of solar plasma with the planets and small bodies strongly influences their environments, first in the upper atmosphere and magnetosphere—either intrinsic or induced—depending on whether a planet possesses a magnetic field. This interaction is referred to as solar-planetary coupling and is substantially dependent on the phase of the 11-year cycle of solar activity. Solar flares, especially CMEs, strongly influence the state of the geomagnetic field and space weather of our planet. Changes in the solar activity result in changes in magnetosphere shape, radiation level, magnetic substorm development, and various upper atmospheric properties of Earth (Fig. 5.12). In particular, the temperature of the Earth's atmosphere in the height range of 200–1,000 km changes by several times, from ~400 K to ~1,500 K, and the mass density changes by one to two orders of magnitude.

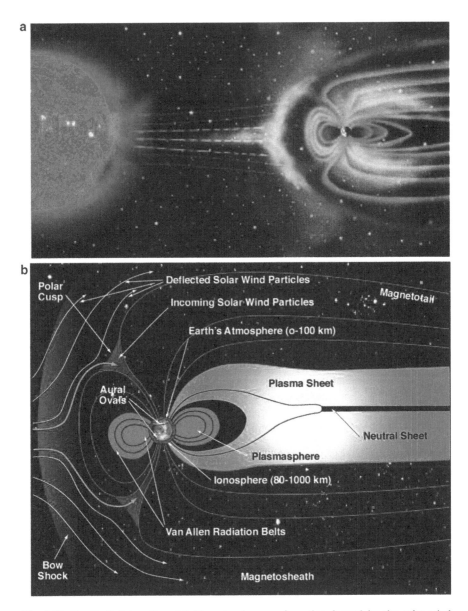

Fig. 5.12 Sun-Earth interaction. (**a**) Magnetosphere configuration formed by the solar wind plasma impact. (**b**) Nomenclature of the main regions of solar wind interaction with Earth (Credit: NASA)

This dramatically impacts the lifetime of artificial satellites. These variations are caused by solar extreme ultraviolet (EUV) and soft X-ray radiation, which correlate well with radio emission of the Sun in decimetric wavelengths. The widely recognized solar activity index is the 10.7 cm radiation, continuously recorded by simple radio antennas over the globe. This index ($F_{10.7}$) changes from about 70 to 180 W/ m^2Hz between solar activity minimum and maximum, respectively, and perfectly reflects the real physical processes depending on the solar energy input. Similarly, indices of geomagnetic activity (A_p, K_p, D_{st}, and some others) recorded on geophysical observatories are used to characterize the Earth's magnetic field disturbance. Effects of solar activity on Earth include auroras at high to moderate latitudes, the disruption of radio communications, breaks in electric power supply, radar operations blocking, and spacecraft electronics damage. Induced electric currents badly influence on oil-gas tube transport systems.

Heliosphere

As we said, the heliosphere is a huge region around the Sun filled with solar plasma and extending to the outer reaches of the solar system. The orbits of the planets and small bodies lie within the heliosphere. For an observer located well outside the solar system (say, at one of the nearest stars) the heliosphere would look like a plasma "bubble" blown into the interstellar medium by the solar wind which prevents the solar system from becoming embedded in this medium (Fig. 5.13). The Sun supplies almost all of the material in the heliosphere, although electrically neutral atoms from interstellar space can pass through the magnetic field lines confining this bubble. Solar protons and electrons dominate the heliosphere composition; the typical proton density in its inner part is 5 cm^{-3}, decreasing as the reciprocal distance squared. The interstellar medium consists mainly of neutral hydrogen and helium atoms. Its density is less than 0.1 cm^{-3} and atoms move through the solar system in the downstream direction with a typical supersonic velocity of ~15–20 km/s.

The heliosphere is asymmetric in configuration and is also slightly distorted by the galactic magnetic fields. It resembles a giant comet of teardrop shape moving toward the apex, with a nose on the leading side facing the Sun's orbital motion through the Galaxy (caused by stellar wind pressure) and a heliotail behind. The solar plasma traveling in this direction at supersonic speed (over a million kilometers per hour) in the solar system eventually meets the very rarefied gas of the interstellar medium which, nonetheless, is capable of slowing it down. Because its pressure is not zero, a termination shock is formed at ~80–100 AU from the Sun (about 11—13.5 billion km). According to theory, this is a standing inwards-facing shock wave elongated in the downstream direction and moving back and forth with a velocity of ~100 km/s. Here the solar plasma speed becomes subsonic (relative to the stars), compressed, heated, and turbulized, and may be affected by the ambient flows of the interstellar medium. Deceleration continues in what is called the heliosheath region where the solar wind and interstellar medium pressures are

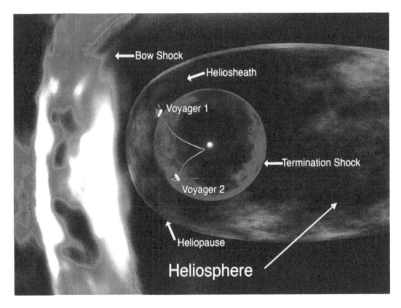

Fig. 5.13 Schematic view of the Heliosphere. The main regions of incoming solar wind on the interstellar medium are shown and asymmetry of heliosphere owing to solar plasma inflow is seen. Voyager 1 and 2 spacecraft positions at heliosheath before leaving the Solar system are indicated (Credit: NASA)

ultimately balanced. In other words, the theoretical boundary (the edge of the heliosphere) is assumed to occur in the heliosheath, where the solar wind's strength is no longer great enough to push back the stellar winds of the surrounding stars. This is called the heliopause. Here, the solar wind velocity drops to zero, the magnetic field intensity doubles, and high-energy electrons from the Galaxy increase 100-fold. We say that here is where the solar wind merges with the interstellar medium.

There is a stagnation zone within the heliosheath which is believed to have been detected by the Voyager 1 spacecraft crossing the termination shock and entering the heliosheath in 2004 at a radial distance of ~ 14 billion km (~94 AU) from the Sun. Voyager 2 crossed the termination shock as well in 2007 at 83.7 AU and brought evidence of denting in the heliosphere probably caused by an interstellar magnetic field. However, the theoretical assumption of the existence of a detached bow shock well ahead of the heliopause, where heliospheric plasma directly hits the interstellar medium, was not confirmed by the results of the Interstellar Boundary Explorer (IBEX) satellite; this could be explained by a slow relative velocity of the solar plasma stream and interstellar medium interaction. Because the pressure of the solar wind plasma varies depending on the solar activity, the boundary close to the heliopause may change within a few tens of astronomical units. This variation seems to have been detected by the Voyager 1 and 2 space vehicles, which appeared to cross this region multiple times and found local small-scale plasma bubbles formed by magnetic reconnection between oppositely oriented sectors of the solar magnetic field as the solar wind slows down (see Fig. 5.13). Moreover, since they crossed the boundary at different positions, it was assumed that the heliosphere may be not only asymmetric, but also irregularly shaped, bulging outwards in the Sun's Northern Hemisphere and pushed inward in the south. Voyager 1 reached helio-pause in 2012. The crossing was confirmed in 2013, and was signaled by a sharp drop in the temperature of charged particles, a change in the direction of the magnetic field, and an increase in the amount of galactic cosmic rays.

Since the fields above and below the solar equator have different polarities point-ing towards and away from the Sun, there exists a thin current layer in the solar equatorial plane, like a ripple, which is called the heliospheric current sheet. It is of magnetohydrodynamic origin and is generated by the electric currents resulting from the interaction of the rotating magnetic field of the Sun and interplanetary plasma charged particles. The heliosphere's current sheet is confined to a surface rather than being spread through a volume in space because of relevant magnetic forces. The current sheet is very thin for its size (its aspect ratio is ~ 100,000:1). Because the Sun's rotation twists the magnetic field lines, at large distances the cur-rent sheet acquires the shape of an Archimedean spiral and resembles a "ballerina skirt." This significantly increases the complexity of the heliosphere's structure and dynamics. We may reasonably assume that other planetary systems have their own heliospheres (stellospheres) of more or less similar configuration and properties.

Chapter 6
Stars: Birth, Lifetime, and Death

General View

Stars are the main astrophysical objects in the universe accessible to observation, and they go through consecutive cycles of evolution from birth to death. Since ancient times people have grouped the most prominent stars into well-known constellations which are clearly distinguished in the night skies, and their configurations gained proper (sometimes curious) names mostly related to mythology. Constellations are only patterns seen in stars that are actually located at completely different distances. A quite definite order is observed in the processes of stellar evolution, in particular, in the birth of stars with various masses in our galaxy, the Milky Way. In the first approximation, this order corresponds to the initial mass function (IMF), while its slope in the region of accreting massive stars corresponds to the well-recognized exponent of the Salpeter function.[1] A typical mass range of stars is from a fraction to tens of solar masses, whereas the diameter range is from only tens of kilometers (pulsars) to hundreds of solar diameters (a supergiant like Betelgeuse in the constellation Orion is 650 times larger than the Sun).

A substantial number of stars are not single but binary or multiple stars, the latter being especially pertinent to massive and hot OB class stars (see below). Detailed observations of many binary star systems collected by astronomers made it possible to determine the masses of stars from computation of the orbital elements which testify to the orbital stability of the systems. Star systems are often organized into hierarchical sets of binary stars and larger groups of star clusters ranging from loose groups of stars (*open clusters*) to enormous *globular clusters* with thousands and millions of stars. Also groups of stars, *stellar associations,* that share a common point of origin in giant molecular clouds have been identified.

---

[1] The Salpeter function is the rate of formation of stars of different masses in the Galaxy inferred from observations of stars of different luminosities and defining the process of stellar evolution. This concept is important in theories of star formation.

© Springer Science+Business Media New York 2015
M.Ya. Marov, *The Fundamentals of Modern Astrophysics*,
DOI 10.1007/978-1-4614-8730-2_6

With the exception of *supernovae*, individual stars have primarily been observed in the visible part of our galaxy and also in our *Local Group* of galaxies, in the M100 galaxy of the Virgo Cluster (about 100 million light years from the Earth), and in the Local Supercluster (see Chap. 10). However, outside the Local Supercluster of galaxies, neither individual stars nor clusters of stars have been observed, even using modern telescopes (the only exception being a faint image of a great cluster containing hundreds of thousands of stars at one billion light years from Earth). However, precise astrometric measurements allow us to measure two components of a star's motion: the radial velocity towards or away from the Sun, and the *proper motion,* the traverse angular movement. Note that because of the relatively vast distances between stars outside the galactic center, collisions between stars are thought to be rare.

By definition, a star is a massive luminous sphere of plasma where self-sustaining nuclear fusion occurs which is sufficient for thermal pressure to balance gravitational pull. This means that at each layer of radial distance r the conditions of hydrostatic equilibrium and mass conservation are satisfied. Similar to the Sun, stars possess a magnetic field that is generated due to stellar intrinsic rotation[2] and convective transfer in the interior by the dynamo mechanism. A mass threshold necessary to ignite thermonuclear reaction (for solar composition), often called the *hydrogen burning limit,* is $M \geq 0.08\ M_O$, where M_O is the mass of the Sun, or ~ 80 M_J, where M_J is the mass of Jupiter. This means that Jupiter lacked only about two orders of magnitude by mass required to ignite nuclear fusion when it formed. If it had ignited, our solar system would have become a binary stellar system. When a star exhausts nuclear fuel it eventually evolves through several stages depending on its mass and ultimately becomes a dead star, a stellar remnant.

Stars are distinguished by their luminosity, color, and spectral energy distribution. The illumination from a star on the Earth's surface (E) is measured in logarithmic scale and expressed in a special unit known as visual stellar magnitude (m). Stars having illuminations E_1 and E_2 differ by one stellar magnitude according to the formula

$$E_1 / E_2 = 2.512^{-(m_1 - m_2)}$$

and because log 2.512 = 0.4, we can also rewrite the above formula as

$$\log E_1 / E_2 = -0.4\left(m_1 - m_2\right).$$

In other words, the illuminations of stars form an infinite decreasing geometrical series with denominator 2.512. Stars of first, second, third, etc., magnitudes are designated as 1^m, 2^m, 3^m, etc., respectively. Obviously, the weaker or fainter an observed star is, the larger its stellar magnitude. In clear night skies one may observe with the naked (sharp) eye stars no fainter than 6^m. For modern telescopes, stars and galaxies

---

[2] In order to determine the rotation rate of a star, methods such as spectroscopic measurements or tracking of the rotation rate of starspots are used.

as weak as 26^m are accessible, which means that their energy flux is more than 10^{20} times (!) less compared to what can be measured from the closest stars. Let us also note that the farther objects are located from us, the farther back in time in the universe we observe them. The reason is that more time is required for light emitted by a distant object in the young expanding universe and propagating in space to reach our eye. In contrast, from the above relationship, the brightest objects have zero or negative stellar magnitude. In particular, the stellar magnitude of the bright star Vega is 0.03^m, that of the planet Venus (at maximum elongation) is—4.4^m, the full Moon has a value of −12.7^m, and the Sun −26.8^m.

The energy distribution in a spectrum gives the most complete information about a star. Besides spectral classification, it also serves to define temperature, the latter being related with the color, though the dependence does not completely fit the standard Planck radiation law. Astronomers usually use the quite objective characteristic called color index. In the international Ultraviolet-Blue-Visual (*U-B-V*) system, *B-V* (main) and *U-B* (ultraviolet) color indices are widely used as the most informative (see Fig. 6.1). In the spectra of stars many differences in the strength

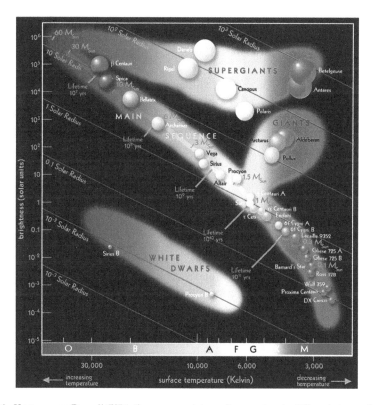

Fig. 6.1 Hertzsprung-Russell (HR) diagram—a picture of many stars in different stages of evolution. Luminosity is normalized to that of the Sun. The Main Sequence (MS) curve corresponds to stars' active lifetime when nuclear fusion occurs. Regions before entering and after leaving MS are shown. Also shown are well known stars in the Milky Way Galaxy Color indeces can be deduced ising spectral band at the bottom (Adapted from Wikipedia)

and number of their absorption lines are distinguished.[3] The dark lines in stellar spectra are caused by the absorption of specific frequencies by the stellar atmosphere, and their identification allows us to reconstruct a star's composition. A typical composition is 71 % hydrogen and 27 % helium (by mass), and the remaining is a small fraction of heavier elements. However, there also exist chemically peculiar stars that show unusual abundances of certain heavy elements in their spectra.

Another important characteristic used in astronomy is the absolute stellar magnitude M, which allows us to compare rather accurately the light fluxes emitted by different stars. We define M as the stellar magnitude m a star would have at the distance 10 pc. Assuming E and E_0 are illuminations from a star at distances r and 10 pc, respectively, we may write

$$\log E_0 / E = 0.4 (m - M).$$

Because illuminations are inversely proportional to distances, $E_0/E = r^2/100$, the following relationship emerges:

$$0.4(m - M) = 2 \log r - 2, \text{ or } M = m + 5 - 5 \log r.$$

In this approach, for the Sun ($m = -26.8^m$, $r = 1$ AU $= 1/206{,}265$ pc) we will have $M = 4.8^m$.

Stars evolve with time, have a finite lifetime, and eventually die. The life cycle of stars at all stages of evolution is traced and classified according to the Hertzsprung-Russell (HR) diagram, a picture of many stars in different stages of evolution (Fig. 6.1). The HR diagram is a plot of a star's surface temperature (related to spectral class O, B, A, F, G, K, M) against its total luminosity L. A convenient mnemonic rule to remember the spectral classes is as follows: "O Be A Fine Girl Kiss Me". Plotted points are not scattered at random, but are confined to well-defined regions. Most stars are located on a line called the main sequence (MS). On the MS a star's luminosity and lifetime are related to its mass, the radius of a star $\boldsymbol{R}$ being proportional to its mass $\boldsymbol{M}$. While stellar temperatures (actually, temperatures of the visible surface, the photosphere) differ by a factor of 10, their radii vary over a very wide range: from hundreds and even thousands of solar radii (R_s) for giants and supergiants, and down to 10^{-2}–$10^{-3} R_s$ for white dwarfs. The following relationship between the luminosity and radius and between the luminosity and mass exists for the MS stars in the mass range $0.1\, M_0 < M < 100\, M_0$:

$$L = R^{5.2} L = M^{3.9}.$$

The lifetimes of massive giant stars (like those of the OB class) are much shorter than those of *dwarf stars* (such as our Sun, which is called a yellow or red dwarf). While for giants the lifetime is only tens of millions of years, for the Sun it is about

---

[3] The surface gravity can influence the appearance of a star's spectrum: the higher surface gravity pertinent to compact stars causes a broadening of the absorption lines, while the opposite is the case for giant pre-main sequence stars having a lower surface gravity.

Fig. 6.2 A photogenic variable star Eta Carinae (bright yellow spot in blue up of center) embedded in the Carina Nebula (Adapted from Wikipedia)

ten billion years, and it is nearly one trillion years for dwarfs with $M < 0.25\ M_O$. Moreover, whereas stars of a solar mass and more massive stars leave the MS at the end of their lifespan when the stored fuel they consume becomes exhausted, no stars under about 0.85 solar mass are expected to have moved off the main sequence.

The most common stars in the Galaxy are those of low-luminosity spectral type M, while our Sun belongs to class G. Note that early stars of less than 2 solar masses are called *T Tauri stars*, while those with greater mass are called *Herbig Ae/Be stars*. The amount of massive OB stars is much less: there are only 10^{-3}–10^{-4} % of 10–100 M_O stars in the whole population, which is fully in accordance with the contemporary theory of star formation. Nonetheless, these massive stars are mostly responsible for supernova explosions, gamma bursts, and black hole formation, and they significantly contribute to galaxy evolution and enrichment of the universe with elements heavier than hydrogen and helium that enter the composition of stars of higher metallicity.[4] Let us also note that binary or multiple systems of stars rather than single stars dominate in the galaxies. It was estimated that more than half of all MS stars and an even larger fraction of pre-MS stars in the Galaxy are binary or multiple stars. There are also stars with periodic or random (sudden) changes in luminosity which are called *variable stars* (Fig. 6.2). Fluctuations of brightness

---

[4] In astronomy all chemical elements heavier than helium are considered to be "metals," and the relative abundance of these elements in a star is called the star's *metallicity* (also designated Z). The metallicity affects its lifetime, magnetic field formation, and stellar wind strength. The younger stars of rather high metallicity and older stars of substantially less metallicity are called Population I and Population II stars, respectively. Population III first generation halo stars are also distinguished in protogalaxies when they began to form and contract. These massive huge stars appeared to consist almost entirely of hydrogen and quickly became supernovae, reionizing the surrounding neutral hydrogen and releasing the first heavy elements into the interstellar medium.

(apparent stellar magnitude), sometimes affecting changes in the spectrum allowing us to identify the luminosity class of a star, may be caused by intrinsic or extrinsic properties. Intrinsic variable stars are thought to be periodically swelling and shrinking stars, while the light emitted by extrinsic variables could be blocked by an orbiting companion at times of eclipse, like, e.g., the eclipsing binary Algol.[5] The intrinsic category includes *Cepheid and Cepheid-like stars*, and long-period variables such as the well-known red giant Mira that pulsates with an 11-month cycle between 2.5 and 11 magnitudes (a 30,000-fold change in luminosity!). Because of the relationship between the period of luminosity variation (fundamental frequency) and the location in space, Cepheids are widely used to scale the position of distant objects for which the parallax method applied for relatively close stars becomes useless. Cepheid variables belong to Population II stars. Among intrinsic variable stars, cataclysmic or explosive variable stars, including novae, supernovae, and symbiotic (interacting binary) variable stars are also distinguished.

Birth

Stars form from the collapse of clouds of molecular gas and dust in the interstellar medium, primarily in the spiral arms of galaxies. The vast majority of stars form in giant molecular clouds (GMCs), and some in smaller diffuse molecular clouds (Fig. 6.3). It is of interest to compare molecular and atmospheric clouds. Molecular gas clouds are composed mostly of hydrogen with about 23–28 % helium, a few percent heavier elements, and dust admixture up to 1 % by mass (Fig. 6.4), whereas the Earth's atmospheric clouds are air and water droplets/ice crystals. The number density of the molecular clouds is much lower than in a vacuum chamber ($n = 10^2$–10^8 cm^{-3}, which is equivalent to a mass density $\rho = 7 \times 10^{-4}$ g/cm^3) and the temperature is only $T = 15$–20 K, as compared to $n \sim 10^{19}$ cm^{-3} and $T = 220$ K in the Earth's clouds. However, their size R and mass M ($R = 1$–50 pc, or 3×10^{18}–1.5×10^{20} cm, $M = 1$–10^6 M_O is enormous as compared to our typical Earth clouds ($R = 10^4$–10^6 cm, $M = 7 \times 10^8$–7×10^{11} g).

The star formation process is rather inefficient, as only a few percent of the mass of the molecular cloud is used to form stars. The peak in the stellar mass spectrum is at about 0.3 M_O; therefore, most stars are less massive than the Sun (1 M_O). Let us note that huge interstellar clouds are called "molecular" clouds because, besides hydrogen and helium, they also contain numerous hydrogen-, carbon-, nitrogen-, and oxygen-bearing molecules, as well as other molecules composed of two, three, four, ten, or even more atoms, including hydrocarbons and deuterated molecules. Emissions of oxygen, hydrogen, and sulfur are clearly seen in the Orion Nebula in Fig. 6.5. Forming massive stars powerfully illuminate the whole region. Another star-forming region in the Orion constellation taken by NASA's Spitzer Infrared

---

[5] Many stars exhibit small variations in luminosity; for example, the energy output of our Sun varies by about 0.1 % over an 11-year solar cycle.

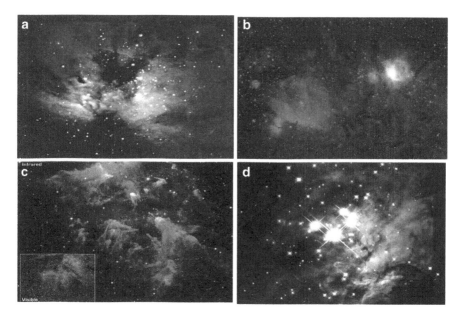

Fig 6.3 Regions of star formation. (**a**) Star-forming region NGC 2024 in the Orion Nebula. The images were obtained with the infrared camera of the Gemini spacecraft in the three near-infrared bands ($\lambda = 1.2$, 1.65, and 2.2 μm—the blue, green, and red colors in the image, respectively—and subsequently processed by superposition of the images. Whereas the central part of the nebula in an ordinary photograph in visible light appears dark due to the absorption of light by dust, a dense cluster of young stars is detected in infrared light. Credit: UCLA and NASA. (**b**) The star-forming region in Orion (Credit: M. Bessell, RSAA, ANU); (**c**) The star-forming region η Carinae. The image was obtained with the Spitzer Space Infrared Telescope Facility (SIRTF). The same sky region in the visible spectral range is shown on the left for comparison (Credit: NASA). (**d**) Birth of stars in the central part of Orion constellation—open cluster TRAPESIUM (Credit: NASA)

Space Telescope is shown in Fig. 6.6. Here, quite unexpectedly, signatures of tiny mineral crystals of olivine around a sun-like embryonic star HOPS-68 were detected. The crystals are in the form of forsterite and belong to the olivine family of silicate minerals. It is assumed that these green mineral crystals were "cooked up" near the surface of the forming star under high temperature (>700 °C), then carried up into the surrounding cloud where temperatures are much colder, and ultimately fell down again like glitter. However, it is more likely that jets of gas blasting away from the embryonic star transported the cooked-up crystals to the chilly outer cloud.

Modes of formation of high-mass stars ($M > 3M_O$) and low-mass stars ($M < 3M_O$) are quite different, although every scenario begins with gravitational instability of a higher density region (caused by fluctuations) within the molecular cloud to satisfy the Jeans instability criterion and collapse. High-mass stars, such as O and B class stars, start thermonuclear fusion while still collapsing, increasing density and converting gravitational energy into heat, and they tend to disrupt the surrounding molecular cloud from which they formed. In contrast, low-mass stars emerge from their cores of dust and gas before hydrogen burning starts, with little disruption of the cloud.

Fig. 6.4 Giant gas-dust
pillars of molecular clouds
(Courtesy of HST Institute
and NASA)

Fig. 6.5 Emissions of oxygen, hydrogen and sulfur are clearly seen in the Orion constellation—an
active stellar nursery containing thousands of young stars and developing protostars (Credit: HST
Institute and NASA)

A protostar forms at the core when gravity and internal pressure are approximately
balanced and the stable condition of hydrostatic equilibrium is achieved. The time to
contract to the MS in the HR diagram is about 30 million years for a low-mass star
of ~ $1 M_O$ and only 10,000 years for a high-mass star of ~ $50 M_O$. Some of the pre-MS
stars are surrounded by a protoplanetary disk (see Chap. 8). Regions of the birth of

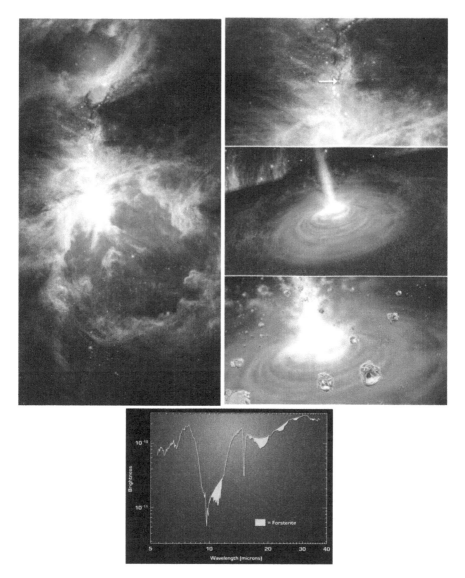

Fig. 6.6 Star-forming region in the Orion constellation taken by NASA's Spitzer Infrared Space Telescope (*left image*). Detection of signatures of tiny mineral crystals – olivine around a sun-like embryonic star HOPS-68 (*arrow* in *upper right image*). The crystals are in the form of the olivine family of silicate minerals—forsterite. They were either cooked up near the surface of the forming star under high temperature and then transported to the much colder surrounding cloud or blasted away from the embryonic star as jets of gas to the chilly outer cloud (right middle and bottom images according to artist's concept). Bottom plate – spectrum recorded by the Spitzer infrared detectors with forsterite signatures (Credit: NASA/JPL-Caltech/University of Toledo)

NGC 3603   HST • WFPC2
PRC99-20 • STScI OPO • June 1, 1999
Wolfgang Brandner (JPL/IPAC), Eva K. Grebel (Univ. Washington),
You-Hua Chu (Univ. Illinois, Urbana-Champaign) and NASA

Fig. 6.7 Region of the birth of high mass star (HST image, NASA Courtesy)

high-mass stars taken by NASA's Hubble Space Telescope (HST) and Spitzer Space Telescope, respectively, are shown in Figs. 6.7 and 6.8. The latter is a spectacular view of a glowing emerald nebula where O class stars—the most massive type of stars known to exist—are formed. Spitzer has found that such bubbles are common and can be found around O stars throughout our Milky Way. Spitzer infrared observations were made in 3.6-, 8-, and 24-μm wavelengths; the image in Fig. 6.8 represents a three-color composite. A couple of giant stars are at the center of this ring whose intense ultraviolet light has carved out the bubble, though they blend in with other stars when viewed in infrared light. The green ring is where dust is being hit by winds and intense light from the massive stars. The green color represents infrared light coming from tiny organic dust grains of polycyclic aromatic hydrocarbons (PAHs). These small grains have been destroyed inside the bubble. The red color inside the ring shows slightly larger, hotter dust grains, heated by the massive stars.

Nuclear fusion in a stellar core is the main source of a star's energy. A young star is composed predominantly of hot, ionized hydrogen and helium, and during a major part of its lifetime (when it is on the MS of the HR diagram) a star releases a great amount of energy due to thermonuclear reactions in its core, burning hydrogen and producing helium. In equilibrium, the thermal energy and gravitational potential energy are in balance (hydrostatic equilibrium). A star's radius is proportional to its mass, $R \sim M$, and the mean density ρ is proportional to the inverse square of its mass, $\rho \sim M^{-2}$. In other words, for stars of low mass density their hydrostatic structure is determined mainly by a balance between gravity and thermal pressure. In such stars the fusion process terminates at helium, according to the nuclear reaction

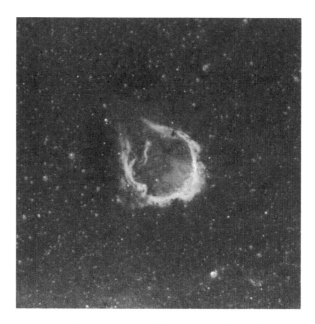

Fig. 6.8 This glowing emerald nebula seen by NASA's Spitzer Space Telescope where O stars—the most massive type of star known to exist – are formed. This is a three-color composite that shows infrared observations from two Spitzer instruments. Blue represents 3.6-micron light and green shows light of 8 microns, both captured by Spitzer's infrared array camera. Red is 24-micron light detected by Spitzer's multiband imaging photometer. At the center of this ring are a couple of giant stars whose intense ultraviolet light has carved out the bubble, though they blend in with other stars when viewed in infrared. The green color represents infrared light coming from tiny organic dust grains—polycyclic aromatic hydrocarbons (PAH). The *red color* inside the ring shows slightly larger, hotter dust grains, heated by the massive stars (Credit: NASA)

chains described for the Sun in Chap. 5, while for high-mass stars the fusion process continues through a few successive stages as the ashes from one fusion cycle become the fuel for the next cycle: from helium through carbon, neon, oxygen, silicon, and finally iron (Figs. 6.9 and 6.10).

In evolved stars with masses between 0.5 and 10 solar masses and temperatures in the cores $T \sim 10^8$ K, helium is transformed into carbon in the triple-alpha process in the following chain of reactions that uses the intermediate element beryllium:

$$^4\text{He} + {}^4\text{He} + 92\,\text{keV} \rightarrow {}^{8*}\text{Be}$$
$$^4\text{He} + {}^{8*}\text{Be} + 67\,\text{keV} \rightarrow {}^{12*}\text{C}$$
$$^{12*}\text{C} \rightarrow {}^{12}\text{C} + \gamma + 7.4\,\text{MeV}$$

And the overall reaction looks like:

$$3\,^4\text{He} \rightarrow {}^{12}\text{C} + \gamma + 7.2\,\text{MeV}$$

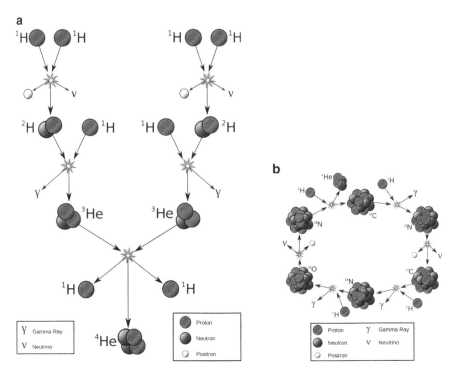

Fig. 6.9 Fusion diagrams. (**a**) Hydrogen-helium burning cycle for low mass stars; (**b**) carbon-nitrogen-oxygen burning cycles for large mass stars (Adapted from Wikipedia)

In more massive stars, heavier elements are burned in a contracting core through the neon, oxygen, and silicon burning processes, resulting ultimately in production of the stable iron isotope ^{56}Fe. Fusion then is terminated, because there is no more internal (endothermic) energy release and it can proceed further only through gravitational collapse. Let us note that the fuel in thermonuclear synthesis is taken from the net mass of the fused atomic nuclei, which is smaller than the sum of the constituents. The lost mass is released as electromagnetic energy, in accordance with Einstein's famous mass-energy equivalence relationship $E = mc^2$. Stellar cores evolve into structures of concentric ("onion-type") shells of elements created during the various stages of thermonuclear fusion (Fig. 6.10). A star's total mass determines how far its core will proceed towards iron, and no fusion can proceed past iron. Elements heavier than iron are formed in supernova explosions. The origin of chemical elements by this process is called *nucleosynthesis*.

We see, therefore, that massive stars with $8\,M_O \leq M \leq 100\,M_O$ go through all of the successive hydrogen, helium, and heavier element burning stages up to the production of iron. When hydrogen is nearly exhausted, the core contracts until the temperature and pressure are sufficient to fuse helium,[6] and the process continues with the

---

[6] During the helium burning phase, very high-mass stars with more than nine solar masses expand to form *red supergiants*.

**1. HYDROGEN BURNING IN THE
CORE OF THE STAR**

**2. HELIUM BURNING IN THE
CORE OF THE STAR**

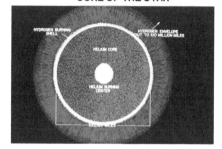

**3. CARBON- OXYGEN BURNING
IN THE CORE OF THE STAR**

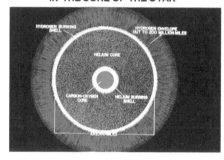

**4. SILICON-IRON BURNING
IN THE CORE OF THE STAR**

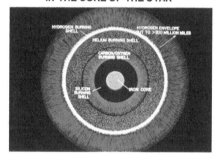

Fig. 6.10 Subsequent cycles of nuclear fusion in star interior (Credit: SAO/G. Fazio)

successive stages of carbon, neon, oxygen, and silicon burning processes along the series of onion-layer shells within the star. Each shell fuses a different element, with the outermost shell fusing subsequently hydrogen, helium, and so forth. The process stops after the formation of tightly bound iron nuclei in the core because contraction of the star can no longer be stopped by the net release of energy through thermonuclear reactions in its central part.[7] The contraction energy is expended on the disintegration of iron nuclei up to the formation of a neutron core accompanied by neutrinos and gamma ray bursts produced by electron capture and inverse beta decay[8]; the central pressure is determined by electron gas degeneracy, while the density is determined by the gas in the atomic core. According to statistical physics, the maximum

---

[7] Some very hot (T ~ $(30–200) \times 10^3$ K) and very massive (over 20 M_o) highly luminous (~$10^6 L_o$) evolved stars are losing mass rapidly (a billion times faster than the Sun!) by means of a very strong stellar wind with a speed about five times more than the average speed of the solar wind. These stars (which also have some specific features in their spectra have characteristic lifespans only in the order of a few million years and this is still sufficient time for their stellar winds to carry away a significant proportion of the total stellar mas - a few tens of solar masses. Such stars are called *Wolf-Rayet (WR) stars* in honor of their discoverers Charles Wolf and Georges Rayet.

[8] Electron capture is a process in which a proton-rich nuclide absorbs an inner atomic electron, thereby changing a nuclear proton to a neutron with the simultaneous emission of an electron neutrino. Inverse beta decay is an alternate decay mode of electron capture for radioactive isotopes with sufficient energy to decay by positron emission.

mass that can be confined by cold electrons is equal to the limiting Chandrasekhar mass $M = 5.75 M / \mu_e$, where μ_e is the number of nucleons per one electron.

The sites of star formation are genetically related to the massive, comparatively flat galactic disk, with the visible matter, a mixture of gas and dust with a nonuniform density distribution, being concentrated in its symmetry plane. Such a multiphase multicomponent medium typically consists of massive cold dense clouds that are in the process of gravitational contraction and that are apparently the early formation stage of star clusters and associations. In disk galaxies, the star formation is active in OB associations with a mass of $\sim 10^7$ M_O and with sizes of the order of the gas disk thickness, so that about 90 % of the cluster breaks up into stars with a mass of $\sim 10^3$ M_O in due course of collapse. Gas condensations are accompanied by contraction and intense heat release due to the conversion of potential energy into kinetic energy, and by angular momentum transport from the forming clump to peripheral regions emerge from this medium when some critical density is reached due to gravitational instability. When the temperature in the interior of a protostar reaches several million degrees, thermonuclear fusion begins and the gravitational forces are balanced by the internal gas pressure (gravitational equilibrium). As a result, the contraction ceases and the star occupies a certain position in the HR diagram dependent on the mass-luminosity relation, where it stays until the reserves of nuclear fuel are exhausted. The energy release through nuclear reactions in the central region of the star is accompanied by radiative and convective heat transport, which leads to active mixing of interior matter (low-mass stars are fully convective) and is responsible for the continual light and heat outflow. Also, every star generates stellar wind from the outer shells into space, similar to the well-known solar wind from the Sun.

Evolution and Death

Stars evolve and eventually die, and the final stage also depends on mass (see Table 6.1). Once the stellar fuel is exhausted, a star collapses. The configurations forming at the final stage of evolution also depend fundamentally on the stellar mass. It follows from the theory of stellar evolution that a low-mass star 0.8–1.4 M_O, after the completion of hydrogen burning, moves off the MS, turning initially into a *red giant* and ultimately into a compact object (a degenerate stellar remnant) called a *white dwarf*. The density in its core increases dramatically, and another source of pressure becomes significant, called the degeneracy pressure. This pressure is produced by electrons occupying higher energy levels than those in normal state atoms where they successively fill up the lowest available energy states, obeying the *Pauli exclusion* principle. Deviation from this principle in compact objects results in a degeneracy pressure domain when the temperature of nuclear fusion increases; in contrast to the ideal gas pressure $\mathbf{p} \sim \rho T$, it scales as $\mathbf{p} \sim \rho^{5/3} T$. In compact objects gravitational energy is balanced by the energy of degenerated particles and the star radius $\mathbf{R} \sim \mathbf{M}^{-1/3}$. In other words, the radius of a white dwarf is inversely proportional to the cube root of its mass: generally, the more massive the star, the more it shrinks.

Table 6.1 Final state of a star depending on mass

Range of mass while a main sequence star	Thermonuclear burning sequence	Evolution	Final state
< 0.08 M_O	None	None	Brown dwarf (T-star)
0.08 M_O–0.5 M_O	Hydrogen	Red giant	White dwarf
0.5 M_O–1.4 M_O	Hydrogen, Helium	Red giant Horizontal branch Planetary nebula	White dwarf
1.4 M_O–8 M_O	Hydrogen, Helium, Carbon	Red giant Horizontal branch Pulsation Supernova	White dwarf or Neutron star
9 M_O–60 M_O	Hydrogen, Helium, Carbon, Oxygen, Neon, Silicon	Red giant Large mass loss Supernova	Neutron star or Black hole

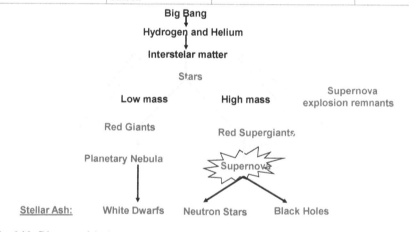

Fig. 6.11 Diagram of the low and high mass stars evolution

As we said in Chap. 5, the Sun will enter the red giant phase in about 5 billion years, when it will expand to a maximum radius of roughly one astronomical unit (1 AU). The released outer shell of a low-mass star is referred to as a planetary nebula and is caused by the enormous increase of radiation pressure on the shell pushing it away. Stars in the lower mass range (0.013–0.8 M_O) evolve to become another type of compact object called a brown dwarf. This is a sort of "intermediate low-mass star," a substellar object with deuterium rather than hydrogen fusion and where electrons serve as degenerate particles. For cool brown dwarfs (when the original inventory of deuterium is burned in fusion) Coulomb pressure dominates. High-mass stars manifest themselves at the final stage of evolution as supernova explosions, leaving behind a neutron star (pulsar) (at 3–8 M_O) or a black hole (at 9–60 M_O) - see Table 6.1 and the diagram in Fig. 6.11.

It follows from the theory of stellar evolution that, after the completion of hydrogen burning, a star with a mass $M \leq M_O$ moves off the MS,, turning initially into a red giant and subsequently evolving into a white (degenerate) dwarf. The existence of such an object was implicitly predicted in 1844 by the famous astronomer and mathematician Friedrich Wilhelm Bessel. Based on the observations of irregularities in the proper motion of Sirius, he hypothesized that it had an invisible companion with a nearly solar mass. As subsequent observations showed, the companion turned out to be the white dwarf Sirius B, and the name itself was given to such degenerate stars later on by American physicist William Fowler.

Low-mass stars, in the case of a relatively slow mass outflow from red giants, leave planetary nebulae with various but fairly ordered shapes (Fig. 6.12) at the stage of their transition to a white dwarf as the star's core shrinks, creating strong radiation pressure at the surface. They contain a strongly turbulized gas, and these shapes are believed to have been acquired in the process of self-organization of the nebular matter. Such order, including a pronounced cylindrical symmetry, can be partly attributed to the existence of a close companion to the star, and to the joint evolution of their envelopes. Extremely complex configurations produced by the

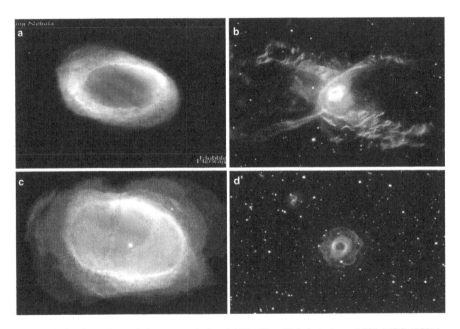

Fig. 6.12 Configurations of planetary nebulae. (**a**) The Ring Nebula in Lyra (M57, NGC 6720) in colors close to the real ones. The *blue*, *green*, and *red* correspond, respectively, to the ionized helium, doubly ionized oxygen, and hydrogen and ionized nitrogen. There is a 15 magnitude star at the final stage of its evolution toward a white dwarf with a surface temperature of ~150,000 K at the center; (**b**) The Tarantula Planetary Nebula. The gas of the ejected envelope is strongly turbulized and has an irregular structure with some ordering; (**c**) A quasi-ring beautiful planetary nebula. It's the dim star, not the bright one, near the center of NGC 3132 galaxy with an inhomogeneous outflow of the glowing gas originated in the outer layers of a star like our Sun; (**d**) Hubble Space Telescope images of another quasi-ring shape planetary nebula (Courtesy of NASA, ESA and HST Institute)

Fig. 6.13 White Dwarf in Stingray Nebula. Image obtained by HST (Credit: HST Institute and NASA)

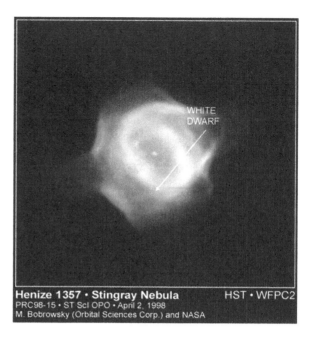

WHITE
DWARF

Henize 1357 · Stingray Nebula   HST · WFPC2
PRC98-15 · ST ScI OPO · April 2, 1998
M. Bobrowsky (Orbital Sciences Corp.) and NASA

interaction of an expanding gas with the interstellar medium, in which a certain order can be distinguished against the background of chaotization, also emerge during supernova explosions. Depending on the extent to which the interstellar galactic medium is enriched with heavy elements produced in nucleosynthesis processes in the interiors of stars and during supernova explosions, the conditions for the formation of first and subsequent generation stars are distinguished. In particular, this determines the metallicity (the ratio of heavier elements to hydrogen) of stellar matter. The cyclic process is permanently maintained on the scale of stellar evolution: ash from decayed stars enters the newborn stars and the disks that surround them, forming planets with a large fraction of heavy elements in their bulk composition. Heavy elements enter the composition of organics and higher forms of evolution: life forms. We human beings are therefore composed of stellar ash!

White dwarfs (Fig. 6.13) are the most characteristic form of the final evolutionary stage for the bulk of the stellar population (up to 95 % of the stars in the Milky Way), with their initial masses lying in the range 0.08–8 M_O. The lower threshold corresponds to the condition for the onset of a classical hydrogen-helium thermonuclear reaction when the star is in thermal equilibrium. For solar-type stars, this is the proton-proton (p-p) cycle that precedes the C-N-O burning cycle in more massive stars in shells following hydrogen and helium burning. Accordingly, carbon-oxygen (C-O) white dwarfs are the final stage of single stars and companions of wide (with semi-major axis $a \geq 300\ R_S$ - star's radius) binaries with $M \leq 8\ M_O$, while degenerate helium white dwarfs are the final stage of companions of close (with $a \leq 300\ R_S$) binary systems with $M \sim 2.5\ M_O$. The electron-degenerate matter that forms inside a white dwarf is no longer a routine plasma.[9] White dwarfs continue to cool, dim, and redden with age for another ~ 4 Gyr, eventually becoming dark *black dwarfs*. Cool degenerate dwarfs oppose their gravitational contraction due to the high density of the degenerate gas of free electrons conserving the kinetic energy even near absolute zero. The amplitude of motions and the degree of degeneracy increase with density, which prevents the collapse of this kind of star at the final stage of its evolution.

Stars evolving to white dwarfs shrink their radius typically to a few thousand kilometers (~8,500 km, close to that of Earth); their typical mass is ~ 0.6 M_O, and the surface gravity is ~ 10^5 times that of Earth. The effective temperatures lie within a wide range, from 150,000 to 4,000 K. A mass of ~ 0.6 M_O is approaching the theoretical upper limit of the mass of a white dwarf, the Chandrasekhar limit equal to 1.38 M_O. Basically, white dwarfs consist of a C-O core and H-He outer shell. The central core density increases proportionally to the growing degenerate electron energies. Cool degenerate white dwarfs oppose their gravitational contraction due to their degeneracy pressure, as we have described above. Let us recall once more that white dwarfs should be distinguished from brown dwarfs, which are quasi-stars with 0.01 $M_O < M < 0.08$ M_O whose interior temperature is not high enough to excite the *p-p* cycle, and, therefore, they are intermediate between low-mass stars and planets.

In some of the white dwarfs, the final evolutionary phase can be accompanied by a nova explosion due to the accretion of the hydrogen envelope from a nearby companion star onto the core. In this envelope, the temperature rises sharply and unstable hydrogen burning takes place in a short C-N-O cycle under conditions of partial electron degeneracy. The final mass of most white dwarfs determined by the "steepness" of the Salpeter power spectrum $M_{WO} \approx 0.6M^{0.4}$ is about 0.6 M_O, and their size as we earlier said is of the order of the size of a terrestrial planet. For this reason, they have a high mean density. The effective temperatures lie within a wide range, typically ~ 4,000 K.

Unlike the low-mass stars, stars with $M >> M_O$ eject their outer envelopes (explode) at the end of their lifetimes. This is observed as a supernova explosion,[10] in which most of the star's matter is blown away leaving behind nebulae composed of an expanding shell of gas and dust and the stellar remnant—a compact *neutron star (pulsar)* or *black hole* as the endpoint of the stellar evolution inside (Figs. 6.14 and 6.15). A beautiful example is the famous Crab Nebula—a plerionic supernova remnant of an explosion in 1054 A.D. containing a neutron star inside (Fig. 6.16). Another example is a pulsar with a white dwarf companion in the globular cluster M4 (Fig. 6.17). Supernovae are so luminous that they often briefly outshine an entire galaxy, before fading from view over several weeks or months. At the supernova explosion the total

---

[9] A degenerate state of matter, as a quantum mechanics entity, is defined as a collection of free, noninteracting particles (such as electrons, neutrons, protons, fermions) with a pressure and other physical characteristics. It arises at the extraordinarily high density in compact stars' interiors (or at extremely low temperatures) and is different from an ideal gas in classical mechanics. When degenerate electrons cannot move to the already filled lower energy levels according to the above-mentioned Pauli exclusion principle, a degeneracy pressure is generated in fermion gas which strongly resists further compression although no thermal energy is extracted.

[10] Basically, supernovae can be triggered by either the sudden reignition of nuclear fusion in a degenerate star or by the collapse of the core of a massive star, and they are difficult to predict.

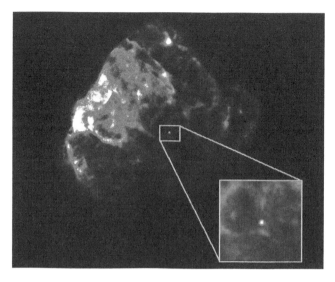

Fig. 6.14 Supernova explosion of high-mass star at the end of its evolution. Image of supernova Puppis A explosion in X-ray (Courtesy of NASA)

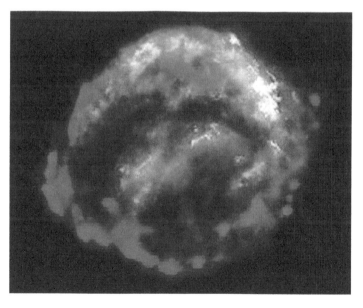

Fig. 6.15 Multiwavelength X-ray, infrared, and optical compilation image of Kepler's supernova remnant (SN 1604) (Credit: HST Institute)

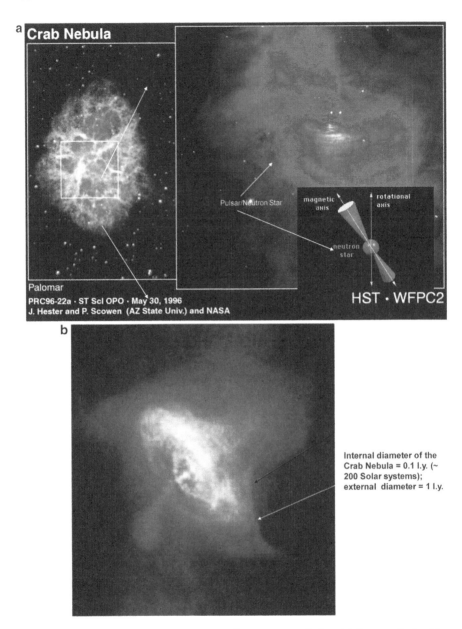

Fig. 6.16 Pulsar in Crab Nebula as remnant of supernova explosion. (**a**) Images obtained by Palomar telescope and HST (Credit: J. Hester, P. Scowen and NASA); (**b**) Pulsar and Crab Nebula in X-ray (Credit: NASA)

gravitational and kinetic energy (mainly in the form of neutrinos) being released reaches 10^{53} erg/s. The visible radiation accounts for about 1 % of this enormous energy release—as much as the Sun is expected to emit over its entire lifespan. This phenomenon can be described in terms of the theory of a strong (powerful) explosion

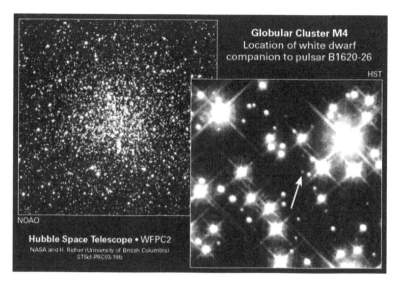

Fig. 6.17 Pulsar B1620-26 with a white dwarf companion in Globular Cluster M4 (Courtesy of HST Institute and NASA)

when much or all of a star's material is expelled at a velocity of up to 30,000 km/s (one-tenth of the speed of light), driving a shock wave propagating through the interstellar gas as a detonation wave.

Two types of supernova remnants of stars with a threshold mass of 8–10 M_O are distinguished. Stars with a mass below this limit evolve in an ordinary way and shine as red supergiants after the explosion phase, lying on the corresponding branch of the HR diagram and becoming the supermassive compact neutron stars known as pulsars (Figs. 6.16 and 6.17). More massive objects with $M \geq 10$–$25\ M_O$ completely lose their hydrogen envelope, and the supernova explosion is accompanied by core collapse followed by a black hole formation (Fig. 6.18). Several mechanisms can be responsible for core collapse. We will mention here the one in which an iron core developed in a massive star becomes much larger than *Chandrasekhar limit* of about 1.38 M_O.[11] Then the core will no longer be able to support itself by the above mentioned *electron degeneracy pressure* and will collapse further to a neutron star or black hole. Astronomers distinguish Type I and Type II supernovae with an adjoined lower case letter depending on spectral features (e.g., supernovae of Type Ia are produced by runaway fusion ignited on degenerate white dwarf progenitors, whereas the spectrally similar Type Ib/c supernovae are produced from the core collapse of massive Wolf-Rayet progenitors). Type II or Ib supernova explosions of massive stars ($M \geq 10\ M_O$) after they have burned nuclear fuel for millions

---

[11] Basically, the maximum (Chandrasekhar) mass is expected to be below ~2.5 M_O depending on the stiffness of the nuclear equation of state (EoS), but it could be lower if phase transitions take place. Observations of large neutron star masses of order ~2.3 M_O would therefore restrict the EoS severely for dense matter.

Fig. 6.18 Massive gas disc associated with Black Hole in the center of active galaxy M87 and its assumed configuration (*bottom right*) (Courtesy of HST Institute and NASA)

of years give rise to neutron star formation; when the iron core in the center of the aging star exceeds its Chandrasekhar limit mass, the star undergoes gravitational collapse in just seconds and suffers a violent death. Gravitational and kinetic energy of the order of ~10^{53} ergs is released mainly by neutrino emission that blows off the outer layers. Only ~1 % of the energy is actually seen in a brilliant burst: the supernova. A recent Type Ia explosion occurred in the galaxy M82 located at ~3.5 Mpc, and a Type II explosion was SN 1987A, located in the Tarantula Nebula in the neighboring galaxy called the Large Magellanic Cloud.

A magnetized rotating neutron star, called a pulsar,[12] emits radio waves recorded on Earth as periodic pulses. Neutron stars are widespread in the universe; at present there are just over 1,000 known radio pulsars with pulse periods ranging from 1.557 ms to more than 8 s. It is estimated that they represent less than 1 % of the active pulsars in the Galaxy lying within a few kiloparsecs, with the others either too faint or too distant to detect. Nonetheless, some very luminous pulsars are observed in the distant reaches of the Milky Way and a few in the Magellanic Clouds. It was estimated that a new pulsar is born in the Galaxy every 50–300 years, which is somewhat lower than the estimated rate of core collapse supernova explosions, with the caveat that some supernovae do not produce active pulsars.

The most stable are the rapidly rotating millisecond pulsars, which rank as the best known clocks in the universe (though it is known that pulsars progressively slow down their rotation). The pulses are accurately measured by taking into account the nonlinear distance variation between the source and observer caused by

---

[12] A pulsar is sometimes also called an X-ray burster. Pulsars with extremely high magnetic fields are called magnetars.

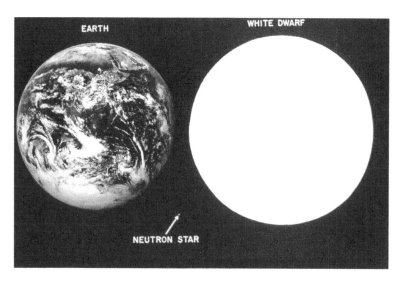

Fig. 6.19 Comparison of White Dwarf and Pulsar sizes with Earth

Earth's motion around the Sun and the Earth's rotation. These variations (basically, the motion of the telescope about the barycenter of the solar system) are then subtracted from the data. A neutron star is composed of degenerated neutrons[13] with a few percent of protons and electrons in the extreme nuclear density. Indeed, the density inside a neutron star having a radius of only $R \simeq 10$–15 km (see Fig. 6.19) is enormous, reaching $\rho_0 = 2.8 \times 10^{14}$ g cm^{-3}, the critical density inside the atomic nucleus, and the matter consists predominantly of free neutrons with a few percent of protons and electrons. This means that on Earth one cubic centimeter of pulsar matter would weigh more than 100 million tons! These huge neutron-rich "nuclei" are bound by gravitation and require a minimum neutron star mass of ~0.1 M_0. Above a maximum (Chandrasekhar) mass of order 2–3 M_0, neutron stars are unstable towards gravitational collapse to black holes. Astrophysicists also speculate about more *exotic compact stars* that have some quantum properties to resist gravitational collapse, rather than the degeneracy pressure of matter composed of electrons, protons, and neutrons. These are stars assumed to be composed of strange matter like very large particles (nucleons) and stars composed of preons, postulated particles conceived to be subcomponents of quarks and leptons (see Chap. 11).

A black hole is formed mainly at $M \geq 30 M_0$ and, unlike a neutron star, it has no surface. The idea of a very massive body preventing even light from escaping was first proposed by geologist John Michell at the end of the eighteenth century, while the term "black hole" was introduced by John Wheeler to describe a very massive entity swallowing up any closely approaching object and not emitting intrinsic radiation. The center of a black hole is described as a gravitational singularity, a

---

[13] Let us recall that lile electrons, neutrons belong to the particles called fermions. They provide *neutron degeneracy pressure* to support a neutron star against collapse. An additional pressure is assumed to be provided by repulsive neutron-neutron interactions.

region where the space-time curvature becomes infinite. There is a mathematically defined surface around a black hole of its gravitational radius called an *event horizon,* which marks the radius of no return (the *Schwarzschild radius* r_g). Basically, this is the defining feature of a black hole, meaning an invisible (imaginary) boundary in space-time through which matter and light can only pass inward towards the black hole center and nothing, light including, can escape.

The event horizon is expressed by a simple formula similar to what is used in celestial mechanics to define circular velocity v around a body of mass M. The formula uses gravity constant G and replaces v by the speed of light c:

$$r_g = 2GM / c^2 \text{ km}$$

It follows from this formula that if the Sun were converted into a black hole its Schwarzschild radius would become only 3 km! In the case of Earth, r_g would be only 0.9 cm! For a black hole of a massive star ($M \sim 10M_O$) $r_g \sim 30$ km, whereas a black hole in the center of a galaxy ($M \sim 10^{10} M_O$) would have $r_g \sim 3 \times 10^{15}$ cm, or ~ 200 AU, which exceeds the distance from the Sun to Pluto by a factor of 5. Interestingly, the critical density would be equal to that of a neutron star for a black hole of stellar mass and less than air density on Earth for a black hole of galactic mass.

As general relativity theory predicts, the presence of a mass deforms space-time in such a way that the paths taken by particles bend towards the mass. At the event horizon of a black hole, this deformation becomes so strong that there are no paths that lead away from the black hole. Moreover, to a distant observer, clocks near a black hole appear to tick more slowly than those further away from the black hole. Due to this effect, known as *gravitational time dilation,* an object falling into a black hole appears to slow down as it approaches the event horizon, taking an infinite time to reach it. At the same time, all processes on this object slow down, for a fixed outside observer, causing emitted light to appear redder and dimmer, an effect known as a gravitational red shift. Eventually, at a point just before it reaches the event horizon, the falling object becomes so dim that it can no longer be seen. Therefore, assuming a spacecraft overcomes r_g (bad luck!) it would be swallowed up and unable to transmit outside any information, though the astronauts themselves falling into the black hole would not notice any of these effects as they crossed the event horizon.[14] In other words, for an observer within a black hole there is no outer boundary and no possibility of transmitting information in electromagnetic wavelengths outward.[15] Let us note that because black holes have only an event horizon

---

[14] Actually, from an outside observer's viewpoint, the spacecraft would never penetrate inside a black hole; whereas for the astronauts it would happen nearly instantaneously and they would see its interior of infinite density (the singularity).

[15] Very recently, the world-renowned British physicist Stephen Hawking, based on quantum theory rather than gravity, suggested that leakage of information from a black hole is possible, proposing that because of the quantum effects of space-time fluctuations in the wide range, no clear horizon boundary around a black hole exists. Instead of an event horizon, he introduced a "visible horizon," a surface where the light leaving a black hole is temporarily retained.

Fig. 6.20 Scheme of binary star system with black hole accreting mass of a nearby star companion

rather than an observable surface, they do not emit radiation and, hence, unlike pulsars or X-ray bursters, cannot be directly detected from the outside. This constraint allows us to distinguish black holes from neutron stars.

However, emission of the enormous radiation energy from a black hole occurs when there is a nearby star companion in a binary star system. The accretion of a companion mass on the black hole allows us to infer its existence (Fig. 6.20). Indeed, numerous stellar black hole candidates have been identified in binary systems.[16] Also, an accretion disk composed of gas and dust may form where matter falling downward is heated by friction and active processes (bursts) occur. Such disks surrounding black holes may emit enormous amounts of energy as electromagnetic radiation in the ambient space. In particular, the most recently discovered powerful bursts of radio waves coming from distances of billions of light years could be the result of such interactions.

It was found that massive galaxies have much more massive black holes which are associated with the nuclei of galaxies, and some are identified as quasars (see Fig. 6.21). They are called supermassive black holes (SMBHs) and contain millions and even billions of solar mass stars. For example, the core of the galaxy M87 is estimated to contain $> 10^9$ black holes. In small galaxies there is evidence of stellar clusters coexisting with a moderate black hole in the center.

In our Milky Way there is also an SMBH named Sgr A* (Sagittarius A-star) containing about 4 million solar mass stars, which was inferred by careful tracking of stars orbiting Sgr A* with the optical Keck telescope in Hawaii. At the same time, with the use of a global network of radio telescopes, the physical size of Sgr A*'s

---

[16]This method allowed us to discover the first strong candidate for a black hole, Cygnus X-1, as early as 1972.

Fig. 6.21 Galaxy-scale phenomena: Black hole eating gas (Credit: NASA)

radio source has been defined as nearly 40 million km (~1/3 AU), and a region of a comparable size has been determined from observations of quiescent X-ray emission of Sgr A* and the rapid variability of X-ray flares based on space telescope observations from the Chandra X-ray Observatory. SMBHs (quasars) accrete galactic gas and release a huge amount of gravitational energy. Some SMBHs periodically absorb ambient stars and gas and throw out a part of the captured matter in the form of hot plasma jets at nearly the speed of light. These relativistic objects provide an ideal laboratory for testing the theories about general relativity and enormous gravity deforming space-time and, at the same time, they place constraints on models for the behavior of accreting material under extreme conditions.

The lifetime of a massive black hole is estimated to exceed the age of the universe, whereas less massive black holes have much shorter lifetimes. The existence of microscopic black holes having masses of only a fraction of a gram is also assumed, because in accordance with general relativity theory, a sufficiently compact mass will deform curved space-time to form the analog of a black hole. Microscopic black holes, however, would evolve on a very short time scale in a mode of "quantum evaporation in space." That is, they would quickly transform their mass into radiation with an efficiency depending on their mass. The branch of astrophysics dealing with populations of black holes is referred to as black hole demography. Interestingly, some theories assume that at the end of the current era of star formation (tens of trillions of years from now) our universe will be populated with cold compact objects like brown dwarfs, white dwarfs, black dwarfs, neutron stars, and black holes, which eventually will either disperse in space due to numerous mutual collisions or fall into central SMBHs.

The gravitational collapse of heavy objects such as stars is commonly accepted as the primary process for black hole formation. However, there could also be more exotic processes leading to the production of black holes, in particular in the early universe shortly after the Big Bang or in the process of high-energy collisions. However, we should emphasize that as yet there is no direct experimental evidence proving the existence of black holes in the contemporary universe, though there is a lot of indirect evidence, e.g., from observations of emission from X-ray binaries presumably caused by accreting matter on the black hole from a nearby compact companion, or from studying the proper motion of stars near the center of our own Milky Way. In any case, new observations, specifically at and nearby the event horizon, are required. Unfortunately, the resolution achieved in all wavelengths (X-ray, optical, infrared, radio) is still insufficient to resolve these mysterious objects and, in particular, the phenomena in their immediate surroundings. Indeed, one may assume that numerous gas clouds circulating around and absorbed by the black hole as well as strong stellar wind from nearby stars would cause periodically powerful outbursts that should make Sgr A* a million times brighter than it actually is. Also, as observations from the Chandra space observatory showed, the Milky Way's central black hole exhibits the curious behavior of surrounding gas which seems never to reach the black hole. Such unexpected behavior is strange and seems to testify in favor of a radiatively inefficient accretion flow model. Also, we do not know why the inherent luminosity of the black hole is orders of magnitude below its theoretical potential and looks like a "shadow" of the SMBH. Answering these major challenging questions is intrinsically related with the development of new generations of both ground-based and space telescopes, in particular those using very-long-baseline interferometry (VLBI) at (sub)millimeter wavelengths capable of ensuring much better resolution. An advancement in this direction was recently made with the Russian Radio Astron satellite, equipped with a ten-meter radio telescope, that was launched into a very high elliptical orbit approaching at apocenter the Moon's orbit (~300,000 km) and establishing a nearly equidistant Earth-Moon baseline. The next breakthrough is planned with the more capable spacecraft project in the millimeter wavelength observations.

Chapter 7
Extrasolar Planets

Brief History

The idea that planetary systems are widespread in the universe and, in particular, in our galaxy, the Milky Way, was in vogue for a long time. It was supported by the observed distribution of protostars of fixed mass by their angular momenta. As a matter of fact, binary and multiple stars are born from protostellar gas clouds if their angular momentum exceeds some threshold, whereas there is a ten times smaller constraint on the angular momentum of main sequence (MS) stars to keep their rotational stability. In an intermediate range is where stars with planetary systems are born; in other words, they transfer excess angular momentum to planets. As in the solar system, in such a case the bulk mass is concentrated in the star and the bulk of the angular momentum is in the formed planetary system, although the precise mechanism of angular momentum transfer from a protostar to a planetary system is not clear. We will return to this problem in Chap. 8.

Several scenarios of planetary system formation are conceivable around protostars, as well as around formed MS stars in the process of their evolution. There is also a correlation between the metallicity of a star and the formation of planets: the probability of having planetary systems sharply increases towards stars of higher metallicity (late spectral classes). Statistically, it has been estimated that 30–40 % of single and close binary stars should possess planets, though their detection until recently was limited by the insufficient power of our astronomical instruments.

Historically, the first distant planets were discovered in the early 1990s by Aleksander Wolszczan and Dale Frail around pulsar PSR B1257+12, which is 1.4 times more massive than the Sun (Fig. 7.1). They found periodic variations in the reduced data of the recorded pulses of this pulsar, which has a 6-ms rotational/pulse period, and these variations were attributed to the presence of companions orbiting the pulsar. It was reported that two planets were orbiting this non-solar class remnant star with periods of 66.54 days and 98.21 days. Their masses M_p (more precisely, $M_p$sini, the product of mass and orbital tilt to the plane of the sky i, see below) were

© Springer Science+Business Media New York 2015
M.Ya. Marov, *The Fundamentals of Modern Astrophysics*,
DOI 10.1007/978-1-4614-8730-2_7

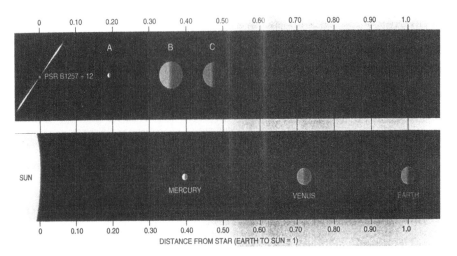

Fig. 7.1 The first distant planets discovered around pulsar PSR B1257-12 (*upper band*) as compared with the position of three terrestrial planets around the Sun (*bottom band*) (Courtesy of A. Wolszan and D. Frail)

estimated as 3.4 M_E and 2.8 M_E, respectively. Also the existence of a third lunar mass object having a period of 25 days was suggested. Later on a much more massive distant planet ($M_p \sim 2.5\ M_J$) was discovered around pulsar PSR B1620-26 at ~23 AU with a period of 191.4 days. However, such discoveries are quite rare because of the small size of the targets and the paucity of planets orbiting stellar remnants.

The first extrasolar planet (ESP) around an MS star was found in 1995 by Michel Mayor and Didier Queloz from Geneva University around the star 51 Pegasus. Two months later this discovery was confirmed by Geoffrey Marcy and his colleagues from the University of California in Berkeley, who soon reported that planets had been found around several other stars. Since then, owing to the remarkable perfection of the methods/techniques and the use of space-borne instruments, great progress has been achieved (Fig. 7.2). Nearly 1,800 exoplanets have been discovered and characterized during less than twenty years including planetary systems around some of stars, and the process is continuously accelerating. A significant portion of planets turned out to be hot bodies as massive as Jupiter and Saturn in tight orbits, and this is why they were found first using the radial velocity (RV) technique. Later on, however, super-Earth and a few Earth-like planets at different distances from the parent stars were also detected with the use of the transit method. Planetary systems with planets in tight orbits, especially those with hot super-Jupiters, form exotic configurations completely different from that of the solar system.

The great breakthrough in ESPs discovery, specifically towards planets of Earth's size and mass, was made with the ESA CoRoT and especially, with the NASA Kepler missions launched in 2006 and 2009, respectively. Of ~1,800 discovered

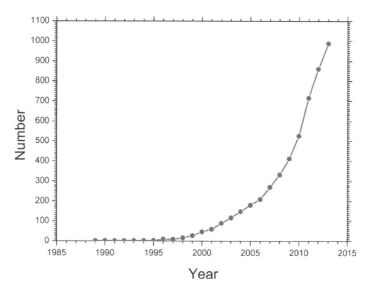

Fig. 7.2 Growth of number of discovered exoplanets since 1990 through Oct. 2013 (Courtesy of Extrasolar Planets Encyclopaedia (http://exoplanet.eu))

exoplanets more than 1,000 form planetary systems[1] with two and more planets. Obviously, our lack of detection of Earth-sized planets is caused by an insufficient capacity of the techniques that are utilized. Nonetheless, by now the surveys are biased away from large mass, very hot, low period planets towards detecting those similar to the terrestrial planets in our solar system. Generally, with no observational selection impact, an inverse proportion between number of planets and their masses was revealed that roughly follows a hyperbolic law $1/M_p$. The data collected gave us extremely valuable information about the amazing diversity of planetary systems, including their characteristics and internal structure. Of the detected planets the one usually considered the largest, TrES-4b, is 19.8 times larger than Earth (1.8 times larger than Jupiter) and is similar to Jupiter in mass, which means that it is a gaseous giant of extremely low density (0.3 g/cm^3 as compared with 1.33 g/cm^3 for Jupiter). The smallest detected planet was reported to be sub-Mercury-sized Kepler-37b, which is only slightly larger than the Moon and is about 1/100 the mass of Earth. Some planets orbit two or even three "suns"—this situation exists in binary or multiple stellar systems. Finding Earth-like planets suitable for habitation in the upcoming decades is a challenging goal for both astrophysics and astrobiology. The discovery of exoplanets has become one of the great domains in astrophysics and has dramatically enriched our study of cosmochemistry, comparative planetology, and astrobiology.

---

[1] As for April 21, 2014, 1,783 exoplanets and 1,100 exosystems has been discovered.

Methods of Detection

Different methods are used to search for exoplanets, and currently all are indirect. They include RV surveys, transit photometry, astrometry, and microlensing. Every method has both advantages and disadvantages, and when we can combine them, we obtain the most efficient results. The effects of observational selection limit the possibility to discover planets around solar-type stars, though the Kepler mission has revolutionized the existing capabilities. Stars of lower mass (and hence lower brightness) are often too faint to enable a detailed study, whereas stars of bigger mass are too bright to detect a perceptible photometric change during transit. Also, the fast rotation of these stars prevents the discovery of planets using RV surveys. By now, ground-based RV and transit surveys and space missions—CoRoT, Spitzer, and especially Kepler—have provided outstanding results.

The method of radial velocity surveys (the RV method) is based on the Doppler shift of lines in a stellar spectrum that occurs due to a target star wobbling towards and away from the observer caused by a planet's orbital motion and gravitational attraction (Fig. 7.3). Actually, the velocity variation of the stellar barycenter is recorded, its amplitude being dependent on the mass of a planet and its radial distance from the parent star. The radial distance also defines the period of a planet's revolution around the star. Obviously, only planets having orbital planes close to our line of sight are detectable, which is also true for pulsars (also, in the latter case the period of the planet's revolution should be less than the timing of the pulses). This also means that only the product of mass and orbital tilt i to the plane of the sky, $M_p \sin i$, can be determined, accounting for an observer motion relative to the barycenter of the solar system. The method of Doppler spectrometry is most sensitive to massive short-period planets orbiting a star at a relatively close distance. The extreme accuracy of the Doppler shift measurements is ~ 1 m/s. For comparison, the velocity variation of the Sun's barycenter under the gravity attraction of Jupiter

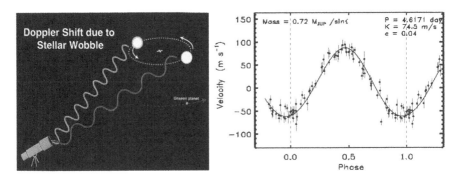

Fig. 7.3 Illustration of the method of radial velocity based on Doppler shift of lines in stellar spectrum due to a target star wobble towards and away from the observer caused by a planet gravity attraction (*left*). Discovery of the first exoplanets Peg 51 by the observed curve of star wobble (*right*) (Courtesy of M. Mayor)

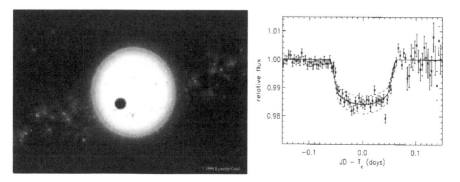

Fig. 7.4 Illustration of the method of transit photometry grounded in the passage of extra solar planet in front of the disc of parent star (Courtesy of L. Couk)

and Saturn at 1 AU is 12.5 m/s and 2.7 m/s, respectively, and it drops down to 0.02 m/s for the Earth. This means that the discovery of Earth-like planets is inaccessible using this technique. It is also poorly applied for hot stars of early spectral classes (O, B, A) because they have fewer features in their spectra compared to the cooler stars like our Sun.

The method of transit photometry is based on the passage of an exoplanet in front of the disk of its parent star (Fig. 7.4). Planets orbiting a star are much fainter than the star, and because they are located rather close to the star they are lost in the star's light and difficult to distinguish. A mutually favorable geometric position of Earth and the orbital plane of a transit planet is required to accommodate the observations, which yield the size and orbital period of the detected planet. The situation is similar to observing transits of the inner planets, Mercury and Venus, in the solar system. We had a chance to observe transits of Venus in front of the Sun's disk in 2004 and 2012, although observations of exoplanet transits are much more complicated. In particular, the real data needs to be distinguished from the background of stellar variability, variations of brightness across the disk (including influence of dust brightness), starspot modulation, and limb darkening, while the detected signal is very small, about 10^{-5} in the visible wavelength. This means that the photometric accuracy of a change in stellar brightness should be up to 0.0005^m, and the excellent modern technology used in the photometry technique has achieved this. This was further advanced by the record of weak changes in a stellar spectrum caused by the gravitational influence of a transit planet. The higher the accuracy, the lower the fractional decrease in the star's apparent luminosity or color that is detected, and thus, smaller planets can be found. A central transit, when the center of a planet blocks the center of the stellar disk light, is the most informative, but the more difficult partial transits, when a planet passes over some part of the stellar disk, are valuable as well. Transit events are quite rare because for many exoplanets the required proper geometry condition, including the position of a planet's orbit normal to the line of sight, is not fulfilled. Nonetheless, since many thousands of stars are observed, the statistics are good enough so that

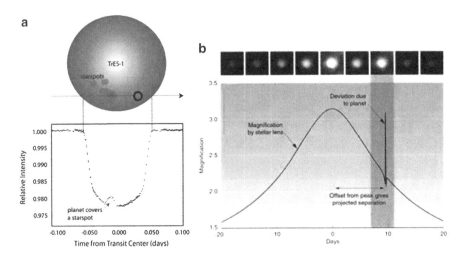

Fig. 7.5 Methods of exoplanets detection by transit photometry and microlensing. (**a**) Curve of brightness of the star TrES-1 Transit Timing Variation (TTV) method. Small peak on the bottom of transit curve is caused by star spot blanking the planet (Courtesy of D. Charbonneau; G. Laughlin, http://oklo.org); (**b**) Modeling photometric observations of distant planet brightness magnification owing to gravity lensing of its light by the star-planet systems intersecting star-observer's light of sight. The secondary narrow peak marks deviation due to planet while offset from peak gives projected separation (Courtesy of PLANET Microlensing Collaboration)

the method has turned out to be very efficient. A modification called the transit timing variation (TTV) method (Fig. 7.5a) is applicable in the case of several planets orbiting a star or for a multiple stellar system.

In addition to the methodological problems pertinent to transit photometry, the ground-based observations are also affected by variability and scintillations of the Earth's atmosphere. This is why a great breakthrough in transit photometry occurred with space-borne instruments installed on board the first satellites used to search for exoplanets: CoRoT and Kepler, with telescopes of 30 cm and 95 cm in diameter, respectively. Currently the most advanced is the Kepler satellite, which resides in a heliocentric orbit, with fixed pointing and a wide field of view (FOV) towards the constellation Cygnus. Kepler is equipped with 42 very sensitive CCD matrices, and is capable of observing 512 targets per minute (17 K targets in 30 min). About 2,500 exoplanet candidates have been observed and about 800 were firmly confirmed by mid-2012.

It is important to note that the most recent observations have been carried out in the near infrared wavelengths, where the ratio of the planet to stellar luminosity is much higher; this allows fainter (smaller) planets to be discovered. It is also very important and useful to combine the complementary methods of transit photometry and RV survey. Since in a transiting system, the position of the orbital plane is known, no assumption about inclination i is required and the mass of the planet M_p directly, rather than $M_p \sin i$, can be defined. Both size and mass allow us to estimate a planet's density. Also, if the time interval between successive transits is not preserved as constant, it means that the planet moves along a perturbed rather than a Keplerian orbit, and such irregularities betray the existence of more unseen companion planet(s).

The method of astrometry deals with wobble measurements in the star motion projected onto the plane of the sky. A star's (barycenter) wobble itself (rather than the Doppler shift of lines in the stellar spectrum it causes) can be measured in two directions in the plane of the sky and, as in the method of transit photometry, orbital tilt to the plane of the sky is determined unambiguously, and provides the most accurate mass M_p evaluation. The method requires high accuracy, especially when observing low-mass planets, and very good stability of the instruments is required in order to reduce the noise. The best precision accomplished with ground-based telescopes employing adaptive optics is ~ 1 ms of arc. Much better accuracy (~20 μs of arc) is obtained using ground interferometers, while space interferometers still under development promise to raise the precision to a few microseconds of arc. This would allow us to detect distant Earth-like planets located at ~ 1 AU from a solar mass star at distances within several parsecs. However, no such planets have yet been detected with the astrometry technique.

The method of microlensing (Fig. 7.5b) is rooted in Einstein's general theory of relativity. It uses the effect of light bending when it passes by a massive object located between the distant source and the observer. A massive object serves as a lens deflecting the light beam, and since the effect is small, it is called microlensing. During the event the brightness of the source can increase several fold, and this allows us to reconstruct the lensing star properties. If the star has a planet, some characteristic patterns appear on the light curve and the star-planet mass ratio can be determined, which allows bodies of Earth and even Moon size to be detectable, especially when different telescopes located at the same site are used. The micro-lensing technique provides an opportunity to find stars with multiple planetary systems as well, though some parameters (such as eccentricity and inclination of orbit) can only be estimated statistically. The method is most advantageous for finding very distant planets too faint to be discovered by other techniques; therefore, it provides a powerful tool to assess the amount and distribution of planetary systems in the Galaxy.

Direct detection and imaging of exoplanets is difficult to perform. The reason is clear if we recall that star brightness and reflected starlight from a regular planet (accounting for its size and radial distance) differ by about a billion times, and a planet is fully obscured in the stellar light. As we have mentioned, the situation is more favorable in the thermal infrared wavelengths because, according to the Planck curve of black-body radiation, the planet emits more and the star less energy, respectively, and consequently the contrast is nearly three orders of magnitude higher. The corona-graph technique, widely used by astronomers in observations of the Sun, where the central star is screened in the telescope's image, is regarded as the most appropriate approach. However, direct imaging with ground-based facilities is additionally complicated by the diffraction of light in telescope optics and atmospheric turbulence, but adaptive optics are playing a critical role in enabling such images to be achieved.

The first successful attempts were most recently undertaken with the Gemini Observatory consisting of two 8.19-m telescopes in Hawaii and Chile, the Keck telescopes in Hawaii, and with the Hubble Space Telescope (HST). Space telescopes are probably the most promising for direct imaging of exoplanets. There have also been reported images of a brown dwarf of ~ 30 M_J at about 30 AU from the 0.6 M_O star Gliese 229 and about ten images of substellar objects not orbiting stars, the closest one being at about 100 ly. These peculiar objects are known as *free-floating giant planets,* and they are also called rogue, stray, or orphan planets and were probably thrown out from planetary systems, especially around binary stars. One may suppose that stray planets can be repelled when gravity forces of a few massive planets "come into conflict." Because these objects are unlighted by stars, they are difficult to detect.

Properties of Exoplanets

By now a few thousands ESP candidates were detected and nearly 1,800 exoplanets reliably discovered, including more than a thousand multiple planetary systems.. This has allowed us to compile more or less rigorous statistics concerning their sizes, masses, orbital, and physical characteristics—which would have been inconceivable only ten years ago. However, we are still far away from being able to order exoplanets in a kind of Mendeleev's periodic table of chemical elements, although some attempts have been undertaken. The majority of stars with planets detected so far have only one planet, although there are some planetary systems that have two, three, four, five, six, and even seven planets.

The most impressive information that came with the first exoplanet discoveries was their size and close proximity to the parent star, which is why they have been detected first with the use of the RV method, as it is most sensitive to massive planets close to a parent star. These exoplanets turned out to be giant gaseous balls similar to Jupiter (some even greatly exceeding it in size and mass), but in tight proximity to the parent star. Thus, these planets, soon after discovery, were called hot super-Jupiters (Fig. 7.6). However, the situation changed dramatically after new, more comprehensive observations utilizing new methods and techniques became available which allowed us to detect much less massive planets and thus to conclude that the range of mass of the discovered planets lies mainly within ~ 0.01–10 M_j, where M_j is the mass of Jupiter, as could be expected. Nonetheless, the known planetary system configurations are completely different from that of the solar system.

Now we know that giant and supergiant exoplanets compose only a part of the whole family and that typical masses of exoplanets are tens and hundreds of times larger than the Earth's mass (Fig. 7.7). Based on the current state of knowledge, the mass distribution of the discovered exoplanets (N) corresponds to an exponential dependence $dN/dM \sim M^{-1.05}$. With no observational selection constraints, the distribution of massive planets roughly satisfies the hyperbolic ($\sim 1/M$) law.

Nonetheless, super-Jupiters per se represent a great interest for planetary science, first of all from the viewpoint of planetary system origin, stability and evolution,

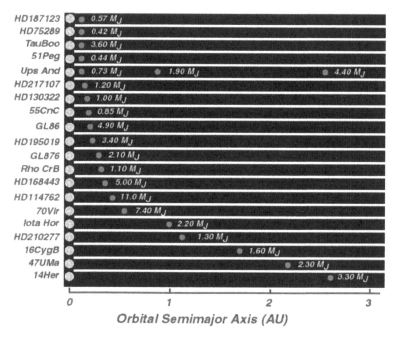

Orbital Semimajor Axis (AU)

Fig. 7.6 The plot of the first discovered massive ESP against distance to the parent star. Majority of exoplanets turned out giant gaseous balls exceeding manifold Jupiter by size and mass in close proximity to the parent star, with period of only a few months that means that they are very hot. This is why such planets were called hot super-Jupiters. The follow on observations with no constraints placed by Doppler shift method discovered many planets of much smaller size at the different distance from parent star. This allowed us to infer much more complete mass/size distribution including numerous lower mass planets (Courtesy of G. Marcy)

Fig. 7.7 Mass distribution of the discovered exoplanets corresponding to exponent distribution dN/dM ~ M$^{-1.05}$ which rather properly reflects the current state of knowledge. Unlike earlier results caused by observation selection indicated, the new evidence are steep decrease of massive planets corresponding roughly to hyperbolic (~1/M) law (Courtesy of G. Marcy)

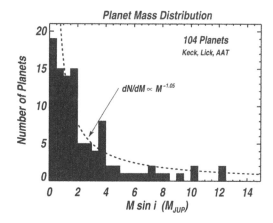

though the main focus is currently on finding terrestrial-type planets analogous to those in the solar system. The radius of giant exoplanets is within 1–3 M_j, i.e., about one-tenth of the radius of the Sun R_s. The characteristic values of the semimajor axes a range between 0.017 and 10 AU, and some have quite large eccentricities. A low a value for some planets means that they orbit their stars at a very small distance (up to ~6 R_s) with periods of only a few months or even weeks and days. Consequently, they are very hot; these are the planets called hot super-Jupiters (Fig. 7.6). Their equilibrium (effective) temperature T can be easily evaluated using a very simple formula defining the equilibrium temperature of a rotating planet:

$$T^4 = L_s(1-A)/(4\sigma d^2(1-g)).$$

Here L_s is stellar luminosity, d is radial distance to the planet, A is the spherical integral (Bond) albedo, and σ is the Stefan-Boltzmann constant. The last term accounts for greenhouse mechanism efficiency g and, therefore, the formula actually defines temperature on the planet's surface. Note that for solar system planets, instead of L_s we use the solar constant at the Earth's orbit, $S = 1,387$ W/cm^2 (measured with high accuracy), which allows us to obtain effective temperatures for Earth and other planets (e.g., for Earth $T = 249$ K (−24 C), for Jupiter $T = 135$ K (−138°C)). Using the above relation for exoplanets we immediately derive their T, and for hot Jupiters it turns out to be 1,200–1,500 K. One may thus assume that these planets are fully composed of refractory elements and compounds to avoid evaporation and survive in close proximity to the parent star. We can further assume that they possess exotic silicate atmospheres and iron clouds. Certainly, this is an extreme case for the parameter generally used to characterize the bulk composition of a planet—the water/rock ratio, which commonly ranges from ~10^{-4} (for Earth) to 0.3–0.5 (for Jupiter's satellites Europa or Ganymede with an assumed water ocean). Another extreme case is the ratio ~1, which might be assumed for a fully oceanic planet.

As we said above, the great breakthrough in ESPs discovery towards planets of Earth's size and mass was made with the ESA CoRoT and the NASA Kepler missions (Fig. 7.8). Among about 1,800 detected exoplanets,, five categories were distinguished depending on the Earth's size (radius R_E): Mercurians (0.02–0.4 R_E), Subterrans (0.4–0.8 R_E), Terrans (0.8–1.25 R_E), Superterrans (1.25–2.6 R_E), and Neptunians (2.6–6 R_E), all also distributed by their orbits (radial distance) and respective effective temperatures (Fig. 7.9). In this classification, for all groups, including Terrans and Subterrans, the domain was primarily in the hot zone close to the parent star. Only a few candidates among the Terrans and Superterrans were found in the warm, habitable "Goldilocks zone" where equilibrium temperature is estimated between roughly 185 and 300 K to allow liquid water (accounting for the greenhouse and generation of internal heat).

Kepler has revealed that planets comparable in size to Earth are abundant in the Galaxy (including those which might have the right temperatures to support life). The majority of planets, however, are of three to four times the Earth's radius, possibly because smaller sized planets are less apt to form. Obviously, of special interest are the detected Venus-like and Earth-like planets, Kepler-20e ($R = 0.87R_E$) and

Fig. 7.8 Kepler spacecraft equipped with 0.9 m telescope for exoplanets search. Artist's concept (Courtesy of NASA)

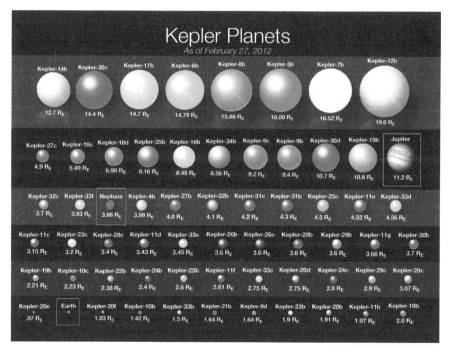

Fig. 7.9 Planets discovered by Kepler mission including Terrans, Superterrans, and Neptunians before February, 2012 (see text) (Courtesy of N. Batalha, NASA Ames Research Center)

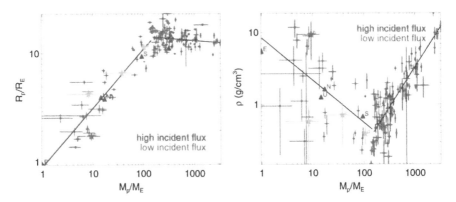

Fig. 7.10 Exoplanets size-mass relationship (*left plot*) and density-mass relationship (*right plot*). In the left mass-radius plot compiled of 138 planets of known mass and size, the sharp break is revealed which corresponds to transition from terrestrial-like and icy giant planets ($M < 150\ M_{Earth}$) to gas giants ($M > 150\ M_{Earth}$). Unlike for small and low mass bodies, radii after $M > 150\ M_{Earth}$ only slightly change whereas exponent of the corresponding isodense curve of the bodies with $M < 150\ M_{Earth}$ is equal (in logarithmic scale) to ~1/2 rather than 1/3. The latter speaks in favor of the growing volatiles contribution in the mass increase. Sharp density increase occurs for very massive $M > 150\ M_{Earth}$ bodies. *Red* and *blue* colors mark planets experiencing either high or low incident flux depending on proximity to star, respectively (Courtesy of L. Weiss)

Kepler-20f ($R = 1.03\ R_E$), respectively, although they are different in other parameters. The closest to an Earth by climate conditions is Kepler-22b ($R = 2.38 \pm 0.13\ R_E$) with an equilibrium temperature of 262 K and an orbital period of 289.86 days, very close to those of Earth, though the planet is more than twice its size. In turn, detection of Earth-sized planets at very close proximity to the host star, like for example Kepler-78b, gives a unique opportunity to use both transit and Doppler shift methods, allowing us to deduce the planet's bulk density. Remarkably, in the case of Kepler-78b, only 80 % percent larger than Earth in size and orbiting a Sun-like star at only two star's radii, the density turned out to be ~5.5 g/cm^3, practically identical to that of Earth. Hence, Kepler-78b is a rocky planet, though generally it is a hellish world.

Other important parameters to characterize exoplanets are size-mass and density-mass relationships. Based on 138 planets of known mass and size synopsis (Fig. 7.10), a sharp break corresponding to the transition from terrestrial-like and icy giant planets ($M < 150\ M_E$) to gas giants ($M > 150\ M_E$) was revealed. Clearly, unlike for small and low-mass bodies, radii after $M > 150\ M_E$ only slightly change because of decayed electronic gas contribution. It was also found that a sharp density increase occurs at the $M \sim 150\ M_E$ threshold in the isodense curve depending on the volatiles contribution in the mass increase: a growing fraction of volatiles results in the density decrease.

Moderate sized exoplanets appear to resemble Earth in bulk composition (mainly O, Mg, Si, and Fe) and interior structure. Also consistent with the Earth's bulk composition are the measured C/Si and C/O ratios. However, they may be of exotic rather than solar composition, such as the C-dominated planet HD 4203 (~3 M_E at 0.9 AU) with C/O = 1.85 and a smaller amount of water. Similarly, the interiors of massive hot super-Jupiters and extrasolar giants with a lower effective temperature are assumed to be generally akin to the giants of our solar system. Their rocky cores appear to contain a few tens of the Earth's masses, and their outer shells are composed mostly of hydrogen and helium enriched with carbon, nitrogen, sulfur, and other heavier elements (see Chap. 3). Their abundance depends on the star's metallicity. Their mantles possibly include ices of H_2O, NH_3, and CH_4. For example, the exoplanet HD 149026b, comparable in size to Saturn, is assumed to have a similar ratio of the heavy and light fractions of elements. In turn, the interiors of exoplanets of moderate size and temperature may be generally intermediate between those of the terrestrial planets and icy giants.

We summarize by saying that until recently more Superterrans (super-Earths) than Terrans (Earth-like planets) have been detected. Some of them orbit their stars at a distance where quite moderate climate conditions should exist and phase transitions of water, including its liquid state, occur. An example is the recently detected planet around dwarf star HD 40307 located at 42 ly from Earth and orbiting its parent star (less bright than the Sun) for about 200 Earth days. Planets like this one could be shrouded by dense clouds of water droplets and/or crystals, but some may even contain water (up to 50 % by mass) and oceans on the surface; they are called ocean-bearing planets. These planets are in contrast to the gaseous hot Jupiters, whose atmospheres probably consist of both gases and evaporated heavy elements and compounds including silicon and hot dust. In turn, the discovered distant planets with upper cloud temperatures of 100–200 K could resemble the solar system giants and possibly have similar systems of icy satellites, while planets having intermediate temperatures of a few hundred degrees may be like Venus, or slightly different if one takes into account a Neptune-like mass.

The most recently discovered planet Kepler-186f around red dwarf Kepler 186 in Cygnus constellation at about 500 l.y. from us caused a sensation. It is one of the system of five planets located rather close to their parent star having 25 times lower luminosity than our Sun. Hence the habitable zone is located much closer, Kepler-186f orbit is at its outer boundary of only 0.4 AU corresponding to Mercury orbit in the solar system. Although spectral composition of light the planet receives, year duration and probably some other characteristics are different, one may assume its climate its suitable for habitation and this is why it was called Earth's cousin.

The challenge is to find constraints for the origin of life on Earth-like and/or super Earth planets in the habitable zone. Unlike Kepler-186f, the planet found closest to us in the constellation Centaurus turned out to be Earth-sized rather than Earth-like ($M_p > 1.1\ M_E$). It circles its parent star Alpha Centauri B in just 3.236 days and, though the star of K1 spectral class is cooler than the Sun (its $T_{eff} = 5{,}214$ K), the planet has a very hot (~1,000 °C) surface temperature inhospitable to harbor water-carbon life, as we know it. However, finding other, more hospitable planets in the same system cannot be ruled out, including those of the closest known star to our Sun, Proxima Centauri (1.32 pc from the Sun).

Dynamics of Exoplanets

When dealing with the dynamics problem one should first distinguish between the planetary systems around single and multiple stars. Tidal effects and migration play a key role in the systems formation and evolution. In particular, they allow us to place some important constraints on hot Jupiters surviving in tight orbits around a single parent star. One may assume that such a massive planet formation occurs at a much larger distance followed by its migration towards the star because of drag in the remaining gas of the protoplanetary disk with intermediate/final resonances set up. In the latter case or, alternatively, due to tidal interactions, migration could be stopped; otherwise, the drift will ultimately cause the star to swallow the planet. The problem, however, is far from being resolved. For Terrans and Superterrans, the mechanism of their origin can be related with the formation of massive planets and is dependent on stellar mass.

Even more complicated is planetary system dynamics around binary and multiple stars, which constitute more than half of the MS stars. Of all the discovered exoplanets about 70 were reported to belong to multiple star systems. Two main configurations in the binary systems are distinguished: a planet orbiting one star (internal, or S-type orbit) or a planet orbiting both stars (external, or P-type orbit). The latter is also called a circumbinary orbit and is of particular interest for the planetary dynamics because such a configuration seems marginal in terms of stability (Fig. 7.11). Stability criteria are specifically addressed in Chap. 8. While the majority of exoplanets reside in internal orbits, many circumbinary planets have also been detected—initially from variation of radial velocity (HW Vir, NN Ser, UZ For, DP Leo, FS Aur, SZ Her) and then from transits by Kepler spacecraft (Kepler-16, 34, 35, 38, and 47). Kepler 47, which includes two planets, is the first discovered multiplanetary binary stellar system.

Unfortunately, because of the constraints of the techniques used in exoplanet detection, it is not yet possible to reproduce the configuration of the entire planetary system.

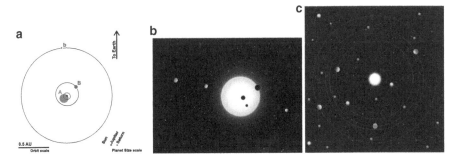

Fig. 7.11 (**a**) Orbital configuration in the circumbinary system Kepler 16. (**b**) Triple transit in the Kepler 11 system (Artist's concept). (**c**) System KOI-730: Four planets orbiting the parent star in close resonance orbits (Image Courtesy of NASA/T. Pyle and Wikimedia Commons)

In the transit method strong constraints are placed by the necessity of having the plane of a planet's orbit transverse the parent star and nicely projected to Earth; the more distant a planet from the star, the more rarely such alignment occurs. Meanwhile, the location of a supermassive planet in the immediate proximity of the star poses the question about stability of the planetary system. One of the above mentioned scenarios assumes that hot Jupiters (and possibly Earth-sized planets as well), which formed quite far from the parent star, later migrated inside the system because of interaction with the residual disk gas through dynamical friction. However, this would severely restrict their lifetime. Nonetheless, giant planet migration and interactions appears to exert a strong influence on orbital configurations as exhibited by large eccentricities, including a possibility for a planet in close approach ($\sim 10\ R_s$) to a massive ($\sim 10\ M_j$) planet to reach parabolic velocity and to leave the planetary system, becoming a free-floating planet. About ten such "stray" planets have been found, though their total number could be comparable with the population of stars in the Galaxy and their velocities traveling in space could be similar to those of stars. Moreover, a category of repelled intergalactic stray planets in the clusters of galaxies is admitted, but they are hard to discover because they are exceedingly faint.

Another result of close interaction and instabilities is the discovery of planets with retrograde orbits in the extrasolar planetary systems, which could be the result of the influence of a highly inclined outer planet on an inner one. About a quarter of exoplanets were assessed to have retrograde orbiting planets. Also the transport of a large amount of material both inwards and outwards (e.g., water) is believed to have occurred in many extrasolar planetary systems, changing the final planets composition, similar to what we assume occurred in our solar system (see Chaps. 2 and 4), with a caveat that the final planetary composition depends on the composition of the host star and possibly the planets' migration history. Let us note that in due course of migration, the thermal regime of a planet would dramatically change as well, though probably slowly enough if we take into account progressive change of the atmospheric opacity and albedo. All these could have important chemical and also biological implications.

An alternative model proceeds from the idea of an important role of massive planetesimals in the close neighborhood of the formed planet. They would exert a significant gravitational influence on the evolution of the initial orbit of the formed planet and, specifically, its migration together with the swarm of planetesimals left behind toward and away from the star, to satisfy the condition for conservation of the orbital energy and angular momentum in the protoplanetary disk. In other words, dynamical instability controls migration, and it would be responsible for the configuration of the forming planetary system. An even more complicated scenario may be assumed in the case when planets are formed in or around binary or multiple stellar systems. Planets found around such dynamically complex systems amount to 20 % of the total number; therefore, planet accumulation in such systems is common, as will be discussed in more detail in the next chapter. Planets may orbit either the companion or the whole system, and these complicated configurations pose the problem of system stability maintenance, at least in dual-star environments, and planetary formation in the presence of planetesimals and dust. Of particular importance

Fig. 7.12 Hypothetical planetary system around the HD 189733b star in comparison with the solar system. Artist's view (Courtesy of NASA)

is to find the difference between dynamical characteristics of planetary systems around single and multiple stars. Among numerous configurations there could be, for example, the one around the star HD 189733, where a massive planet, HD 189733b, was detected. According to an artist's view, the system could be as shown in Fig. 7.12 in comparison with the solar system. The question is whether such a planetary system with a giant planet close to the parent star would be stable and for how long.

Further Study

Extrasolar planets (or exoplanets), specifically the search for terrestrial type planets, are a hot area in modern astrophysics. This topic draws progressively growing attention in the astronomical community and space agencies. Amazingly, great progress has been accomplished only twenty years after the first exoplanets were discovered. The planets closest to us were found in the Alpha Centauri binary stellar system at a distance of 4.4 ly (1.34 pc), see Fig. 7.13, while the farthest planets SWEEPS-04 and SWEEPS-11 were detected by the HST in the framework of the program Sagittarius Window Eclipsing Extrasolar Planet Search using the transit technique (Fig. 7.14). These planets are located at nearly 30,000 ly (8.5 kpc) in the direction towards the Galaxy's dense, spheroidal central bulge, which is composed mostly of old stars. Both planets are super-Jupiters (Mp/Mj <3.8 and 9.7 for SWEEPS-04 and

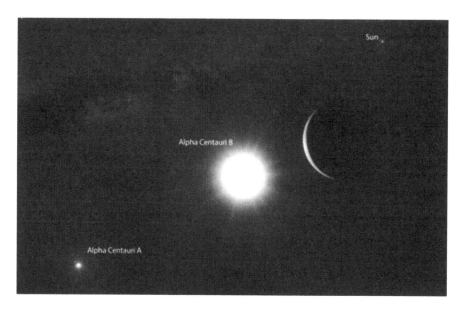

Fig. 7.13 Artist's concept of Alpha Cen A, B, planet alpha Cen Bb and the Sun (Courtesy of ESO)

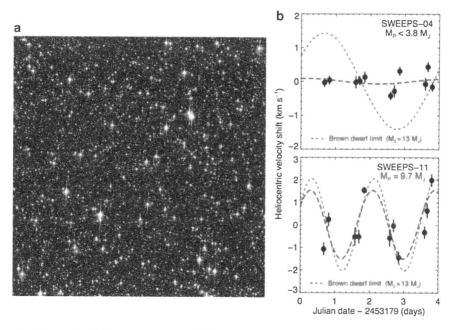

Fig. 7.14 (**a**) Star field where the planets SWEEPS were discovered in the framework of Sagittarius Window Eclipsing Extrasolar Planet Search project. The observed SWEEPS field is 202×202 angular seconds where 245,000 stars of stellar magnitude m < 30 and 180,000 stars of m < 26 are located. Systems with the planets SWEEPS-04 and SWEEPS-11 are located towards Galactic center at the distance 8.5 Kpc. (**b**) Radial Velocity (RV) curves of the detected SWEEPS-04 и 11. Jupiter-mass planets' detection around very dim stars with m ~ 26 are at the extreme of HST (Courtesy of K. Sahu)

SWEEPS-11, respectively, and are nearly similar to Jupiter in size, orbiting their parent stars (which are similar to the Sun) for only 4.2 and 1.8 days. Again, this tight proximity to the star suggests a very high surface temperature for these planets.

There are numerous programs of exoplanet observations by powerful ground telescopes and spacecraft. As we have seen, the first success came with ESA's CoRoT and NASA's Kepler spacecraft, and they have paved the road for future space missions targeted to Earth-like planet discovery in the habitable zones around millions of stars in our Milky Way. Some stars are especially distinguished, specifically those within 8 ly from Earth. The new program of observations is focused on the analysis of planetary atmospheric spectra to detect first of all traces of O_2, CO_2, and CH_4, which are intrinsically related to signs of life. The first images of Earth-like exoplanets will be obtained someday soon.

Future projects of special interest are the NASA James Webb Space Telescope (JWST), Transiting Exoplanet Survey Satellite (TESS), and two Terrestrial Planet Finder (TPF) spacecraft, and ESA Darwin space missions. JWST will carry infrared space telescope of new generation with a mirror of 6.5 m in diameter, operational at 0.6–28 μm wavelength and cooled to <50 K. It will be equipped with the perfect set of instruments, much superior to those of the HST. It is a leap forward in technology which will deepen and broaden astronomy, in particular probing stars and galaxies across cosmic time and searching for exoplanets.[2] TESS will observe the entire sky to search for favorable exoplanets. Both the JWST and TESS capabilities will be greatly amplified by the use, in addition to very high sensitivity optics, of ultra-high precision Doppler spectroscopy to measure exoplanetary sizes and masses. Darwin represents three platforms equipped with telescopes with 3.5-m mirrors and a solar shield of 7.5 m across each that will be positioned along a circle of 100 m in diameter, composing a precise system with laser-controlled mutual positions equivalent to a large-sized telescope. The spacecraft will be placed in the L2 Lagrange point at 1.5 million km from Earth and conduct observations in the far infrared (thermal) wavelengths, where the planet's brightness is a larger fraction of that of the star, allowing it to be more easily distinguished. Very sensitive receivers of thermal radiation will be cooled down 30 K which provides sensitivity by an order of magnitude higher than that of JWST. To help maintain the telescopes' mutual extremely accurate positioning, a special navigation satellite is to be launched together with astronomical vehicles as a part of the overall mission. These projects will utilize the transit method together with spectroscopic measurements, while other new missions will exploit the astrometry technique. In particular the European cornerstone mission Global Astrometric Interferometer for Astrophysics (GAIA), which launched a spacecraft in 2013, will improve the accuracy of former astrometric satellite Hipparcos by two orders of magnitude. Also, the Exoplanet Characterization Observatory (EChO) project to trace back exoplanets' formation history based on their chemical composition, the CHaracterizing ExOPlanet Satellite (CHEOPS) to search for transits by means of ultrahigh precision photometry, as well microlensing

---

[2] A concurrent project is the planned Japanese SPace Infrared 3.2-m Telescope for Cosmology and Astrophysics (SPICA).

planet search programs with the WFIRST and EUCLID projects are several ESA and NASA mission candidates.

All these missions will provide great breakthroughs in our knowledge about other worlds—their natural conditions and possibly the potential for the existence of extraterrestrial life and peculiarities of their forms. It is truly a great challenge to explore the potential for life among the ~10 billion Earth-like planets that statistics predict exist in our Milky Way alone. This will produce invaluable philosophical meaning and manifest a great triumph of science. So let's prepare ourselves for the deluge of new astonishing data and real surprises!

Chapter 8
Planetary Systems: Origin and Evolution

Basic Remarks

Problems of the formation of planetary systems and, in particular, our solar system belong to the most challenging and intriguing fields of modern science and are addressed as the product of a long trail of cosmic evolution. This important branch of advanced astrophysics is also called stellar-planetary cosmogony. Being interdisciplinary by its content, it is based on the fundamental theoretical concepts and observational data available involving the processes of stellar origin and evolution. These fundamentals, alongside with the properties of protoplanetary accretion disk structures around young stars of late spectral classes, and mechanical and cosmochemical properties of the solar system, place important constraints on plausible scenarios of how planets are formed and evolve.

The first attempts to understand how the solar system is structured and how planets were formed were undertaken in the Middle Ages. In the sixteenth century, the Italian monk, doctor of theology, and author, Giordano Bruno, voiced against the church dogma that the Earth is the center of the World, arguing instead for a configuration of the solar system with the Earth orbiting the Sun. But the truth is never free, and it is often necessary to pay a high price for personal conviction, sometimes with one's life. This is what happened to Giordano Bruno, who was sentenced by the Inquisition to be burned at the stake. Nicolaus Copernicus, who revolutionized the World system concept, had a more fortunate fate, and we refer to his concept as a real breakthrough in astronomy and philosophy in general. In 1755, Immanuel Kant, father of classical German philosophy, published the book *General History of Nature and Theory of the Heavens,* based on a hypothesis proposed in 1749 by Swedish mystic author Emanuel Swedenborg, who suggested that stars are formed in the eddy motions of space nebula matter, "as he was told by angels." Kant hypothesized that planets formed from dusty cloud that he associated with original Chaos. Laplace independently put forward a nearly analogous idea and gave it mathematical support.

© Springer Science+Business Media New York 2015
M.Ya. Marov, *The Fundamentals of Modern Astrophysics*,
DOI 10.1007/978-1-4614-8730-2_8

Basically, these ideas have been preserved until now and underlie the principal concepts of the solar system's origin.

Indeed, the Kant-Laplace hypotheses proposed in the eighteenth century about the simultaneous formation of the Sun and the protoplanetary cloud, along with the idea of rotational instability responsible for the successive separation of plane concentric rings from the cloud periphery, underlie our current views. The solar system is currently believed to have formed through the gravitational collapse of a dense fragment (core) of an interstellar molecular cloud with a density $\rho > 10^{-20}$ g·cm^{-3}, a temperature $T \sim 10$ K, a mass larger than the solar one by 10–30 %, and a dust mass fraction of ~1 %. It is also believed that after the central compressed core of a molecular cloud collapses, giving birth to the central star, material from the outer cloud regions continues to accrete onto the disk, causing strong turbulization of the gas-dust medium due to the difference between the specific angular momentum of the falling matter and the disk particulate matter involved in the Keplerian rotation. Observations backed the starting concept that a certain part of the material from the parent cloud (nebula), with an appreciable angular momentum, remains in orbit around the central clump and is incorporated into the protoplanetary disk in the process of stellar collapse. Concurrently, disk matter continues to accrete on the protostar for 1–10 million years and during this time the mass flow decreases by two to three orders of magnitude.

The discovery of circumstellar disks through high-resolution visual, infrared, and submillimeter observations, as well as extrasolar planets (also called exoplanets), has extended considerably our views of the properties of planetary systems and revealed a great variety of unusual configurations involving hot super-Jupiters in orbits very close to the parent star. These were the first detected exoplanets because of the technique constraints at that time. Although remarkable progress has been achieved since then in detecting less massive planets (super-Earths and even Earth-like planets), it is not yet possible to reproduce the configuration of an entire planetary system. Nonetheless, the peculiarity of supermassive planets located in the immediate proximity of a star poses the question about the key processes of planetary system formation and their stability.

It is generally accepted that, like other planetary systems, our solar system formed 4.567 billion years ago from an original molecular cloud (a protoplanetary nebula) consisting mostly of hydrogen and helium with a rather small admixture of heavier elements. In turn, the new protostar and gas-dust disk formation gave birth to planets in the process of continuing evolution. The process starts with a collapse of the molecular cloud (see Chap. 6) fragment with much of its mass concentrated in the center while the rest flattens out into a disk, the whole system continuing to rotate due to conservation of angular momentum. As we will further see, the age of the solar system origin was deduced from the analysis of radioisotopes preserved in the matter of ancient meteorites. The origin is related with a nearby supernova explosion which is assumed to implant stable daughter nuclei of short-lived isotopes, such as ^{26}Al and ^{60}Fe.

Basically, the continuing evolution involves the gaseous disk compression, probably influenced by the explosion shock wave. Its density increases and heats up, the inner core collapses under its own gravity, and pressure/heat increases, triggering nuclear fusion within the newborn star. This is followed by the development of gravity instabilities within the disk material that results in primary dust solid grains and dust cluster (original blobs of matter) formation. It is further assumed that eventually, in the follow-up process of their numerous collisions and growth, planet embryos—planetesimals—formed which led to the origin of planets and small bodies, and thus, of the existing solar system. A schematic view of the solar system formation from a collapsed fragment of molecular cloud, followed by the formation of the proto-Sun and protoplanetary disk and its breakup into individual ring clumps of solid particles, giving birth to planetesimals and ultimately planets through collisional interactions, is shown in Fig. 8.1. A more detailed diagram of the protoplanetary nebula evolution according to Russian scientist Otto Schmidt involving the sequence of transformations of the original gas-dust disk in blobs growing into rocks and coalescing into clumps of planetesimals is shown in Fig. 8.2. The respective time spans for every phase and the overall process of solar system formation are indicated.

However, this concept needs more rigorous and augmented support. The theory of condensation, postulating the successive emergence of high temperature and low

Fig. 8.1 Scheme for the formation of the solar system from the collapse of a molecular cloud fragment through the formation of the proto-Sun and protoplanetary disk (*1*, *2*), then its breakup into individual ring clumps of solid particles giving eventually birth to planetesimals (*3*, *4*). Continuing collisional interactions of planetesimals ultimately leads to the formation of planets (*5*) (Adapted from Wikipedia)

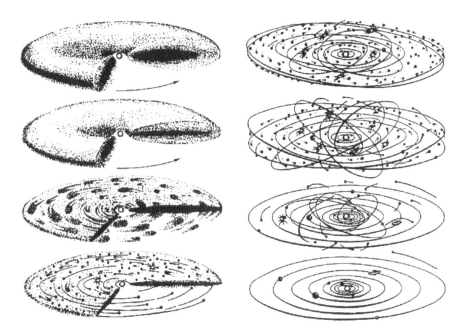

Fig. 8.2 Evolution of the protoplanetary nebula according to O. Schmidt concept. *Left side*: Sequence of transformations of the original gas-dust disk in blobs growing into rocks and coalescing in clumps of planetesimals. The time span is ~ 10^4–10^5 years. *Right side*: These embryos of planets continue to grow through mutual collisions to become eventually protoplanets and ultimately planetary system, here attributed to the solar system. The time span is ~ 10^8 years (Courtesy of B. Levin)

temperature condensates from the protoplanetary disk matter depending on radial distance from the Sun, may be recognized invoking some geochemical and dynamical constraints. This fractionation is believed to be responsible for the rocky inner planets close to the Sun and the gaseous-icy outer planets far away. Migration and collisional processes throughout the solar system history and matter transport appear to play a crucial role in the follow-up planetary evolution. Indeed, as we discussed in Chap. 2, the surfaces of the terrestrial planets have probably been painted with a veneer of volatiles and organic compounds made from life-forming elements which under certain conditions transformed into a biological infestation, at least on Earth.

Generally, the overall scenario satisfies our current understanding of the mechanism of solar system origin and evolution and is in accordance with the available experimental data on the Earth and planets. Infrared telescopes such as Spitzer and Herschel are providing an exciting picture of how all the ingredients of the "cosmic stew" that makes planetary systems from a protostar nebula are blended together. Yet many uncertainties of this scenario still exist and need to be clarified. The discovery of planets and planetary systems around other stars have greatly contributed to the current theories of the solar system's origin and at the same time have posed many new important questions to be resolved. Here we shall address only the main scenario of planetary system formation: star origin together with a compact accretion disk from an original protostar nebula, although there are also other scenarios, e.g., a capture of matter when a star passes through a molecular cloud, late protoplanetary disk appearance around main sequence (MS) stars or white dwarfs, and disk formation in specific binary systems at different stages of evolution, specifically, due to dynamical destruction of one of the companions. These scenarios, however, received much less augmented support.

Prerequisites

Before addressing the problems of planetary origin in more detail, let us summarize the most important issues considered in previous chapters in the context of stellar-planetary cosmogony, as prerequisites of the relevant concepts subject to further discussion.

1. Planets form in the common process of stellar origin/evolution and can be viewed as a more or less routine by-product of star formation. Stars surrounded by a disk are quite a common phenomenon of star formation regions (see Figs. 6.3, 6.6, and 8.3). Astronomical observations have shown that about 20–30 % of newborn

Fig. 8.3 The star-forming region NGC 1333 from Spitzer Space Infrared Telescope Facility (SIRTF) observations. The surrounding disks are clearly seen around several stars (Courtesy of JPL/NASA)

stars have disk-shaped configurations around them, but they do not all appear to evolve towards planet formation. The probability and intrinsic process of planet formation strongly depends on the mass of a star and its position on the Hertzsprung-Russell (HR) diagram. As we discussed in Chap. 6, the mass constraint for a body to become a star (to ignite regular nuclear fusion reactions in the interior) is $M \geq 0.08\ M_O$. Bodies with $M < 0.01\ M_O$ are regarded as planets (this threshold is ten times more than the mass of Jupiter), while bodies in the intermediate range of mass ($0.01\ M_O \leq M \leq 0.08\ M_O$) are brown dwarfs. The most relevant stars to possess planets are those of the late spectral classes (G, K, M). We see, therefore, that not every star-disk system gives birth to planets and a lack of sufficient mass results in different evolution scenarios with the emergence of such objects as brown dwarfs, or possibly free-floating planets. We can try to speculate on the scenario proceeding from a part of the HR diagram (see Fig. 8.4).

Inside stars primordial hydrogen atoms are cooked into heavier rock-forming elements such as silicon and metals, as well as biologically the most important element, carbon. Stellar explosions are responsible for these heavier elements returning to the interstellar medium, where they become incorporated into

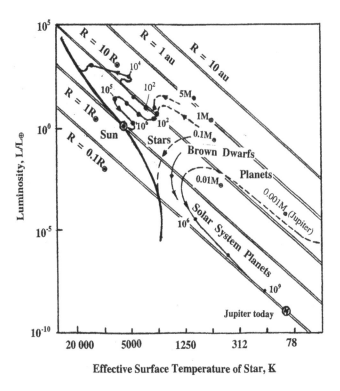

Effective Surface Temperature of Star, K

Fig. 8.4 A part of Hertzsprung-Russell diagram. Diagonal lines running from *upper left* to *bottom right* are disc radii from 10 AU to 0.1_O. Solid curve is the Main Sequence. Tracks for protostars of the solar, larger and lower masses are shown. Objects having $M << M_O$ do not reside the MS (Credit: the Author)

huge molecular clouds from which succeeding generations of stars born, some with surrounding disks. The chemical elements which entered the rocky planets (which are heavier than the cosmically most abundant hydrogen and helium) were produced in stellar interiors during stellar evolution by nucleosynthesis. These products are sprayed out into the interstellar medium in the final phase of star evolution; specifically, they enter the original nebula from which new stars and presumably planets may form. We assume this was the case of the solar system origin, complemented also by a nearby supernova explosion. The latter would serve as the main source of elements heavier than iron.

2. The scenario of the solar system and other planetary systems origin proceeds from a cold protoplanetary nebula that gives rise to a gas-dust accretion disk. The main parameter defining the fate of a collapsing nebula is its specific angular momentum J_N which results in a single or binary star formation. Fast rotating single stars with protoplanetary disks are formed only from a nebula having some restricted range of J_N. Historically, protoplanetary nebulae have been associated with giant gas-dust cloud formation in space stretching out for tens of parsecs. These great chunks of matter are called molecular clouds (see Chap. 6) because numerous rather complicated organic molecules have been found there using spectroscopic techniques. Basically, fragmentation of a molecular cloud caused by local fluctuations gives rise to a more compact rotating nebula that steadily flattens down, resulting in gas-dust disk formation around the collapsed central core to become a protostar, where the pressure and temperature progressively grow until thermonuclear reactions are ignited (Fig. 8.5). This process is

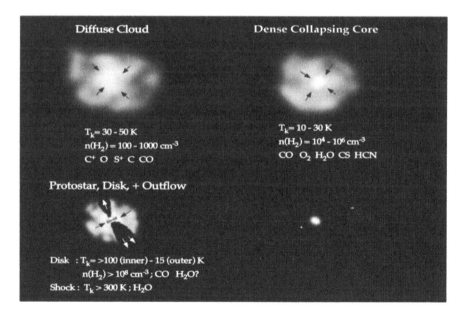

Fig. 8.5 Sequence of protostar-disc formation from an original diffuse cloud. Changes of the parameters involved in the process of evolution are shown (Courtesy of Smithsonian Astrophysical Observatory/G. Fazio)

accompanied by accretion of matter from the nebula on the turbulent disk, most of it being absorbed by the central star. The remaining matter experiences expansion with compression/flattening and angular momentum transfer from the collapsing star to the accretion disk, followed by a relatively dense dusty subdisk formation around the central plane at a very early stage of the disk evolution. In the subdisk, a gravitational (Jeans) instability[1] develops and is responsible for the appearance of primary dust clusters in the predominantly ring-shaped subdisk configuration. The clusters contain grains composed of condensates emerging under different temperatures—from refractory compounds in the proximity of the protostar to ices farther away. Their permanent mutual collisions are assumed to lead to further particles growing with the additional involvement of processes of coagulation/coalescence and the formation of rather fluffy structures. In our solar system, accretion of growing small particles/original dust clusters of presumably fluffy structure and their further collisions gave rise to the formation of denser planetesimals (planet embryos) within the first few million years, and these planetesimals ultimately formed the planets. These bodies formed before the protoplanetary accretion disk fully dissipated for about 4–5 million years, and it was during this period that the process of gas accretion on the giant planets occurred. A longer time (about 30–100 million years) is required for terrestrial planet accumulation of primary solid bodies through runaway (oligarchic) growth with "sweeping out" of residual planetesimals. Therefore, the overall process of our solar system formation involved the overlap of formed giant planets and still forming terrestrial planets, a period that occupied roughly 10^8 years. Asteroids and comets are regarded as the remnants of this process. Similar processes appear to occur in extrasolar planetary system formation involving not only single but also binary and multiple stars.

3. Key mechanical properties place severe constraints on scenarios of the solar system origin and evolution. All planets orbit the Sun in the same prograde (anticlockwise) direction, in coincidence with the Sun's rotation around its axis. The orbits are nearly circular and have a very small inclination to the ecliptic, the imaginary plane containing the Earth's circumsolar orbit. Similarly, all planets (except Venus and Uranus) rotate in a prograde direction, and the same is true for most satellites inside the planetary system. This helps confirm the idea that planets and their satellites formed in a unified process from the same original disk matter. Satellites are locked in resonance with the planet's intrinsic rotation and therefore they face the planet on the same side, similar to our Moon. The outermost satellites behave more randomly, exhibiting both prograde and retrograde orbits and rotations, and they are regarded as small bodies captured later on by the planet's gravity field. There is a peculiar mass and angular momentum distribution in the solar system: while the Sun comprises 99.8 % of the whole solar system mass, the planets comprise nearly 98 % of its angular momentum.

---

[1] Jeans instability occurs when the internal gas pressure is not strong enough to prevent gravitational collapse of a region filled with gas-dust matter.

This resulted from the process of disk evolution and planet formation, though it is not yet clear how the angular momentum redistribution in early solar system history occurred.

4. Chemical constraints involve similar cosmic abundance of chemical elements in the Sun and the most primitive of meteorites, which are viewed as original pristine substances and remnants of planetesimals. There is some evidence that the inner (terrestrial) planets were formed of matter resembling that of chondrite meteorite composition, while the gaseous-icy giant planets preserved their composition essentially unmodified since their origin. There exists an obvious correlation of the planetary bulk composition with their distance from the Sun, in support of the above-mentioned condensation theory that favors emergence of different substances from the hot gas disk depending on radial temperature distribution and, thus, on the distance from the Sun. This allows us to explain the rocky composition of the terrestrial planets containing refractory elements/compounds and the mostly gaseous and icy composition of the giant planets. Note that the composition of asteroids in the main asteroid belt between Mars and Jupiter is intermediate between the silicate/metal-rich inner planets and the volatile-rich outer planets, which also supports the condensation theory. In turn, comets are mainly composed of water ice and other frozen volatiles, and these bodies retain the most pristine matter from which the solar system formed.

5. The discovery of protoplanetary accretion disks and exoplanets places some additional constraints on the basic theory of solar system origin. Numerous accretion disks around young stars where planets could be born, and about two thousands exoplanets, have been discovered within thousands of light years from Earth. However, as was shown in the previous chapter, most of these planets (and planetary systems) turned out to be very different from those of our solar system; many of them are equal or even exceed the Jupiter-Saturn-Neptune mass and occupy eccentric orbits in very close proximity to the parent star. This means that the effective temperature of such close planets amounts to more than 1,000 K and their atmospheres could be very exotic in composition. The peculiar orbits of most exoplanets poses the question of their stability and, therefore, how long they might survive in close vicinity of the star. Only a small number of planets close to the Earth's size have been found because of the limits placed by the observational techniques utilized. Nonetheless, these discoveries call into question whether there are other planetary systems with an identical scenario of origin to our own. One may doubt whether another configuration similar to that of the solar system exists in the Galaxy. In other words, is the solar system unique? Even the nature of the Earth-like exoplanets, not to mention the hot super-Jupiters, is poorly consistent with the intriguing idea of the origin of life. However, new breakthroughs in finding exoplanets are expected, specifically those similar to Earth in natural conditions, which may extend a possibility to meet extraterrestrial life.

Let us now discuss these fundamental issues in more detail, drawing attention to the key processes of protoplanetary disk evolution and ultimately, planetary body formation.

Protoplanetary Accretion Disks

Disk Formation: Observations, Scenario, Model

The conditions for the formation of the Sun, the circumsolar protoplanetary disk, and the solar system were probably similar to the present-day conditions for the formation of solar-type stars with their protoplanetary disks and extrasolar planetary systems.

The fact that a significant fraction of young stars are surrounded by disks, which are assumed to be similar to the disk around the young Sun, had already become obvious by the early 1990s, though historically the first disks were discovered around stars more massive than the Sun, such as Vega (α Lyra). The observations of young protostellar and stellar objects are currently performed in a very wide wavelength range: from X-rays to the radio band. One of the most informative methods for studying these objects is to analyze their spectral energy distribution. The studies of the infrared, submillimeter, and millimeter spectra have revealed gas-dust disks with Keplerian rotation around hundreds of T Tauri stars (early classic stars of less than two solar masses, see Chap. 6) with ages from 10^5 to 10^7 years. Gas-dust disks have been discovered around most of the observed T Tauri stars with ages $\leq 10^6$ years and around 20–30 % of the stars with ages $\leq 10^7$ years with a mean lifetime of 3–6 Myr. The disk masses are ~ 0.01–0.2 M_S and they extend up to ~ 100–100 AU. Quite a few stars with gas-dust disks of ~ 10 Myr old were also observed. Among them, there is the gas-dust disk around the star TW Hydrae with a quite recently discovered young planet.

Based on these and other available data, the sequence of processes for the formation of stars with protoplanetary disks (young solar-type stars, i.e., single stars with a mass close to the Sun's mass) appears to be as follows. Rapid core collapse lasting $\sim 10^4$ years begins as a result of an increase in the density of the molecular cloud core in the course of its own evolution or, more likely, is due to an increase in pressure under an external force. Several external factors that could cause the core contraction and collapse are known, e.g., a nearby supernova explosion and the consequent supernova ejecta and shock, the contraction of the molecular cloud when passing through a galactic spiral arm, or the gas flows from massive stars forming nearby or an expanding ionized hydrogen (H II) region.

Before or during the collapse, the rotating core of the molecular cloud can break up into fragments, which will give rise to a single, binary, or multiple star. An important factor contributing to the stability and counteracting the fragmentation of the protostellar core (or the collapsing protostellar object) is the magnetic field. If there was no breakup into fragments, then the development of the collapse, which is faster in the central, denser region, gives rise to a clump. Growing and becoming denser, it turns into a single protostar in hydrostatic equilibrium. The material from the surrounding accretion envelope with a gradually decreasing mass falls (accretes) onto this protostar and the disk forming around it. Such an early protostellar object incorporating the protostar with the embryonic disk and the accretion envelope with a mass larger than that of the protostar is classified by its spectral energy distribution as belonging to the class 0 (zero—not to be confused with the spectral type O of stars).[2] The disk at this stage is not yet clearly revealed spectroscopically, but its existence and the accretion of matter from it onto the protostar manifest themselves in observed powerful gas outflows from the protostellar object with an extent of 0.1–1 pc. The outflows entrain the matter from the neighboring regions of the molecular cloud, reducing their velocity, and manifest themselves as CO radio emission. For this reason, they are called molecular outflows. The outflow velocities reach ~10–100 km/s.

Both young (10^6–10^7 years) and older (10^7–10^8 years) stars possess disks either before entering or already residing on the MS (Fig. 8.4). However, young stars of T Tauri type are grouped in the regions of star formation (such as the Orion Nebula) and contain more gas and dust than the older ones. Clearly, this happens because the process of planet formation has occurred in the older disks and, respectively, a major part of the matter has entered the solid bodies. Dust in these disks appears to have a secondary origin as the result of numerous collisions and debris left behind during planetary formation and is partially depleted by sedimentation on growing bodies.

However, not all stars having gas-dust disks evolve following this scenario, and the number of stars that are known to have low-mass dust debris disks is greater than that of stars known to harbor planets. Many stars have a wide variety of debris disks of both axisymmetric and asymmetric structure. Some disks created as the result of destruction of planetesimals may resemble the main asteroid belt and/or the Kuiper Belt in the solar system. We emphasize that the study of debris disks plays a vital role in our understanding of the formation and evolution of extrasolar planetary systems.[3] In particular, the mass and angular momentum content of the disk is likely to be comparable to that in the planets. Also, disk-planet interactions naturally produce orbital migration and circularization through the action of tidal torques, which in turn may lead to orbital resonances in multiplanet systems, which we will discuss below.

Once the major part of the accretion envelope has accreted onto the disk (but not by 100 %), the protostar turns into a young star surrounded by a gas-dust disk that can be observed in the visible and shorter wavelength parts of the spectrum. At the same time, the accretion rate of matter from the disk onto the star either strongly decelerates (less than half remains in the envelope) or continues. Such stars with disks are classified by their spectroscopic characteristics as belonging to the class I

---

[2] Depending on the spectral energy distribution in the protostar and surrounding disk due to a change in the distribution of masses, velocities, and temperatures, classes I and II are also distinguished (see below).

[3] An example of a stellar system with a debris disk and with no observable planets is Tau Ceti, while a system with both a debris disk and cold planets is Epsilon Eridani. Our solar system belongs to the systems with gaseous cold giants at r >> 0.1 AU, in contrast to the systems with hot gaseous giants at r < 0.1 AU (see Chap. 7).

or class II types, respectively.[4] Because of perturbations in the accretion process, these variable stars with a mass $M \sim 0.25$–$1.1\ M_\odot$ surrounded by accretion gas-dust disks exhibit irregular brightness variations, as is the case for the classical T Tauri stars.

Numerous observations based on the described scenarios have led to various models that aim to answer questions regarding the basic physicochemical processes involved. They include the sequence of changes in the aggregate state of the main protoplanetary matter components, the location of the condensation-evaporation fronts depending on the thermodynamic parameters of the disk, the role of particle sublimation and coagulation in the two-phase medium, the relative contribution of radiation and turbulence to heat and mass transport, and the mechanisms for the development of hydrodynamic and gravitational instabilities with allowance made for the shear stresses in boundary layers and polydisperse suspended dust particles responsible for turbulence appearance. Because terrestrial planets form close to the Sun, the focus in modeling is especially narrowed at the poorly resolvable inner disk regions within several astronomical units, where matter actively accretes onto the young star. This causes dust/gas ratio, optical opacity, and thermal regime changes, as well as a significant contribution of photochemical processes in transformation of the matter composition. The study is closely related with the general problems of disk evolution in the immediate vicinity of the parent solar-type star leading to the formation of some peculiar configurations of extrasolar planetary systems.

Many important constraints on the models of protoplanetary disk evolution have been determined in the last decades owing to the great progress in observations in the wide spectrum range from the ground and space. Infrared gas and dust emission measurements and high-angular-resolution submillimeter observations made the most significant contribution to the study of the disk chemistry. Basically, a differentially rotating gas-dust disk is composed of 98 % hydrogen and helium and 2 % other relatively minor species in either gaseous or solid states depending on temperature. Many transitions of molecular components have been revealed and their isotopic composition identified. Numerous molecules found in protoplanetary disks (such as H_2O, CO, N_2, H_2CO, HCN, etc.) are probably genetically related to the volatiles contained in the frozen granules of primordial accreting material. They are assumed to be subsequently subjected to substantial chemical and thermal processing. Some of these molecules are apparently ionized or in a nonequilibrium state due to the photolysis processes attributable to hard ultraviolet and X-ray radiation from a young star. Observations carried out with the Spitzer Space Telescope allowed the discovery of several young stars surrounded by disks in a fairly small star-forming region that contributed to an understanding of the structure and evolution of the circumstellar disks (see Fig. 8.6). The extension of some disks is comparable with Neptune's orbit in the solar system (Fig. 8.7). Note that the Kuiper Belt could be regarded as analogous to the dusty debris disks observed around many MS stars.

---

[4] Note that the accretion rate from the disk onto the protostar ($\dot{M}$) can be a factor of 3–4 lower than that from the envelope onto the disk ($\dot{M}_d$) due to the mass loss in the protostellar flows/winds.

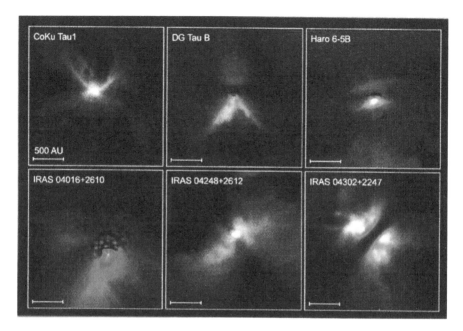

Fig. 8.6 Gas-dust disks (the dark bands between the bright regions) (Courtesy of NASA)

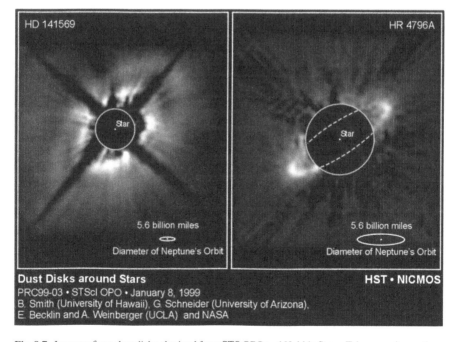

Fig. 8.7 Images of gas-dust disks obtained from STS OPO and Hubble Space Telescope observations (Courtesy of the University of Hawaii, UCLA and NASA)

Dust in Disk

No less important for understanding the evolution of the protoplanetary accretion disk, in particular its inhomogeneous structure, thermal regime, and the dynamics of its inner regions (Fig. 8.8), are the data on dust composition and properties. Observations were carried out in the optical, near, and thermal infrared spectral ranges and included measurements of emission spectra with long baseline interferometry in the millimeter wavelength. An efficient technique was developed for solving the problem of inverse light scattering by taking into account the spectral dependence of particle properties. The physical characteristics and mineralogy of dust particles were reconstructed incorporating as an analog meteor particles in the Earth's atmosphere. Interestingly, the disk particles turned out to be much larger than micrometer-size dust in the diffuse interstellar medium: the largest are millimeters to centimeters across and resemble sand or even pebbles. Moreover, they follow a height stratification, with the smaller micrometer-size particles concentrated near the disk surface. This stratification is thought to persist for millions of years. Naturally, the content and size spectrum of solid particles (granules) affects the disk medium opacity and turbulence flow patterns and strongly influences the disk thermal regime, viscous properties, chemical transformations in a gaseous

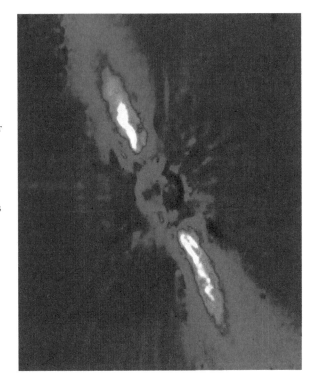

Fig. 8.8 Disk around the star Beta Pictoris. Its extent in every direction from the star is 25 AU; the clearly distinguishable inhomogeneous structure of circumstellar gas–dust disk is attributable to turbulent processes in the gas–dust medium on which gravitational perturbations from the planets forming inside the disk may be superimposed (Courtesy of European Observatory. Image obtained at the 3.6-m European Observatory telescope with adaptive optics)

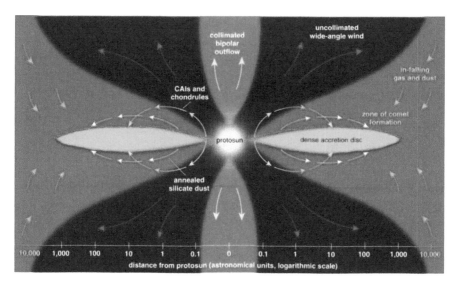

Fig. 8.9 Scheme for the formation of a gas–dust accretion disk and a subdisk. The proto-Sun onto which matter from the protoplanetary nebula (the *red color*) continues to accrete is at the center. The *green color* indicates the forming dust subdisk near which the outflow of gas and dust, including the formed high-temperature condensates in the inner zone, such as refractory CAIs, takes place. The *blue color* indicates the bipolar flows of matter attributable to the solar magnetic field (Courtesy of ISU)

medium and, in the end, its evolution including the dependence of the processes on the radial distance from the protostar and the early subdisk formation (see Fig. 8.9). The dust distribution may also signal distribution patterns of larger bodies, such as planetesimals (planetesimal belt) and/or forming planets.

It was reasonably assumed that the disk's solid particles are roughly similar in origin and composition to those located in the interstellar medium and found in meteorites. They appear to incorporate such compounds as carbon-bearing insoluble organic matter and annealed glass with embedded metal and sulfides. Submicrometer particles can be identified with the crystalline component of the predominantly amorphous interstellar dust silicates enriched with magnesium or, rather, with condensates retained in comets and related meteor fluxes as well as in chondrite meteorites. It may further be assumed by their genesis that these particles were formed inside the circumstellar disk very close to the protostar and may have undergone the follow-up processes of evaporation-crystallization during radial motion in the disk. The particles may also experience heating by shock waves in the accretion zone and subsequent rapid cooling. In accordance with this concept, we find the same type of quite refractory crystals in comets and in the frigid outskirts of our solar system. As we discussed in Chap. 4, comets are born in regions beyond the snow line where water is frozen. These regions are much colder than the searing temperatures needed to form such crystals (approximately 700 °C). The most probable scenario we may

Fig. 8.10 Artist's view of asteroid-size bodies, comets and dust around the young star Beta Pictoris (B-Pic), as seen from the outer edge of its disk. Six likely comets around distant stars were recently discovered by astronomers at the University of California, Berkley and Clarion University by weak absorption in their outcome features varying from night to night, which were attributed to large clouds of gas emanating from the nuclei of comets as they neared their central stars. It was suggested that exocomets are just as common in other stellar systems with planets and they are addressed as a sort of the missing link in current planetary formation studies (Courtesy of NASA and L. Cook)

assume is that comets acquired the crystal materials in the young solar system mingled together in a planet-forming disk. The materials forming near the Sun eventually migrated out to the outer, cooler regions of the solar system. Also, jets might have lifted crystals into the collapsing cloud of gas surrounding the early Sun before raining onto the outer regions of the forming solar system where the crystals would have been frozen into comets. Water signatures in gas-dust disk spectra encircling a young star at the later stage of evolution can also be associated with comets streaking by (Fig. 8.10). Generaly, fluffy dusty aggregates of fractal nature are assumed to enter the structure of an original circumstellar disk (see Fig. 8.11).

What are our state-of-the-art views on the mechanism of the solar system setup? We currently believe that the circumsolar gas-dust disk that developed from the protosolar nebula was 98 % molecular hydrogen and helium by mass in the ratio 70.5:27.5 %, which corresponds to a ratio of ~ 10:1 by the number of particles, while the remaining 2 % was accounted for by other elements and compounds, including hydrogen compounds, specifically water. Some of them were also in a gaseous state (from 0.5 % to 1.5 %, depending on the temperature), while others were in the form

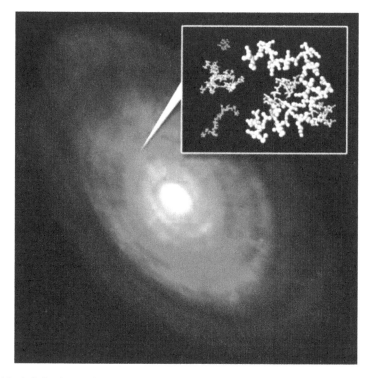

Fig. 8.11 Artist's view of fluffy aggregates of fractal nature in a circumstellar disk (Courtesy of SOKENDAI/NAOJ)

of ices (from 1.5 % to 0.5 %, respectively). The above-mentioned idea that the protosolar nebula was produced by a supernova explosion in the vicinity of a compact gas cloud initially formed through the fragmentation of a more massive gas cluster is acknowledged when addressing the solar system origin. A basis for support of this hypothesis was given by the discovery of enrichment of the matter in the Allende meteorite of ^{26}Mg and a comparison to its cosmic abundance. This stable isotope is the product of radioactive decay of the short-lived radionuclide ^{26}Al; therefore, ^{26}Mg is the daughter isotope of ^{26}Al having a half-life time decay of only 0.74 Ma. Provisionally it was implanted into the protosolar nebula's particles as the result of supernova matter injection. This idea supports general views on the important role of short-lived radionuclides in the early solar system evolution. The concept of a supernova explosion that triggered molecular cloud fragmentation is also favored by the modeling results, which suggest that an excess pressure was needed to cause the gravitational collapse of a diffuse cloud similar to the parent cloud of the solar system and the disk separation. In principle, such an excess pressure, along with the process-accelerating turbulization of the interstellar medium, could be provided by the shock waves generated by a supernova explosion.

Chronology of Disk Evolution and Cosmochemical Constraints

Of primary importance is the problem of the protoplanetary gas-dust disk evolution chronology and cosmochemical constraints placed on the thermal history of its evolution.

The study of meteorites is the main tool used to reconstruct the original matter's appearance and transformations. Clearly, the time of differentiated iron and stony meteorite formation is to be in coincidence with the time of crystallization of cores and shells of their parent bodies. The time sequence was defined from radio isotope analysis of the meteorite matter based on the measurements of ratios of long- and short-lived isotopes and products of their decay (isochrons). The main isotopic chains used in the study were the following: ^{208}Pb-^{207}Pb-^{206}Pb, ^{26}Al-^{26}Mg, ^{53}Mn-^{53}Cr, ^{87}Rb-^{86}Sr, ^{129}I-^{129}Xe, ^{182}Hf-^{182}W. The refractory inclusions of micrometer-millimeter sizes enriched with refractory elements Al and Ca called calcium-aluminum inclusions (CAIs) were found in the chondrite meteorites Allende and Efremovka and thoroughly analyzed. They were assumed to belong to the ancient solid material that entered the primordial disk composition, enabling a determination of the absolute age of the solar system, from the time of first condensed dust until now. The value was deduced to be $4{,}567.22 \pm 0.21$ million years (Myr). Concurrently, the absolute age of iron meteorites was defined to be $4{,}567.5 \pm 0.5$ Myr, i.e., accounting for referred margins, the time of the solar system origin is determined with an accuracy ~ 1 Myr, or 0.002%. The age of the Moon's oldest anorthosite rocks and the Earth's zircons is only slightly younger and is estimated to be ~ 4.4 Gyr. In turn, the absolute age of stony meteorites turned out to be $4{,}564.91 \pm 2.58$ Myr. The difference $\Delta = 3.64 \pm 1.52$ Myr can be admitted as an estimate of the total time of accumulation and thermal evolution (differentiation) of the parent bodies of these ancient meteorites.

In addition, the origin of submillimeter chondrules (*spherules*) embedded in stony meteorites and composed of ferromagnetic silicates (see Chap. 4) is also dated between 1.7 and 2.0 Myr after the origin of CAIs, close to the timeframe for the formation of chondrites of different petrological classes. It may therefore be assumed that primordial parent bodies of about 100 km in size formed in the very first few million years since the solar system origin. This size was sufficient for the body to experience differentiation due to intense heating by short-lived isotopes (mostly ^{26}Al, meaning that in ~ 5 Myr this isotope would have been fully exhausted) with an iron core emergence. A following core and silicon shell fragmentation caused by numerous collisions was probably responsible for the above iron and stony meteorites found in lab analysis. It is also believed that several million years later, when very efficient heat sources such as the short-lived ^{26}Al (and also ^{60}Fe) nuclides were exhausted, undifferentiated chondrites that experienced no melting were formed and eventually became the parent bodies of planetesimals. This type of scenario is supported by the study of Mn-Cr isochrons, refractory elements of CAI age not melted by ^{26}Al–^{60}Fe decay, which are assumed to be inherited from nebular dust and accreted as chondrules, refractory inclusions in carbonaceous chondrite parent bodies between 1.5 and 5 Myr after the solar system origin.

Basically, this time scale is in accordance with the results of computer modeling which argue that accretion from the disk on the proto-Sun terminated at 1–2.5 Myr from the system formation, while the dust subdisk composed of 1–10 cm particles formed much earlier, at 0.01–0.1 Myr at radial distance r ~ 1 AU, where critical density was achieved and gravitational instability developed. Evidently, this time was sufficient for accumulation and thermal evolution of the first solid bodies. Assuming that the mass of the protoplanetary cloud M_{cl} was ~ 0.1 M_O and that ~ 0.1 M_{cl} ultimately entered the planets, we may estimate that ~ 10^9 original bodies of ~ 100 km size were born followed by subsequent collisional evolution.

Now let us address the thermal history of the disk evolution. The view of "cold" accumulation of solar system bodies had dominated in cosmogony over a long period of time. However, a gradual transition to the "hot" protoplanetary disk model, based on the accumulated experimental data on the discrepancy between the abundances of many elements on the Sun, in undifferentiated meteorites, and on Earth, has been occurring since the early 1970s. In particular, it was established that all chondrites (except CI) and our planet are depleted in moderately volatile (Na, K, Rb, Sn, etc.) and highly volatile (Cs, Pb, etc.) elements relative to the solar abundances coincident with their abundances in carbonaceous CI chondrites. The depletion is most pronounced for such elements as Bi, Cd, Cs, Hg, In, Pb, Se, Te, Tl, Zn, S, etc. Subsequently, the depletion in these elements was found to be typical not only for various types of chondrites but also for the bulk composition of the terrestrial planets and some large planetesimals (for example, the parent bodies of eucrites).[5] Hence, it was concluded that the differentiation of moderately and highly volatile elements was an important large-scale process at the early evolutionary stages of the solar nebula and the protoplanetary disk.

Basically, the observed depletion could result from either partial evaporation or incomplete condensation of the protomatter of planets and the parent bodies of chondrites, because it was noticed that the higher the volatility of an element, the greater the depletion. Thermodynamic calculations showed that temperatures of no lower than 1,200–900 K are required in both cases. However, the fractional evaporation mechanism did not meet either theoretical or experimental confirmation because partial evaporation of moderately volatile elements requires heating the matter to temperatures at which the highly volatile elements are completely lost. Meanwhile, an experimental study of the fractional evaporation of the material of CI chondrites aimed at obtaining the material of chondrites of other types showed that, irrespective of the redox evaporation conditions, the residue obtained through heating differs radically from the actual chondritic material in the abundances of highly and moderately volatile elements. These differences are particularly true for the abundances of such pairs of elements as Zn and Se, Sn and Pb, Rb and Cs, etc. with similar degrees of depletion relative to the abundances in chondrites.

Therefore, the condensation mechanism for the differentiation of moderately and highly volatile elements accompanied by their incomplete accumulation in the

---

[5] Eucrites are a group of achondrite class meteorites enriched with calcium; they are also referred to as basaltic achondrites.

protomatter of planets and the parent bodies of chondrites seems more justified. For most elements, there is a clear correlation between the degree of depletion and the temperature of its condensation (~ 50 %) from a solar-composition gas. It is most likely attributable to the influence of the kinetic constraints on the heterogeneous reactions in a gas-solid system associated with the reduction of the reaction surface of small dust particles during their accumulation and some other factors. This allowed us to conclude that much of the chondritic material formed through the condensation of the gas phase of the protoplanetary disk and that the maximum temperatures in the circumsolar disk in the formation zone of the chondritic material ($r \sim 1$–2.5 AU) reached $T \sim 1,200$–900 K.

Apart from mineral phases containing moderately volatile elements, there are crystalline Mg- and Fe-bearing silicates in the material of chondrites in support of the "hot" protoplanetary disk formation model. The crystalline Fe-Mg silicates most likely form in high-temperature processes in the envelopes of giant stars of some classes. This idea was favored by recent studies of the composition of interstellar dust, the solid component of interstellar molecular clouds and circumstellar disks, and the material of comets and interplanetary dust, although a significant fraction of these minerals might be modified by secondary processes in the parent bodies and decrystallization under the influence of various external factors. They could also enter into amorphous olivine and pyroxene forms of dust composition, as observations of protoplanetary disks around T Tauri stars showed. The condensational origin of the Fe-Mg silicates seems even better confirmed from the observations of disks around the above-mentioned Herbig Ae/Be stars with slightly higher solar masses.

We may therefore assume that the crystalline Fe-Mg silicates formed as a result of the condensation processes in the inner zone of the circumstellar disks and that, subsequently, part of the dust was carried out into their outer zones. This idea is particularly supported by the detection of signatures of olivine crystals around a sun-like embryonic star HOPS-68 in the Orion constellation (see Chap. 6 and Fig. 6.6). Extending this conclusion to the conditions in the circumsolar protoplanetary disk, the amorphous silicates that entered into the composition of the interstellar molecular cloud from the fragment of which the solar system formed is believed to have also evaporated in its inner zone ($r \leq 1$–1.5 AU) at the early evolutionary stages. The crystalline silicates that condensed during the subsequent cooling of the disk gas phase were partially carried out into regions farther from the Sun as a result of the radial drift, where they entered into the composition of the parent bodies of chondrites as refractory chondrules, whose formation temperatures are estimated to be $T \sim 1,700$–2,100 K.[6] One may suggest that the inclusions enriched in the most refractory elements (such as Ca and Al composing CAIs, as well as the rare Hf, Sc, Lu, etc.) were the earliest condensates formed near the Sun ($r < 0.5$ AU) at $T \sim 2,000$–1,700 K and partially carried outward through the radial drift up to the formation zone of the parent bodies of chondrites.

These elements entered into the composition of the first condensed phases forming primordial bodies when a solar-composition gas cooled down. A small fraction

---

[6] Some chondrules were likely formed during local heating caused by impact processes.

of them reached $r \sim 5$–10 AU, as evidenced by the findings of olivine crystals in the material of Comets Halley and Hale-Bopp as well as in the Comet 9P/Tempel 1 nucleus investigated in the Deep Impact experiment (see Chap. 4). All this led to the conclusion that the temperature in the inner part of the protoplanetary disk at $r \leq 1$–1.5 AU reached $\sim 1{,}500$–1,300 K. The temperatures in the circumsolar disk at greater heliocentric distances ($r \sim 5$–10 AU) at the initial formation stage of Jupiter and Saturn did not exceed the water condensation temperature; this follows from the fact that the presence of water ice is a necessary condition for the formation of Jupiter's solid embryo and the planet as a whole.

Planetary System Formation

Planet formation is a widespread although very complex process, believed to be the succession of several stages affected by different mechanisms of physical interactions, chemical transformations, and numerous perturbations in the gas-dust disk. The statistics based on the results of discovered planets supports an earlier mentioned estimate that at least about one-third of the stars in the Galaxy possess planets, which means that their total amount is comparable to or even exceeds the number of stars! This estimate could probably be extended to other galaxies.

Scenarios and the model for the origin of protoplanetary nebulae and evolution are generally backed by observational data. The mechanical, physical, and cosmochemical characteristics of the solar system serve as the starting concept for the formation of planets around stars. The existing patterns in the system of planets and satellites definitely point to a unified process of their formation, while the data on the surface properties and matter composition for planets and small bodies, when compared with the samples of material from their embryos and "debris" (meteorites), provide insights into the probable sources, paths, and chronology of this process.

Model of Primary Body Formation

According to modern views, the planets around a solar-type star are formed after loss of gravitational stability in the dense dust subdisk produced out of the dust component settled down to the equatorial (central) disk plane perpendicular to the rotation axis of the turbulized accretion gas-dust disk. The sequence and time scale of a protoplanetary gas-dust accretion disk and dense dust subdisk formation followed by subdisk breakup due to gravitational instability and the formation of primary dust clusters and solid bodies (planetesimals) are shown schematically in Figs. 8.1, 8.2 and 8.9. The process is thought to occupy less than 10^5–10^6 years. Observations of the spectra for T Tauri stars allowed the accretion rate (total mass flux) from the disk onto the central star to be estimated: for most stars, it lies within the range $\dot{M} \sim 10^{-9}\text{--}10^{-7}$ M_S/yr with a mean value of $\sim 10^{-8}$ M_S/yr; in the range of stellar ages from 10^5 to 10^7, there is a tendency for the flux to decrease from 10^{-7} to 10^{-9} M_S/yr. Let us note that the drift of dust particles in directions radial and orthogonal to the central disk plane and the phase transitions during particle evaporation and/or condensation depending on temperature stratification in the disk are of key importance in influencing the subdisk origin. Turbulence generated at the boundaries of the protoplanetary disk layers and caused by shear flows corresponds in character to the parameters of a boundary (Ekman) layer. It significantly affects the disk dynamics and largely determines the subdisk structure and evolution, irrespective of whether the protoplanetary disk is formed around a single star or in a close binary system, as will be further discussed. It is important to emphasize that generation and maintenance of shear turbulence at various evolutionary stages of the disk involves a two-phase (gas-dust) medium with a differential angular velocity of rotation, and different relative contents of dust particles, their size distribution, and coagulation processes. In particular, invoking the shear turbulence mechanism backs the views about the possibility of ring-like contraction of a flat protoplanetary cloud and the formation of planets from initially "porous" (fluffy) gas-dust clumps filling the main part of their sphere of attraction (Hill's sphere) and slowly contracting due to internal gravitational forces.

The idea of the planets forming from the original dust clusters emerging in a dense subdisk breakup due to gravitational instability and eventually growing to medium-size bodies (planetesimals) is generally recognized, and both terrestrial planets and the cores of giants are thought to form by an accretion process within a gas-dust disk. Giants continue to grow, accumulating on their cores gas and/or ices abundant beyond the snow line beyond which water ice becomes thermodynamically stable. However, many details of this scenario, especially its earliest stages, remain unclear. This first concerns the mechanism by which particles grow from original micrometer-millimeter to centimeter-meter and then even to hundred meter-kilometer sized bodies. The process of particle mutual collisions is usually invoked as the factor which presumably gave rise to integration of small particles to pebble/boulder-size solid clumps. However, a higher probability of destruction rather than combination of particles and larger size bodies may occur under significant collisional relative velocities until gravitational attraction becomes dominant for bodies of about kilometer size. As a compromise, collisional integration was assumed to occur with original micrometer-sized particles of fluffy structure that eventually resulted in fluffy dust clusters of a fractal nature (Fig. 8.10). The bottom line is therefore to address fluffy dust aggregates with an account of their self-gravitation as a mechanism of coagulation to form primordial seeds of planetesimals. It was also assumed that particles and dust clusters could easily assemble in turbulent eddies. Particles coalesced more easily into "rings" of matter in vortices which could play an important role in the embryo accretion process. Nonetheless, the problem is still far from resolved, especially when we bear in mind that the global qualitative effect of disk gravity further increases collisional/impact velocities and adds additional jitter to the orbital evolution of the primary bodies.

Further Evolution

The next stage of how bodies grow to planetesimals seems quite conceivable and much easier to implement, though it occupies a much longer time ($\sim 10^8$ years). In the first approximation, the planetesimals' mass distribution obeys the Smoluchowski coagulation equation[7] with an account for gravity mutual attraction and collisional fragmentation. These interactions, jointly with dust sedimentation on such bodies residing on intersected orbits (with chaotic velocities superimposed on quasi-circular orbital velocities), result in further growth of these protoplanetary embryos followed by the gradual scooping up of smaller bodies in due course of the swarm evolution. Due to braking in the remaining gas, the rotational velocity of these bodies becomes lower than the Keplerian one, which must contribute to the acceleration of this process, as well as the mechanism of original matter exchange in the radial direction. A significant role could be played by migration of primordial bodies and planetesimals, as well as resonances in an original planetary system (in particular, the solar system) at the different stages of formation and evolution. An efficiency of the various evolution scenarios imposes certain constraints on the models attempting to describe the early phases of original body and planetesimal formation in the disk (Figs. 8.10 and 8.12). In this regard let us recall (see Chap. 3) the popular idea of Neptune and Uranus formation in the feeding zone of Jupiter and Saturn. Indeed, computer calculations showed that the embryos of Uranus and Neptune could increase their semi-major axes from ~ 10 AU to their present values, moving permanently in orbits with small eccentricities due to gravitational interactions with the planetesimals; the latter migrating outward were ultimately ejected into hyperbolic orbits.

The matter of the protoplanetary gas-dust disk is a complex multiphase unstable system with regions of different densities, temperatures, and degrees of ionization. Basically, it is an inhomogeneous dispersal medium composed of multicomponent gas and dust particles of various sizes. This matter, which is generally dusty plasma, is magnetized and is in a state of strong turbulization. Note that flattening of the rotating protoplanetary cloud results from an opposition of the two main dynamical forces, the gravitational and centrifugal ones. When these forces are in balance, weaker factors, such as the thermal and viscous processes, disk self-gravity, and electromagnetic phenomena, dominate the disk's evolution. They decisively affect the condensation of volatiles, including water, and have a significant effect on the relative content of gas and solid particles, as well as on disk energetic and angular momentum transport.

The thermal regime in the primordial disk stipulated by radiative transfer from the proto-Sun with the account of disk matter opacity variation and turbulent energy release was responsible for different conditions in the formation of the terrestrial and giant planets. Under the high temperatures close to the Sun, gases could not be retained and were exiled outward leaving behind bodies of relatively small mass

---

[7] This is regarded as a sort of Boltzmann kinetic equation when dealing with coagulation processes.

Fig. 8.12 Gas-dust disc at the later stage of evolution with concentric rings of matter encircling a young star with primary solid bodies (planetesimals) emerging inside as comets streak by (Adapted from Wikipedia)

depleted with volatiles which accumulated in the giants. An interesting exercise is to estimate what mass the Earth would have had it not lost volatiles. In order to do this, let us scale the cosmic ratios of H and He to Si based on their cosmic abundance. For hydrogen we have $H/Si = 2.6 \times 10^4$ by number, or 940 by mass. For helium we have $He/Si = 1.8 \times 10^3$ by number, or 250 by mass. Let us recall that Earth contains 6×10^{26} g of Si (10 % of the total Earth's mass). Then, with the incorporated volatiles, the mass of Earth would be $(1 + 940 + 250) \times 6 \times 10^{26} = 7.1 \times 10^{29}$ g. This is about 120 times the mass of the contemporary Earth (6×10^{27} g), or more than the mass of Saturn!

Problem of Angular Momentum Transfer

One of the major problems in the formation of a planetary system is that related to the mechanism for the transport of angular momentum J_N from a collapsing star to the protoplanetary disk. The constraints on the initial angular momentum of the circumsolar protoplanetary disk are set with an allowance for the physical processes at the early evolutionary stage of the solar system. For a uniform-density protosolar nebula, the angular momentum lies within the range $10^{52} < J_N < 10^{53}$ g cm^2 s^{-1} This range of angular momentum is typical of both single young solar-mass stars with disks and binary stars and is limited by a value that is approximately an order of magnitude higher for a disk with a mass concentrating to the center. For the solar system we estimate $J_N = (1–4) \times 10^{52}$ g cm^2 s^{-1}.

During the accretion of matter from the disk, an angular momentum is transferred to the star, accelerating its rotation. If the entire matter from the disk were absorbed by the protostar, then it would lose its stability while continuously accelerating its rotation. This probably did not occur due to the formation of two ionized gas flows, the protostellar and/or disk wind, in the inner disk near the stellar surface. Rotating around the polar axis, these flows are ejected with a high velocity (>100 km/s) under the action of a magnetic field in both directions from the disk inside a cone with a larger or smaller opening angle enclosing the rotation axis of the disk and star. Such rotating gas flows carry away an excess angular momentum, causing the rotational velocity of the protostar to remain well below the instability threshold and satisfying the condition for conservation of the angular momentum in the protostar-protoplanetary disk system. Because of the collimation of the flows along the rotation axis, they have the form of jets streaming far from the star. When the flows interact with the molecular cloud, small patches of nebulosity known as bright Herbig-Haro objects[8] and bipolar molecular outflows are formed, which may reduce the angular momentum of the collapsing star.

An even more complex situation of angular momentum transfer is in the planetary systems around binary and multiple stars, where the angular momentum is redistributed between planetary components and those of the stellar system. It is believed that due to the viscous forces of friction (arising from the relative displacement of gas suspension elements during their orbital motion), the disk matter drifts toward the protostar along a very gently sloping spiral trajectory as its angular momentum is transported outward, from the inner disk regions to the outer ones. This transport is most likely due to turbulent viscosity in a rotating, convectively unstable gas disk, which determined the time scale of its expansion. Subsequently, the turbulent vortices through which the particles accelerated and coalesced more easily into "rings" of matter could play a no less important role in the embryo accretion process.

Alternative possibilities for the loss of angular momentum by the Sun at an early stage of evolution are associated with shear motions during Keplerian disk rotation and, in the presence of a partially ionized medium, with the action of electromagnetic forces or the emergence of local shear instabilities in a poloidal magnetic field. Small-scale magnetic fields along with the assumption about disk turbulent viscosity were suggested as the most important mechanism of the angular momentum transport in the accretion disk. Under conditions of significant disk matter ionization at a

---

[8] It was also supposed that these jets, in combination with emitted radiation and/or radiation from nearby massive stars, may help to drive away the surrounding cloud from which the star was formed.

certain ratio of the disk thickness to the radial distance H/r (called the *aspect ratio*), this approach allows one to limit the restrictions on the possibility of the generation of hydrodynamic turbulence in a purely Keplerian flow. In other words, the magnetic (Maxwellian) stresses rather than the dynamics of the flow itself turn out to be dominant in this case. The possibility that the excess angular momentum is carried away at the contraction stage to which a magnetic field contributes cannot be ruled out either, although the magnetic fields in stars are relatively weak. Subsequently, the angular momentum of matter can be carried away by the stellar wind, and since the sum of the plasma angular momentum per unit mass and the angular momentum related to magnetic stresses remains constant, the angular momentum is transported via magnetic stresses. This causes the angular velocity of the star to gradually decrease. However, an adequate mechanism for the appearance of an effective viscosity cannot yet be found in any of the above models, including turbulent convection, to explain the transport of angular momentum J_N in the radial direction away from the central core.

In any case, the turbulent and electromagnetic mechanisms should be considered to be justified from a physical viewpoint and probable for explaining this phenomenon. Indeed, the protoplanetary accretion disks have a significant viscosity, which, incidentally, in combination with the differential rotation of matter, leads to the presence of a constant "intrinsic" source of internal thermal energy. According to the present views, shear turbulence and chaotic magnetic fields having energy comparable to the energy of hydrodynamic turbulence are the most likely causes of viscosity in the differentially rotating disks. Chaotic (local) magnetic fields stretching with accreting plasma, mixed due to the differential rotation of the disk, and experiencing reconnection at the boundaries between chaotic cells should also contribute significantly to the viscosity in the inner disk region and the outer layers of its atmosphere where a sufficient degree of matter ionization is attained. In turn, large-scale magnetic fields can play an important role in both angular momentum transport and physical mechanism of accretion.

Planetary System Dynamics

The dynamics is of key importance in terms of planetary system formation and stability maintenance. Special attention must be given to resonances defining the system's dynamical structure and configuration at all stages of origin and evolution. Resonances of mean motions (which we have briefly discussed in Chaps. 1 and 4) and secular resonances are distinguished. The former are caused by commensurabilities in the mutual orbital motions of planetary bodies around a single star or between (proto)planets and a parent star in multiple systems, while for the latter commensurabilities in orbit precession are responsible.

Capture in different kinds of resonances is stipulated by migration of planetesimals and forming planets in the protoplanetary disk in the course of planetary system evolution. In multiplanet systems, orbital resonances also appear to arise during interaction of formed planets with a gas-dust disk. Such interactions naturally affect orbital migration and circularization through the action of tidal torques which in turn may lead to an orbital resonance. The most common are 2:1 and 3:1 resonances of mean motions resulting in some dynamical similarities in the system configurations. In the solar system, typical examples are the earlier mentioned Uranus-Neptune (resonance 2:1) and Saturn-Uranus (resonance 3:1), as well as numerous resonances in the main asteroid belt and Kuiper Belt. In exoplanetary systems, the respective resonances were found in Gliese 876 and HD 82943 (2:1) and 55 Cancri (3:1). Let us note that 2:1 resonance is regarded as the natural result of dynamical evolution (migration) of original planetary bodies in the protoplanetary disk. At the same time, interaction of mean motion resonances (or subresonances) in a planetary system may cause its chaotic behavior, Kepler 36 serving as an example. In such a case the form and orientation of the planetary orbit experience both slow variations and precession.

Resonances are intrinsically related with the problem of disk/planetary system stability. In celestial mechanics, nonchaotic (stable) behavior of a planetary system is defined by stability criteria. The most fundamental is the analytical Hill criterion defining the zone of stability—the Hill radius (or Hill sphere, as we called it earlier)—i.e., the region of stable orbits around a planetary satellite or a smaller mass companion in the binary system. Also widely used is the Chirikov criterion of overlapping resonance zones and its modifications, in particular the Wisdom criterion defining a "locked" resonant region around a planet for other bodies to enter. Some other criteria are applied for multiple stellar systems, specifically for planetesimal dynamics in a circumbinary disk. This approach allows us to establish chaotic regions and, in particular, to evaluate Lyapunov's time (defining the dynamical system's predictability) of a planet's motion in an unstable (chaotic) zone. A typical example is the planet Kepler-16b, which is in a circumbinary planetary system and is located nearby a dangerous chaotic zone. As was shown by numerical modeling, it survives owing to its position close to orbital resonance 11:2 with the central binary star—a situation that resembles Pluto and plutinos surviving due to 3:2 resonance with Neptune.

Note also that primordial body dynamics is closely related with the above-mentioned debris disks found around many stars known to harbor planets. The number of stars having debris disks is greater than that of stars having planets. These disks are thought to form because dust is created in the destruction of planetesimals embedded in the disk, much in the same way that dust is produced in the main asteroid belt and Kuiper Belt in the solar system. For the nearest stars, direct imaging of debris disks has revealed a wide variety of structures which could be interpreted in terms of small body dynamics in the planetary systems, the observed dust distribution providing information on the distribution of larger objects, such as planetesimals and planets.

Of particular interest are configurations with supermassive planets in the immediate proximity of a star. One of the scenarios assumes that hot Jupiters (and possibly Earth-sized planets as well) which formed quite far from the mother star then migrated inside the system because of interaction with the residual disk gas through dynamical friction. Indeed, orbital migration and circularization in a multiplanet system should naturally result from disk-planet interaction and tidal torques, which in turn may lead to orbital resonances. However, this mechanism may restrict the planet's lifetime. In due course of migration, the thermal regime of a planet should be dramatically changed depending on a change of the atmospheric gas opacity and albedo.

An alternative model proceeds from the idea that massive planetesimals left in the close neighborhood of the formed planet play an important role. This would exert a significant gravitational influence on the evolution of the formed planet's initial orbit and specifically, its migration together with the swarm of planetesimals left behind toward and away from the star, to satisfy the condition for conservation of the orbital energy and angular momentum in the protoplanetary disk. In other words, in such a case dynamical instability controls migration and is responsible for configuration of the forming planetary system. In a multiplanet system, disk-planet interaction naturally produces orbital migration and circularization through the action of tidal torques, which in turn may lead to a general first order orbital resonance, as a direct consequence of the conservation of energy and angular momentum in a two-planet system.

In this regard let us recall that the migration mechanism probably affected the evolution of planetary orbits in outer regions of the early solar system. Existing scenarios admit Saturn migrating inwards and being captured in 2:3 resonance with Jupiter, which could halt and reverse the inward migration of Jupiter, presumably by a few AU. This process would inevitably exert an influence on planetary embryos in the inner region, terrestrial planet formation and the configuration of their orbits, as well as the position and storage of remnant pristine bodies in the main asteroid belt. Besides, early evolution could be responsible for the movement of Uranus and Neptune from their initial formation close to the Jupiter-Saturn zone to farther away from the Sun, as well as for the formation of the Kuiper Belt, which was discussed in Chap. 4. This scenario is justified by evaluating the time which would have been required for the accretion of Uranus and Neptune in their present-day orbits: modeling has shown that such a time would exceed the age of the solar system. The model, which is sometimes referred to as reconfiguration of giant planet orbits (the Nice model), suggests the existence of a primordial disk of several tens of Earth masses, made up of comet-like objects, residing just outside the initial orbits of the giant planets. The disk was thought to be scattered throughout the solar system by gravitational interactions between the giant planets leading also to the planets' migration. Let us note that the Nice model is linked in time to the extended time of Late Heavy Bombardment of the Moon and terrestrial planets (see Chap. 2) and the Moon crater chronology. It is also supported by a number of cosmochemical considerations.

Multiples

The discovery of exoplanets, which was addressed in the previous chapter, generated great breakthroughs in the concept of protoplanetary accretion disk evolution and planetary systems formation, stability, and dynamics. This has posed many intriguing questions about the principal mechanisms responsible for the origin of multiple configurations different from that of the solar system. It became evident that planets may form not only near single stars with angular momentum excess but also around binary and multiple stars and their components. As we said in Chap. 7, about 20 % of all known exoplanets have been found to inhabit multiple stellar systems. Most are in binary systems, but some have also been discovered in triple stellar systems, like the planet in the triple system HD 132563. The total number of multiplanetary systems is estimated to be about 130. Examples of outstanding multiplanetary systems are 55 Cancri with five planets, Upsilon Andromedae with three planets, Kepler 11 (KOI-157)[9] with six transit planets, and KOI-730 with four transit planets. Note that although our solar system is unlike those with exoplanets, among the numerous detected multiplanetary systems there are Terrans and Superterrans located close to habitable zones, such as those in the systems Gliese[10] 581, 47 Ursae Majoris, and Mu Arae (HD 160691). In particular, Superterrans Gliese 581c, d, and g are located basically within a habitable zone, whereas the configurations of 47 UMa and μ Arae (HD 160691) resemble that of the solar system.

Unfortunately, the particular scenarios involving numerous physicochemical processes in gas-dust disks around young stars and especially around binary and multiple systems in due course of original gas-dust cloud accretion and/or matter exchange between the system components are still difficult to comprehend and reconstruct. Nonetheless, the data available give important insights into the general problems of the early evolution of stellar systems and place additional constraints on the developed models. The investigation of extrasolar planetary systems contributed substantially to planetary dynamics, whereas the key answers about origin can be taken from cosmochemistry encapsulated in the primordial bodies and preserved pristine matter that it is not yet available.

The above discussed problem of angular momentum transport is easier to solve in the model framework of a disk forming and evolving around a close binary and multiple star system. Disks were assumed to form within these systems, and original matter entering planets has indeed been indirectly observed around one or both components of some young binaries. Similar to the solar system's formation within the gas-dust disk, an accretion process could be common in a binary star system as well, though the circumprimary disk is truncated by the companion. Indeed, some of the systems within dual-star environments contain stellar companions in moderately close orbits, implying a dynamically complex evolution of such systems that is different from the planetary systems around single stars. Alternatively, one of the components of a binary system can break up into a disk near the more massive component. Studying how such scenarios came about is of great interest. The specific angular momentum of the original gas-dust cloud is known to be the main factor determining the difference between the early evolutionary stages of binary and single stars and planetary system stability. Thus, close binaries can serve as a test bench for planet formation models and reveal some crucial parameters pushing the

---

[9] KOI is the acronym for Kepler Object of Interest.

[10] Gliese means that a star is taken from the Gliese Catalogue listing stars within 25 pc from Earth.

system to extreme positions. For example, one of the planets in the binary system HD 196885 has an orbit located at 2.6 AU in a highly perturbed region—just at the limit for orbital stability. Its formation stage is arguably the most sensitive to binary perturbations of primary planetesimals causing induced planetesimal shattering. Obviously, both ultimate planet accumulation and orbital stability achievements pose a clear challenge to planet formation scenarios.

More in-depth study has allowed clarification of some additional details. It was found from the analysis of close binary star distribution by their angular momentum that for approximately one-third of protostars the angular momentum is insufficient for gas-dust disk formation near the close binary system. Numerical simulation of the evolution scenario as a function of parameters for the original binary star showed that different configurations can be obtained depending on the donor-to-accretor mass ratio and on the extent to which they fill their Roche lobe.[11] The range of the parameters involved place constraints on the scenarios of either binary star coalescence or mass exchange resulting in different discs/planetary systems formation. Of special interest are the cases of protoplanetary systems with a central star and expanding excretion disk origin or, alternatively, an extended gas arm of spiral shape breaking up into individual clouds (condensations) with masses of the order of giant planets. The latter occurs in the case of binary star coalescence and occupies about one hundred orbital periods of the original binary system. The most massive clouds were obtained in orbits with semimajor axes in the range $1 \, \text{AU} < a < 3 \, \text{AU}$, which can be reduced through tidal dissipation and friction when interacting with the disk, providing the angular momentum conservation condition is fulfilled. In the framework of the numerical model for a circumstellar gas accretion-decretion disk, it was independently shown that the conservation of angular momentum requires the development of an extended decretion part of the disk that accumulates its "excess" angular momentum on the periphery. Accordingly, accretion goes into a ring having a radius of only several stellar radii, while the disk itself expands to tens or even hundreds of stellar radii.

The Problems to Be Solved

The significant advancements in the field of stellar-planetary cosmogony were accomplished due to in-depth theoretical and experimental studies performed mainly during the two most recent decades. The new data largely supported the well-known scenario of stellar-planetary formation caused by original fragmentation of a molecular cloud into primary bodies embedded in the disk structure, as is nicely depicted by an artist in Fig. 8.13. This involves several stages of compression of a protoplanetary nebula and ultimately collapse of its inner core to produce a fusion ignition/star birth with an accretion disk left behind, its continuing evolution

---

[11]The Roche lobe is the radial distance from a planet where tidal forces exceed internal forces holding a solid body together, and it can be destroyed within this limit.

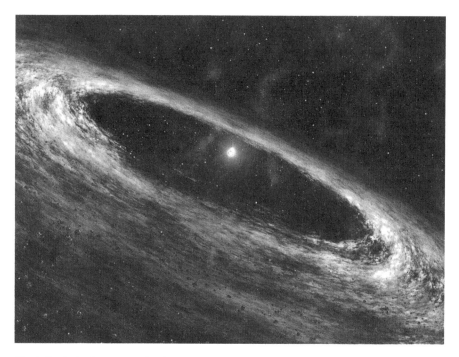

Fig. 8.13 Artists' concepts of a planetary system formation. Original gas-dust disc around central protostar where collisional processes occur and primary duct clusters forms (Courtesy of NASA)

resulting eventually in planetesimals/protoplanets and then planets. Planets are therefore regarded as a by-product of star formation. Currently, this scenario is generally accepted. Let us emphasize again that reasonably assuming that about 30 % of stars possess planetary systems, the total number of planets in the universe is estimated to be roughly equal to or even exceeds the number of stars.

The most important advancement can be summarized as follows. Observational results of the accretion of gas-dust disks around stars of different spectral classes in different wavelengths and at the various phases of evolution allow us to resolve, in quite a lot of detail, the structure of disks and their dynamics. The discovery of exoplanets around stars predominantly of late spectral classes and high metallicity (G,K,M) has introduced several planetary systems that seem to be strongly different (at least, as we know it) from the solar system in terms of body size, location, and overall system configuration. The solar system's age was firmly established from radioisotope dating of the chondrite meteorites available, with CAI dating as the starting point $(4,567.5 \pm 0.5$ million years). The chronology of these meteorites was intimately related with formation and heating of their parent solid bodies, and shed light on the earliest stage of the protoplanetary disk evolution for the first ~ 3–5 million years. Significant progress has been made in the theoretical study and computer modeling of protoplanetary accretion disk origin/evolution including its thermal regime, evaporation/condensation depending on the radial distance, coagulation/clustering, and mechanism of collisions and dynamics, including stricter observational constraints. Finally, the first successful attempts to discover Earth-like planets were undertaken, and they have mostly focused on finding evidence of the existence of a favorable natural environment suitable for the origin of life and possibly even evidence for signs of life itself.

However, these breakthroughs are still insufficient to clarify the many problems intrinsically related to planetary cosmogony. Moreover, very important new questions have been posed which are waiting to be answered. The following problems (to mention a few) are of principal interest:

- How concurrent were the evolutionary processes of the dust subdisk in the first ~0.1–0.2 million years, its fragmentation into clusters and solid bodies of ~ 100 km in size formation in the next ~ 1–3 million years at the background of continuing accretion of gas and dust on the disk and from the disk on the proto-Sun which are estimated to be ~ 3–5 million years long depending on the accretion rate?
- How are the above estimate and also similar observational constraints on the disks accretion phase (<10 million years) compatible with the hypothesis of the Earth-Moon formation of a single clump of matter in the ring-shape compression at the Earth's orbit, which is assessed to have occurred ~ 50–70 years later based on the age of the ancient lunar rocks? In contrast to the giant impact scenario, this hypothesis is supported by the radioisotopes shift, an absence of the lunar volatiles isotopic fractionation and remaining water in glasses of the Apollo samples which otherwise, together with other volatiles, would be exiled at the giant impact scenario? In short, how the Earth-Moon system originated?
- Are interactions through collision the most conceivable process for dust particles to grow from original micrometer-millimeter to centimeter-meter and then even to hundred meter-kilometer size bodies for which gravity begins to affect the processes? Could the problem be solved by invoking original fluffy dust clusters of a fractal nature assembled in fluffy dust aggregates which under the influence of self-gravitation and ambient gas pressure would coagulate, evolving to primordial seeds of planetesimals? How important is the role of turbulent eddies in assembling particles and dust clusters? In other words, how efficient are the proposed scenarios with the involvement of primary fluffy particles and particle assembly in turbulent eddies to alleviate the problem?
- What were the roles of turbulence and the processes of self-organization at the different stages of the protoplanetary gas-dust disk evolution? What was the concurrence of turbulence and magnetic braking in the angular momentum transfer?
- What was the protoplanetary disk chemistry, and how can we reconstruct the main processes involved to obtain the matter petrology conserved in the composition of the different classes of meteorites? How would it be possible to incorporate chemical kinetics in the mass and energy conservation equations, specifically in the equation of heterogeneous mechanics and magneto-hydrodynamics?
- What was the original configuration of the solar system and what underlies a planetary system configuration? How did the solar system evolve from an

original multiple of 100-km bodies (estimated number ~ 10^9 based on an assessed subdisk mass) to larger planet-sized bodies and ultimately, to its contemporary configuration?
– What mechanisms are responsible for planetary system stability? Which role did migration of primordial bodies and planetesimals and resonances in original planetary systems and, in particular, the solar system play at the different stages of formation and evolution?

Only a small part of these problems were addressed in this chapter, although essentially all topics discussed in the other chapters are in some way related to the general concept of the origin and evolution of the solar/planetary system. New observations of the evolved disks around stars and exoplanetary systems will allow us to place stricter constraints on the developed models and to reconstruct the processes of planetary system formation in a more coherent scenario. Further progress in the field will be provided by the synergy of astrophysics (stellar-galactic astronomy) and planetary sciences. This, in turn, will tremendously advance the problems of the origin of life on planets within the circumstellar habitable zone and astrobiology as a whole, to unify them into a single concept encompassing the entire universe. Undoubtedly, as the most intriguing goal, we are driven to understand the process by which we came to this world and to learn about our place in space. So stay tuned!

Chapter 9
Astrobiology: Basic Concepts

Some Historical Highlights

The intriguing problem of the origin of life on Earth, and possibly elsewhere, remains unresolved to this day. Obviously, it is intrinsically related with a planet's formation and its outer shells, the upper crust coupled with the interior through tectonic-volcanic activity, and the existence (if any) of an atmosphere-hydrosphere. On Earth, these original processes resulted in the formation of a biosphere followed by the progressive development of biological evolution. Therefore, we think of the origin of life as a combination of cosmic, geological, and biogenic processes. However, we do not know whether such an approach is universal and justified in terms of searching for the same life features elsewhere in the solar system and in the universe at large or whether an abstract logical construct must be adapted as more appropriate. In any case, this is still only a challenge because of the lack of any signs proving the existence of extraterrestrial life, but at the same time we address it as a legitimate scientific question and an alluring intellectual endeavor.

Life requires liquid water, the presence of biogenic elements, and available sources of free energy. Among the fundamental properties of life that distinguish living from non-living matter are the consumption of energy and natural substances, replication (reproduction), secretion of wastes, active biomineral exchange, and evolution (Fig. 9.1). The basic question we are addressing concerns the origin of life—the origin of the transition from prebiotic chemistry to the processes of metabolism, replication, and transmission of genetic information—since life in the modern sense has to be defined as a functional system, capable of processing and transmitting information on the molecular level. Biogeochemical functions underlie basic cyclical mass-energy exchange processes on Earth, in conformity with the fundamental laws of thermodynamics, and they are regarded as being responsible for life's origin, support, and proliferation. However, we do not know what triggered the possibility of abiogenesis (the "spontaneous generation" of life) in earlier geological epochs on our planet as a special form acquired by matter at a certain

© Springer Science+Business Media New York 2015
M.Ya. Marov, *The Fundamentals of Modern Astrophysics*,
DOI 10.1007/978-1-4614-8730-2_9

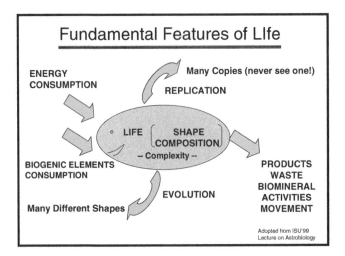

Fig. 9.1 Schematic representation of the fundamental features of life (Courtesy of ISU)

stage of its development. Neither can we reject or confine the role of an external (cosmogenic) source of life's origin (panspermia), first proposed at the end of the nineteenth century by the renowned Swedish scientist Svante Arrhenius. Extending this hypothesis further and recalling what we discussed earlier about migration and matter transport in the solar system, one may assume that living microorganisms, when they fell to a planet and found favorable conditions conducive to and necessary for abiogenesis, were able to establish themselves there.

Basically, we pursue the study of life with a geochemical approach, which was put forward by the outstanding Russian scientist Vladimir Vernadsky at the first half of the last century. He believed that prebiological evolution occurred very rapidly and that already in the earliest geological epoch, the Archean, millions of open systems could have emerged on the basis of diverse primordial macromolecules— precurses of high molecular weight protein and nucleotide compounds. These systems generally would have been capable of remaining for a certain time in a state of dynamic equilibrium and unchanged as indicated by the character and the paragenesis of minerals. The high degree of internal organization of some of these systems led to the appearance and persistence of metabolic processes and primitive replication, which served as the foundation of the incipient biosphere. The formation of the biosphere, in turn, launched the process of evolution and the creation of "morphologically different hereditary lines." Evidently, the physicochemical state of the biosphere changed in very close connection with the evolution of living forms from bacteria or calcareous algae in the Precambrian time and skeletal organisms in the Cambrian time followed eventually by plants, reptiles, mammals, and man. Obviously this earliest stage of the biosphere included the abiogenetic synthesis of organic compounds and the matrix synthesis of macromolecules, followed by formation of the properties of metabolism, the mechanism of replication, and eventually the development of prokaryotes.

Evidently, chemical processes involving a huge variety of isomer formation underlie the world of macromolecules, starting with segregation of their structural variants in the process of prebiotic evolution to enforce their role as polymeric catalysts and information carriers. Therefore, the abiogenetic appearance of diverse life forms from inorganic substances, represented by the totality of many species, belonging morphologically to various sharply divided classes of organisms is considered as completely expected. This means that biogenesis must have developed immediately, although the subsequent evolutionary process was prolonged. Possibly, the most primitive organisms, the eobionts, as Vernadsky called them, appeared on Earth about 4 billion years ago, while the emergence of photosynthesis in the prokaryotic protobionts—the first living organisms in an abiogenic medium—dates from about 3.5 billion years ago.[1] This implies that the Earth's biosphere features took shape through an evolutionary process over the subsequent billions of years, in which the biogenic cycles of atoms played a decisive role. The great breakthrough occurred only 0.5 billion years ago with the appearance of eukaryotes, microorganisms morphologically identical to prokaryotes though functionally superior to them and capable of performing intercellular biomineral exchange. This led to the appearance of an internal skeleton function, which in turn paved the way to the formation of progressively complicated organisms. The main point is that, unlike prokaryotes, which used biomineral cycles outside a cell (making them vulnerable to the environmental conditions), metabolic processes inside the eukaryote's cells made the skeleton formation independent of the environment, which changed the cyclic processing of key elements pertinent to life, such as calcium, silicon, etc.

This concept gives us a better understanding of the peculiarities of life emergence and the way in which organisms act on their environment, and also allows us to formulate the conditions necessary for life to appear. This, in turn, imposes limits upon our conceptual models of forms in which either abiogenesis or the introduction of life from outer space (probably unicellular organisms as ancestors of more complex forms) might have occurred. In any case, the structure and properties of the space occupied by life (the biosphere, as distinct from other geospheres) had to have changed, and diverse special biogeochemical functions must have appeared. The latter were brought about by living organisms. They are the functions of a single, indivisible set of organisms, a set comprising the numerous morphologically diverse forms that cause the complexity of life and the fundamental distinction between living matter and inert matter, as well as those reflecting the progress of life through time, which is directly linked with irreversible thermodynamics and Prigogine's notable principle of "the arrow of time."

---

[1] Microbially induced sedimentary structures (MISS) are found in many modern environments. Some of the oldest well-preserved sedimentary rocks in shelves, lagoons, river shores, and lakes provide a geological record extending to the early Archean period (3.48 billion years ago). This means that complex mat-forming microbial communities likely existed almost 3.5 billion years ago.

Prerequisites and Constraints

Astrobiology came into being as a framework for the attempt to uncover the relationships of life and outer space, to understand the phenomenon of life and how it arose on our planet, and ultimately to detect signs of life in the solar system and beyond. As the most important segment of the origin of life, we address the chemical evolution of matter in outer space, which is the subject of astrochemistry, as well as the intriguing problem of "habitability zone" in terms of the entire universe. Indeed, a star's position in a galaxy in terms of heavy element abundance and safety and circumstellar habitable zone (sometimes called an "ecosphere") is key in assessing the possibility of the origin of life in the stellar neighborhood. Let us note that only about 5 % of the overall stars in the Milky Way are located in a circular zone at about 7–9 pc from its center. The solar system is nicely located in this region of the Milky Way, making it a suitable place for life to originate. In contrast to the Galaxy's edges, this region is populated with stars of higher metallicity, containing the higher abundance of heavy elements that favors the appearance of life. It is also distant enough from the Galaxy center to avoid the influence of harmful radiation. (On the other hand, supernova explosions in this region may be either promoting or fatal for life's origin and sustainable development.) Obviously, because the average age of stars in this zone is 4–8 billion years, many could be much older than the solar system, and advanced complicated life forms could have originated in the course of their evolution.

The next step is to find prerequisites for the origin of life within the solar system, at least on one of the planets, and constraints (if any) which prevented its appearance on other planets. The critical point is the position of a planet relative to the Sun, which determines its climate conditions; this position depends, in turn, on the processes of dynamical protoplanetary body evolution in the protoplanetary disk. Life is also vulnerable to asteroid and comets impacts, in particular by those perturbed by nearby stars. Earth perfectly fits the position inside the solar system in terms of natural conditions necessary for the origin of life, at least for the well-known form based on carbon and liquid water. This is the so-called habitable zone in the solar system (Fig. 9.2), within which a planet could theoretically support a

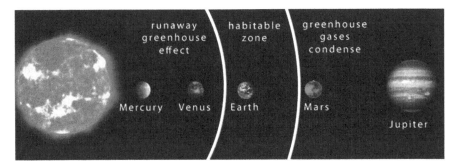

Fig. 9.2   The habitable zone in the vicinity of Earth (Adapted from Wikipedia)

climate favorable to the emergence and continued existence of life. This zone is just around Earth's orbit, falling far short of the orbit of Venus, and only approaching the orbit of Mars. The existence of water, carbon, and other volatiles on an Earth-like planet (or their delivery through migration mechanisms, as discussed in Chap. 4) is key for the origin of life. Are there life forms based on other types of chemistry and principles somewhere in the universe? Though the answer to this is unknown, carbon and water entering organic compounds appears to be the most appropriate combination for this process, and the potential to make these organics and then "dump" them on any planet can probably be regarded as a universal process.

Organic compounds are commonly found in meteorite and comet samples, although their origin presents a mystery. Indeed, the dust component of interstellar molecular clouds contains organic compounds in significant amounts (~34–38 wt. %), mainly in the form of refractory polycyclic aromatic hydrocarbons (PAHs). Having entered the circumsolar disk, the organic matter was apparently preserved virtually without any changes in the cometary matter, where its mass reaches 50 % of the dust mass. In addition, the organic compounds were partially preserved in the material of carbonaceous chondrites, particularly the most primitive ones (CI), where their content reaches ~3.5 wt. %. The maximum temperature at which these organic compounds remain stable is estimated to be ≤ 450–600 K. This range can be taken as an upper temperature limit at radial distances $r \sim 3.5$ AU, the most probable formation region of the parent bodies of CI chondrites.

It was revealed that organic synthesis, a process that takes no more than a thousand years, occurs in the interstellar medium. Synthesis is particularly efficient in interstellar molecular clouds of gas and dust (see Chap. 6), where it is fostered by the turbulence and evaporation of particles in the cloud. More than 200 fairly complex organic molecules have been found in molecular clouds, including a large quantity of hydrocarbons (building blocks of the PAHs), the simplest of which is benzene. About 70 amino acids were discovered in the Murchison and Murray meteorites, a finding which supports models of the extraterrestrial origin of the precursors of biomolecules.

The origin of some organic matter is probably caused by chemical processes triggered by the energetic ultraviolet (UV) radiation to which simple ices in space are easily exposed. This process is called irradiation. Numerous attempts to reproduce the situation expected in the solar nebula or in the Earth's primary atmosphere were undertaken beginning in the middle of the twentieth century. In laboratory experiments, glass vessels filled with a mix of gases (such as methane, ammonia, carbon dioxide, and water) were subjected to UV radiation and electric discharges, resulting in different organic outputs. A rich mixture of organics came out the chemical processes that occur when high-energy UV radiation bombards simple ices like those seen in space. These include molecules of biological interest, including amino acids, nucleobases, and amphiphiles, the building blocks of proteins, RNA and DNA, and cellular membranes, respectively. These are the sorts of molecules which could form during the formation of the solar system.

It is no less encouraging to look at computer simulations of the early gas-dust disk linked to the laboratory experiments. They showed that complex organic compounds, including many important to life on Earth, could be readily produced under conditions that likely prevailed in the turbulent environment of the primordial solar system and probably other planetary systems as well. In particular, it was found that although every dust particle within the nebula behaved differently, they all experienced the conditions needed for organics to form over a simulated million-year period. A rich mixture of organics of biological interest was found in irradiated ices made under moderate temperature. The particle dynamics in a turbulent medium also resulted in amino acids, nucleobases, and amphiphiles. However, how important a role these compounds may have played in causing life to originate and whether they were sufficient to initiate biological processes on a planetary surface, remain poorly understood.

In discussing the origin of and search for life, the geochemical conditions and biological mechanism of life on Earth is naturally our primary point of reference. Clearly, the natural conditions on the planet that were necessary for prebiotic evolution and the origin of life are of paramount importance. Life as we know it can exist only in a very limited range of natural conditions. In other words, from the outset there are fairly strict limitations on the mechanical and thermodynamic parameters of a celestial body on which life might come into being. A planet suitable for habitation must meet well-defined criteria, including size and mass (since a large planet accretes material until it becomes a gas giant, while a small planet loses its atmosphere); temperature and pressure allowing for the presence of liquid water; the existence of an atmosphere with a suitable chemical composition, excluding aggressive impurities; a radial distance from the parent star that makes favorable climatic conditions possible; and an optimal distance from the parent star, because a planet that is too close is locked in tidal resonance, which is not favorable for the development of life (Fig. 9.3). Meanwhile, based on our terrestrial experience, we should also keep in mind a number of favorable circumstances for the origin, support, proliferation, and detection of life. Indeed, with respect to metabolism (respiration, alimentation) life has great variety and adaptability, and living organisms are able to withstand extremely harsh environmental conditions, e.g., a wide range of temperatures, a vacuum, radiation, and low pH[2] (see Figs. 9.4 and 9.5), and the ingredients necessary for life are widely distributed. There is even a branch of microbiology that studies microbes known as extremophiles, which are capable of adapting to extreme environments.

Nonetheless, addressing Figs. 9.2 and 9.3, we are aware of the quite narrow range of natural conditions suitable for life's origin and support. Unfortunately, we cannot yet answer the question of what distinguished Earth from the other planets in the solar system, making the emergence of the biosphere possible here. As we discussed in Chap. 2, on Venus this possibility is excluded by the runaway greenhouse effect, which has raised its surface temperature to 475 °C and its surface pressure to 90 atmospheres. At the same time, there is reason to believe that in the early Noachian era, favorable climatic conditions for life to arise existed on Mars, including an ancient water ocean. The climate changed catastrophically about 3.6

---

[2] Bacteria inside the camera taken by the Apollo astronauts from the Surveyor spacecraft survived lunar conditions for over 6 years.

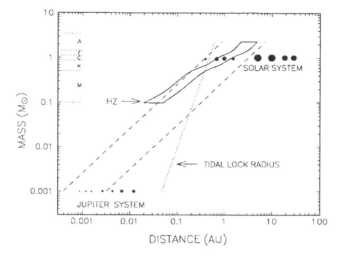

Fig. 9.3 A habitable zone for planets in the vicinity of a mother star (the distribution of the sphere of the habitable zone). The *vertical axis* indicates the spectral class and the mass of the star relative to the mass of the Sun. The *horizontal axis* gives distance in astronomical units. The *dashed lines* show the boundary limits for the planets depending upon the star class and the radial distance, and the *dotted line* shows the tidal lock radius. Theoretically, three planets of our Solar System exist within the boundary of the habitable zone: Earth, Venus, and Mars (Courtesy of J.F. Kasting, D.P. Whitmire, and R.T. Reynolds)

billion years ago, leaving a waterless desert surface and a rarefied atmosphere, though traces of primitive Martian life may have survived.

It is not impossible that life may exist in what are assumed to be oceans of liquid water under the icy surfaces of two of the Galilean tidally heating moons of Jupiter: Europa and Ganymede (see Chap. 3). The evolution of organic material on Enceladus and Titan, satellites of Saturn, is a question of great interest in terms of prebiotic evolution. Jets of gas and solids emitted by very small Enceladus from fractures in its crust imply that a pocket of liquid water (probably in the form of a water-ammonia eutectic) lies beneath its surface. Moreover, organic molecules and salts were observed to be present in the moon's plumes. Titan, 1,000 times larger than Enceladus by volume with a dense nitrogen-methane atmosphere, UV/charged-particle driven photochemistry, and lakes of apparently liquid methane and its product ethane, looks even more intriguing. Numerous organic species in the atmosphere and on the surface could be produced and serve as primitive matter for an exotic biochemistry with no oxygen involvement. On the other hand, in a water ocean beneath Titan's surface at a depth of ~50 km, as suggested based on the Cassini-Huygens mission results, more routine life could originate.

Recently, researchers' attention has been increasingly attracted to exoplanets, especially the Earth-like planets that have already been discovered by the Kepler mission in orbit around other stars (see Chap. 7), and also to the prospect of finding suitable climatic conditions or even life on these planets, the more so since the impact of life on the environment is rather noticeable and lends itself to external observation.

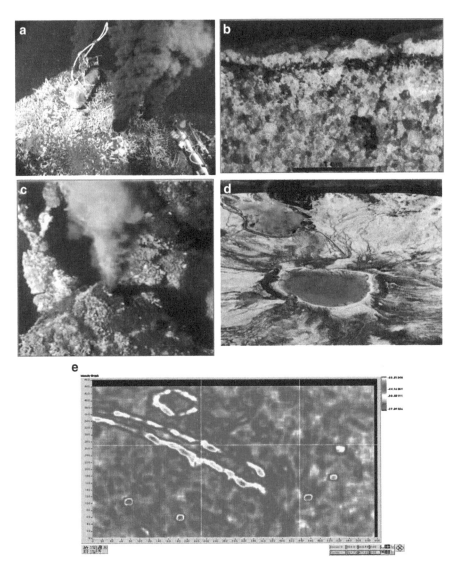

Fig. 9.4 Life is hardy. Microbial life (extremophiles) are found near undersea volcanic vents, in deep underground aquifers (**a**), within rocks (**b**), or in hot (120 °C) bottom ocean springs (**c**) and acid (very low PH) lakes (**d**). Cyanobacteria fossils from 650 million years ago (**e**). The existence of these bacteria suggests that life needs only water, a source of energy, and cosmically abundant elements (Courtesy of NOAA PMEL Vents Program, NPS)

Connection of the biochemical evolution of matter with cosmic factors merits special attention. Here we have a choice between alternative models of the origin of life: starting directly on Earth, or with an external cosmogenic source playing a part. The most challenging is the concept of live seed transfer between the stars; the probability for this is low because of the high relative star velocities and the low stellar

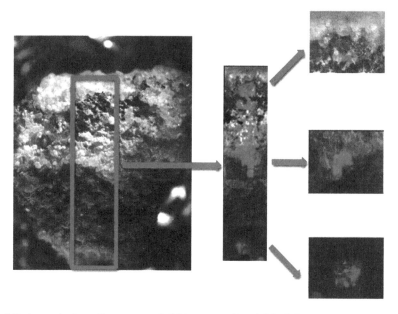

Fig. 9.5 Antarctic dry valley cryptoendolithic community, visible light and deep UV (224 nm) images (Courtesy of Center for Life Detection, JPL/CIT)

densities. A more promising idea is that of interstellar transport within the stellar clusters where the majority of stars are born and relative velocities are smaller. In Chap. 4 we also discussed the important role of matter transport due to migration and collision processes in the solar system, in which the key role is played by comets and asteroids with a carbonaceous chondrite composition and specifically interplanetary/interstellar dust. This allows us to consider these small bodies as likely carriers of prebiotic or even biotic matter from the primary asteroid belt and from the trans-Neptunian Kuiper Belt.

Carbonaceous chondrites are the key to finding extraterrestrial sources of organic matter: they contain chemically bound water, in hydroxyl (OH) form, and their parent bodies (hydrosilicates) were probably formed in a water environment. Moreover, complex biomolecules (such as amino acids of chiral proteins, and chains of hydrocarbons and kerogens) and even fossilized primitive microorganisms (microfossils) were reported to be found in carbonate meteorites of CI1 (Alais, Orgueil, Ivuna) and CM2 (Mighei, Murchinson) classes. Obviously, these claims need to be confirmed by more elaborative studies, first of all, in order to exclude biological contamination of these meteorites.[3] These microfossils are identified with remnants of most

---

[3] To disprove arguments of contamination, the authors argued that only three of the five nucleotides necessary for life and only 8 of the 20 protein amino acids commonly present in a living cell were found in the carbonaceous chondrites under consideration. It is therefore unlikely that only a part of these molecules pertinent to terrestrial organisms would penetrate the meteorite matter; thus, their extraterrestrial origin is more plausible.

primitive one-cell prokaryotes (viruses, cyanobacterias) and more complicated multicell eukaryotes. Comets, probable fragments of ancient large icy-stony bodies beyond the snow line, enriched with water and carbon and experiencing a complex thermal evolution, are even more favorable "cradles" and prolific carriers of the seeds of life. Indeed, the ratio between the carbon in comets and the carbon in carbonaceous chondrites is 10:1, although the meteorites' volatile organics might have been lost at later stages during asteroid impacts.

As we discussed in Chap. 4, given the key role of water in the origin of life, we should note that modeling has indicated that the Earth could have received a large influx of volatile matter from comet and asteroid bombardment: a quantity of water that could be comparable to the volume of our planet's oceans. Biological molecules could constitute a significant share of this input. Tiny water-filled vesicles may form in grains due to the reaction of hydrogen ions in the solar wind with oxygen atoms in dust particles. This mechanism of water formation almost certainly occurs in the solar and other planetary systems with potential implications for the origin of life and its proliferation throughout space. Dust grains floating through our solar system have been found to contain tiny pockets of water that form when they are hit by charged particles from the Sun, a phenomenon created in laboratories but previously unconfirmed inside actual stardust. Organic compounds have previously been detected in stardust as well, and the discovery of water in the dust grains suggests that they contain the basic ingredients needed for life. Let us also note that particles are considered as important carriers of the seeds of life, because they survive entry to a planet under lower temperatures than large bodies. We can assume that dust is probably common in solar systems all over the universe, which also suggests that life might be widespread throughout the cosmos.

Tentative Models of the Origin of Life

The question of how life originated is of paramount interest, along with its implication to search for life in the solar system and beyond. When we talk about the origin of life familiar to us, we are dealing not only with the formation of chains of nucleotides and amino acids (nucleic acids and peptides), which constitute the informational (DNA and RNA)[4] and the functional (proteins) basis of life, respectively, but also with the formation of the first ecosystem. Among the various conceptual approaches to the origin of life, we address here the most noteworthy and well-founded (in our view): the hypotheses of an ancient RNA world and of a sequential ordering process.[5] In each of these approaches, processes of biochemical evolution

---

[4] DNA and RNA are heteropolymers - chains consisting of monomers. Note that only about 5 % of the double DNA spiral is used for coding. The remainder contains information on how the sequence of genes is to be ordered.

[5] We will follow here the basic approach developed by the distinguished Russian scientists Alexander Spirin and Eric Galimov.

Ancient RNA world as the precursor of life origin on Earth

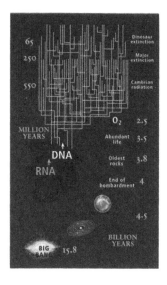

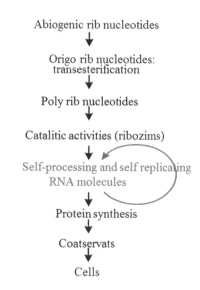

Abiogenic rib nucleotides
↓
Origo rib nucleotides:
transesterification
↓
Poly rib nucleotides
↓
Catalitic activities (ribozims)
↓
Self-processing and self replicating
RNA molecules
↓
Protein synthesis
↓
Coatservats
↓
Cells

[AU1]   **Fig. 9.6** *Left*: a chronology of events in the course of the evolution of the biosphere. *Right*: a schematic depiction of the evolution of life from its origin in the ancient RNA world (Courtesy of J.F. Atkins and R.F. Gesteland; A.S. Spirin)

are crucial. As for Darwinism, it has an important role with regard to the stages of biological evolution, but not at the early stages of life's origin and the development of the molecular mechanisms of biological systems. From this perspective, molecular genetics, biochemistry, and Darwinism are complementary, and constitute the foundation of modern evolutionary theory.

The concept of an ancient RNA world (Fig. 9.6), as the basis for the evolution of the primordial biosphere, is favored by the unique properties of the RNA molecule (a three-dimensional heteropolymer) defined by the sequences of RNA bases along the strands and the character of the coiling. Indeed, ensembles of RNA molecules carry out the functions of assimilation, metabolism, and replication. It is important to emphasize that RNA may contain genetic information or serve as a temporary copy of genetic information. For this purpose it uses a short-lived intermediate molecule (mRNA taken to the cell's protein synthesis machinery, called the ribosome), which carries the initial information for production of a specific protein and copies the cell genome which is the catalog of genetic material in a living system. Thus, RNA has the ability to perform many of the basic functions of DNA, participating in the ribosome's process of protein synthesis. These include encoding (programming the synthesis of biopolymers by a linear sequence of polynucleotides), replication (strict copying of genetic material), the self-folding of linear polynucleotides into unique compact configurations (3D structures), recognition (specific interaction with other macromolecules), and catalytic functions. To this list we should add the fact that an RNA molecule has transfer properties (tRNA); that is, it transports other molecules that are necessary for a number of biological reactions and for protein synthesis. Each of the 20 existing tRNA molecules can attach to only one of the existing 20 amino acids, which it transports to a certain ribosome and then integrates into the chain of the protein being synthesized, in accordance with the specifications contained in the intermediate mRNA molecule.

There are also catalytic (driving reactions) RNA molecules (ribozymes), which are involved in protein synthesis, along with standard protein catalysts (enzymes). They ensure the selection of specific intermolecular reactions and reduce the amount of energy they require. In addition, ribozymes provide the correct arrangement of nucleotide bonds in the chain during splicing (an ability to "cut and paste" links between nucleotides in a strand) of the mRNA molecules; only after this process will they be read correctly by the ribosome in protein synthesis. Thus, molecules of ribosomal RNA (rRNA) play a very important role in protein synthesis, because they form the structural core of the ribosome, consisting of more than 50 different proteins and several rRNA molecules. The ribosome, in a sense, "relies on" the catalytic functions of the rRNA during protein synthesis, and by reading the information encoded in the mRNA, it "knows" which protein to make. However, the extremely complex mechanism by which the genetic information of nucleic acids is decoded into the structural parameters of proteins, and how this mechanism was formed in the process of evolution, is not yet fully understood.

It follows from what we have said here that RNA, as the working instrument of cellular production, could have been the prototype of living systems. One may say that RNA is a working part of celular machinery. However, the emergence of an RNA world and its evolution up to the point of the first highly organized organisms—bacteria—over an extremely short period of time (about the first half-billion years in the Earth's history) is unlikely, as advocates of this concept themselves admit. This difficulty may be eliminated by adducing a hypothesis that ensembles of RNA molecules originated and underwent their initial evolution in the environment of outer space, especially on small bodies such as comets, which bombarded the Earth and other planets intensively about 4 billion years ago. Therefore, the idea of an ancient RNA world is linked with the possibility of the extraterrestrial origin of life.

An alternative to the conception of an ancient RNA world is that of a sequential ordering of the processes of the origin and early evolution of organic matter as the chemical basis of life. This approach is consonant with the above-mentioned ideas about processes of abiogenesis in open dissipative systems that have a high degree of internal organization and are capable of remaining in a state of dynamic equilibrium for some time, and about the organized nature of the biosphere, based on the biogenic cycles of the atoms of chemical elements, which preclude a chaotic state. As part of this concept, in which the basic functions of RNA molecules also play an important role, the origin of life is conceived of as a continuous ordering process in an open stationary system, which, in contrast to a conservative (Hamiltonian) system, which conserves energy, is an open dissipative system that exchanges energy with the environment. Generally, one deals with a system migration between many metastable states with an irregular energy release rather than remaining in the state of single equilibrium. This kind of approach seems pertinent to progressively ordering matter in vital systems and is applicable to both individual organisms (ontogenesis) and the whole history of life.

From the biologic viewpoint, a system under consideration would consist of pre-biotic organic compounds (the above-mentioned macromolecules) that had emerged in the processes of chemical evolution, possibly having originated in outer space, and then triggered on the Earth's surface. Conjugated chemical reactions occur in the system, producing not only positive but also negative entropy, which is a necessary condition for structural organization (ordering) in a chaotic environment. The energy is thereby maintained above a certain minimum level, as long as Prigogine's minimum entropy production conditions are met. Chemical ordering (limitation of the number of partners in a reaction, and the number of mechanisms and interaction paths) is implemented efficiently by selective catalysis employing biochemical catalysts known as enzymes, which are peptide chains (proteins) folded into three-dimensional structures; these are highly active and they efficiently accomplish the ordering by means of selective catalysis. The adenosine triphosphate (ATP) molecule, which consists of adenine, ribose, and a phosphate group, could play a key role in these processes. It absorbs solar energy and transfers it to the conjugated chemical system, and the universal mediator for coupling is water through a hydrolysis process followed by reactions of polymerization. An appealing factor here is that ATP is synthesized from simple molecules, hydrogen cyanide (HCN) and formaldehyde (HCHO), which are widespread in outer space.

The processes of increasing complexity in the above concept and the accumulation of changes appear to occur in a highly nonlinear system, which leads to instability, bifurcations, and successive transformations of the system into a qualitatively new state. In mathematical language, such a process corresponds to the branching (qualitative change) of solutions under certain (critical) parameter values. For each new state (self-organization) of the system there is a different corresponding set of interactions of the molecular complexes. In other words, the increasing ordering of the original (chaotic) system takes the form of a sequence of bifurcations, from the appearance of primitive polymer structures and the development of the universal catalytic function of peptides, to the emergence of the nucleotide sequences involved in protein synthesis, and ultimately, to the genetic code in which the general plan of organism development as well as its numerous individual peculiarities are recorded.

A sequential ordering concept is of great interest from the standpoint of stochastic dynamics. Basically, the events under consideration are nothing other than the outcome and consequence of local instability in a nonlinear chaotic dissipative system with many degrees of freedom, while the sequence of changes in state (evolution) of the system leads to self-organization. The sequential ordering model furthermore requires, as an important property, that there be feedback for the transition to a new level of organization. A reducing medium is also required under conditions of the separate existence of an atmosphere and a hydrosphere, as well as the accessibility and mobility of phosphates, which generally is not inconsistent with current ideas about the natural conditions on Earth at the time of the appearance of the first primitive forms of life.

Let us emphasize again that the capability of ordering through selective catalysis and the capability of self-reproduction are the two most important properties of bio-organic compounds that are necessary for the origin and evolution of life. The initial ordering is created by nucleotide chains and amino acid chains (peptides). Chains of amino acids form the universal design of biological structures capable of infinite variety, and chains of nucleotides provide for self-reproduction (replication) as a fundamental property of living matter. In other words, nature has divided up these two capabilities, the tendency toward ordering through selective catalysis and the capacity for self-reproduction, between these two classes of organic compounds. We may further speculate that even an insignificant impact on a planet's environment like that exerted by primitive life forms could, under favorable conditions, excite an initially stable system with the transition (bifurcation) of the natural medium to a new stable state because of a huge gain inherent in the biological systems. In turn, when passing through early stages of evolution involving genetic code formation, life eventually acquires a very high adaptation capability.

Note in particular that, in the world of organic compounds, ordering is effected by the unique properties of carbon compounds. While we cannot answer the question of whether or not life has ever taken hold as a widespread process elsewhere in the solar system and beyond, we are confident that only on the basis of carbon can complex biopolymer structures be created, and ordering through selective (enzymatic) catalysis and replication (self-reproduction) take place. This statement should be considered as the main paradigm of the origin of life. Therefore, the discussions sometimes encountered about the possibility of conceivable life existing on the basis of silicon, for example, are groundless. If there is life somewhere in the universe, its molecular construction is probably analogous to that of life on Earth: based on carbon and its compounds, and on principles that allow a protein-nucleotide form of functioning.

Evolution

We shall now briefly touch upon the issue of biological evolution. The formation of biopolymers capable of catalysis and replication includes the appearance of an intermediary between peptides and nucleotides, such as the above-mentioned transfer RNA (tRNA); it also includes the formation of the genetic code. The emergence of the genetic code completes the stage of prebiotic evolution, and biological evolution itself (the evolution of life) begins. One of the peculiarities, and a fundamental property of life forms, is dissymmetry, or chirality. Indeed, in contrast to classical symmetry, life is characterized by the preponderance of left-handed or right-handed enantiomers. This phenomenon, which was discovered by Louis Pasteur and substantiated by Pierre Curie, is exclusively the property of living organisms and is absent in non-living nature. It was discovered that compounds concentrated in an egg or seed rotate the plane of polarization of light in a particular direction and such orientation is also present during crystallization of these compounds, as well as in

Life Origin/Evolution on Earth

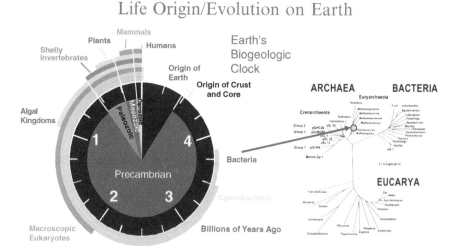

Fig. 9.7 The evolution of life on Earth ("The biological clock of the Earth") (Courtesy of D. Des Martis, NASA Ames Research Center)

organisms' ingestion of similarly oriented enantiomers and avoidance of different ones. Chirality is manifested by the unidirectionality of biological molecules (left-handed L-amino acids and right-handed D-sugars). Dissymmetry is regarded as a powerful factor in the selectivity and stability of life, and it is thought that its genesis from inert matter, abiogenesis, could occur only in the peculiar environment of Earth.

Biological evolution is understood as cumulative changes over time. Through a continuously increasing state of order (including RNA precursors), we believe that the first living organisms appeared on Earth approximately 3.8 billion years ago (see Fig. 9.7). These were bacteria with complex molecular apparatuses for heredity, protein synthesis, energy supply, and metabolism. The emergence of the first living systems (prokaryotes, eukaryotes) was accompanied by evolution on the level of cells,[6] organisms, and ecosystems, and the formation of the biosphere. The most suitable sites for the origin of life were probably volcanoes and deep-seated ocean hydrothermal vents, where various anaerobes could develop. Their feeding chains were discovered by the distinguished Russian microbiologist George Zavarzin, who called them autonomous chemosynthetic communities. These first relatively simple microbial communities, capable of adapting to extreme environmental conditions in the absence of oxygen (anaerobes), led to progressively complicated living organisms that released oxygen, eventually transforming the Earth's atmosphere from a reducing to an oxidizing one. This dramatically changed the structure and hierarchy

---

[6] Here we should emphasize again the striking self-organization of living species on the cellular level, involving a well-controlled and coherent sequence mechanism of turning on and off specific groups of genes in the different cells.

of populations including internal trophic connections, diversity, and biochemistry. As this occurred, the ordering processes were inevitably accompanied by processes of disorder and chaos. Obviously, in the competitive processes of ordering/disordering (degradation), Darwinian natural selection played a decisive role. We note that natural selection was responsible for the elimination of the dominant part of mutations harmful to life; their carriers fail to survive or produce offspring. Thus, we emphasize again the important role of Darwinism in biological evolution, but not at the early stages of the establishment of life and the development of the molecular self-organization mechanisms of biological systems. From this perspective, molecular biology, biochemical genetics, and Darwinism are not contradictory, but rather complementary and quite coherent foundations of modern evolutionary theory. Darwinism may be further developed through the concept of "covariant reduplication,"[7] which is based on the idea of matrix reproduction and replication of different variants of genetic texts, including those which have undergone mutation; these versions are then "offered" to nature to choose from. This concept is closely related with the previously mentioned ideas about the matrix synthesis of organic macromolecules during the evolution of the biosphere. Accordingly, the matrix mechanism of variation and heredity is associated with natural selection and the theory of evolution.

Earlier in this chapter and in Chap. 7 we emphasized the important role of migration and collisional processes in the solar system, implying an existence of permanent matter transport in the system. We also addressed small bodies (including those coming from beyond the solar system) as potential carriers of prebiotic matter which could serve as seeds for the origin of an early microbial biosphere. Alternatively, such bodies might bear destructive functions when impacting planets (Fig. 9.8) and, in particular, result in the destruction of living organisms. Earth experienced numerous catastrophic events caused by large asteroid/comet impacts throughout its half-billion year history, leading to great biosphere mass extinctions. This is recorded in Fig. 9.9 (top). The most well known is Chicxulub, a devastating event caused by an asteroid of about 10 km across that fell about 65 million years ago in Mexico, killing about 90 % of biosphere species. Other events could be related to both asteroids and comets. There have been attempts to find regularity in the mass extinctions, connecting them with comet showers caused by periodic wobbling of the solar system plane relative to the Galactic plane (see Fig. 9.9, bottom). This might have serious implications for the Earth's biosphere existence and maintenance.

Of particular interest are speculations concerning the evolution of life through progressively complicated levels and especially towards its highest forms: mammals and humanoids. As an example, dinosaurs could possibly have had a chance to become more intelligent had they not been killed in the Chicxulub event; in that case it would have been harder for apes and humans to emerge. The emergence of intellectual life can then be regarded as an even much less probable and more vulnerable pathway in the overall problem of life's evolution. The challenging goal of primitive life detection on extrasolar planets has become an agenda of the coming

---

[7] This idea was proposed by the highly regarded Russian scientist N.V. Timofeyev-Resovsky.

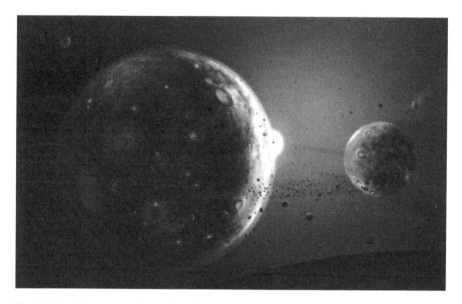

Fig. 9.8 Artist's concept of impacting planet by a large body. Earth experienced numerous devastating events throughout its history, especially during the Late Heavy Bombardment (LHB) at about 3.9–4 billion years ago, just before the biosphere set up (Adapted from Wikipedia)

decades with our advanced technologies. The first potential candidates in habitable zones were discovered with the Kepler mission. The probability of finding life features is certainly non-zero, because, bearing in mind matter transport in space, one may assume that once seeds are planted on a planet with a not too extreme natural environment, primitive life forms can survive. In a more favorable environment, they could proliferate and probably multiply with the enormous amplification coefficient pertinent to microorganisms.

We can still only speculate if life, once developed, could easily evolve to a high level of organization including intellectual capacity. For the moment, such a probability seems vanishingly low. We are yet unable to communicate with other life forms, although the first attempts have been undertaken. The international program Search for Extraterrestrial Intelligence (SETI) has been active since the 1960s and has progressively developed with robust observation of the universe in the most reasonable wavelengths with the hope of being rewarded with detection of artificial signals from other worlds. The discovery of extrasolar planets and continuing efforts in this direction is an extremely important step forward and promises further progress in astrobiology.

However, finding a civilization capable of manifesting itself and establishing communication is a much more difficult endeavor than just a life signs search. SETI is a challenging problem involving deliberative strategy and the most advanced technology, but unfortunately, the numerous attempts undertaken until now have not been successful. To estimate the number of potentially existing civilizations capable

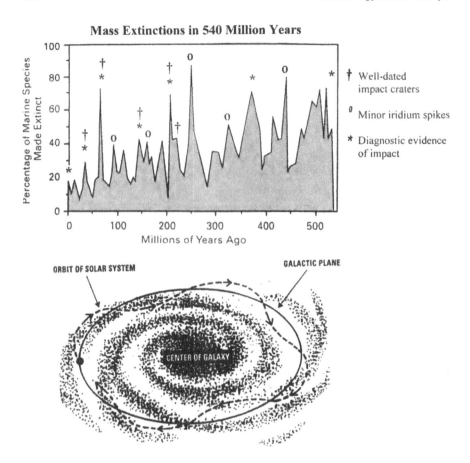

Fig. 9.9 *Above*: the mass extinctions of living organisms on Earth during the last 540 million years. The events correspond to impact craters, enriched with iridium and containing other signs of falling cosmic bodies. *Bottom*: schematic view of the movement of the solar system through the plane of our Galaxy (Adapted from Wikipedia)

of communicating, American radio astronomer Frank Drake suggested a simple formula including both astronomical and biological-social factors. This relation became more accurate after extrasolar planetary systems including Earth-like planets (Terrans) were discovered, but it is still somewhat ambiguous, especially in an attempt to assess a lifetime of technologically developed and well-advanced civilization. The crucial question is: Would a technologically developed civilization, possibly even hundreds to millions of years more advanced than us, be not just capable but also willing to communicate? And if so, was this civilization able to overcome social-economical problems (like those human civilization is facing these last decades with the threat of nuclear weaponry) and avoid destroying itself? Also, is this the key factor that may dramatically limit communication chances in the SETI program with the use of radio telescopes and capable analyzers? In particular, based on our own experience, one may argue that resource-ecological, economical, and politico-social problems might and still may crucially impact the evolution and existence of human civilization. And even accepting the concept of favorable evolution of extraterrestrial intelligence throughout thousands of years, we should keep in mind the enormous size of our Milky Way galaxy (~100,000 ly across) in considering the chances to hear a signal sent out there during our lifetime. We may conclude this section with the remarkable words of Arthur Clark: "After 2000 years of imagining life beyond the solar system, Earthlings are poised to use science for their search! What is potential level of their adaptation to the environment and how different these environments could be?"

Chapter 10
The Structure of the Universe

View of the Universe

We shal now address our space environment at large, streching from the solar system to frontiers of the universe. Stars, the most familiar and numerous objects in the universe, which we discussed in Chap. 6, are grouped in great stellar systems of thin disk shape called galaxies, which appear as lighter colored bands in the night sky. Galaxies are the main "building blocks" of the visible universe, and they are essentially islands of matter in a predominantly empty space, which we must consider when pondering the origin of the universe. Galaxies range from dwarfs with as few as ten million stars to giants with about 100 trillion stars, each orbiting their galaxy's own center of mass. Most galaxies in the universe appear to be dwarf galaxies, many of them orbiting a single larger galaxy. Historically, different apparent shapes of galaxies have been categorized related to their morphology, the most prominent being elliptical, spiral, and irregular forms (Fig. 10.1). The largest galaxies are giant ellipticals, some of them are thought to form due to the interaction of galaxies resulting in a collision, in which they merge and grow to enormous sizes, as compared to spiral galaxies. Stardust galaxies are also distinguished (and are possibly also the result of collisions leading to the emergence of elliptical galaxies) where star formation at an exceptional rate has been observed. Stardust galaxies are thought to have been more common in the early history of the universe, but currently they contribute significantly to the total star production rate. In spiral galaxies (Fig. 10.2), visible stars take on a spiraling pinwheel shape in a rotating disk, probably within an invisible sphere—an enigmatic dark matter halo.

The spiral arms are thought to be areas of high-density matter or "density waves" and have a pattern of approximate logarithmic spirals, possibly resulting from disturbances in a uniformly rotating mass of stars. In the spiral arms are located dense molecular clouds of interstellar hydrogen, with their potential for new generations of star production. Generally, most star formation occurs in smaller galaxies where cool hydrogen is less depleted. In contrast, the big elliptical galaxies are largely devoid of this gas, which has already been converted into

© Springer Science+Business Media New York 2015
M.Ya. Marov, *The Fundamentals of Modern Astrophysics*,
DOI 10.1007/978-1-4614-8730-2_10

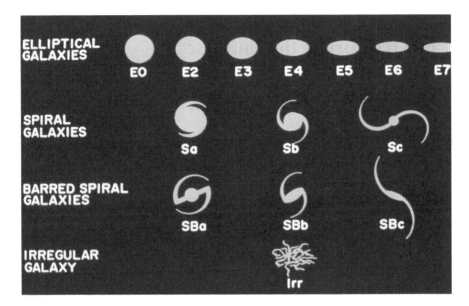

Fig. 10.1 Classification of galaxies (Courtesy of Smithsonian Astrophysical Observatory (SAO))

heavier elements; thus in these galaxies new star formation has essentially terminated. A majority of spiral galaxies have a linear, bar-shaped band of stars in the central part that extends outward to either side of the core and then merges into the spiral arm structure. Some of the barred spiral galaxies exhibit activity, possibly as a result of gas density waves being channeled into the core along the arms or due to tidal interaction with another galaxy. The latter is especially pertinent to peculiar (ring-shaped) galaxies, which are thought to form when a smaller galaxy passes through the core of a spiral galaxy.

A typical galaxy contains hundreds of billions of stars of different age, mass, luminosity, and chemical composition that are at different evolutionary stages, interstellar clouds and interstellar medium filled with the most cosmically abundant hydrogen and helium gas, as well as dust and cosmic rays. There are about 100 billion ($\sim 1.7 \times 10^{11}$) galaxies in the observable universe with nearly 100 billion stars in each moving with relative velocities reaching $\sim 1,000$ km/s. The total amount of stars in the universe is estimated to be as large as 3×10^{23} (300 sextillion). Although stars exist within galaxies, intergalactic stars have been discovered as well. Galaxies represent the main state of matter in the visible universe; however, as we will see further on, the invisible matter in a galaxy (referred to as dark matter, see Chap. 11) exceeds the visible matter by nearly an order of magnitude, and it is clearly responsible for spiral galaxies avoiding the spurious dissipation of angular momentum.

Our galaxy, the Milky Way (Fig. 10.3), is a large barred spiral galaxy representing a thin revolving disk system; it resembles the Andromeda Galaxy (Fig. 10.2).

Fig. 10.2 Examples of spiral galaxies: Andromeda (*up*) and NGC 4414 (*bottom*) (Courtesy of NASA)

It consists of nearly 400 billion stars[1] with a central bulge (active galactic nucleus), out of which wind several spiral arms interspersed with clouds of gas and dust. Unlike Andromeda and many other galaxies, we can view our Milky Way only edge-on, and it exhibits different shapes when observed in different wavelengths

---

[1]The idea that the bright band in the night sky known as the Milky Way might consist of distant stars was first proposed by the Greek philosopher Democritus (450–370 BC). Actual proof that the Milky Way consists of a huge number of faint stars came in 1610 with the telescopic observations of Galileo Galilei.

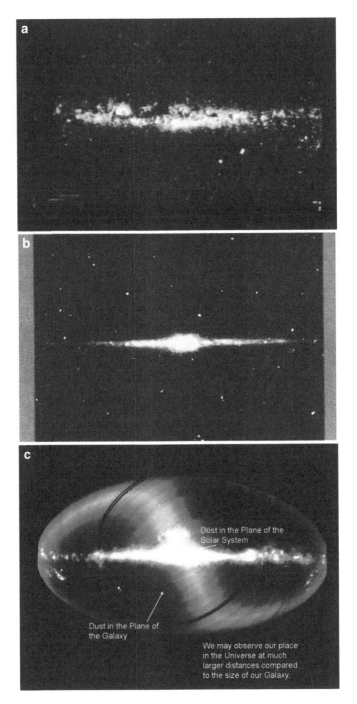

Fig. 10.3 Edge-on view of our Galaxy Milky Way in visible (**a**), near infrared (**b**) and far infrared (**c**) wavelength. In far infrared Zodiacal Light caused by interplanetary dust in the plane of ecliptic is observed indicating its position relative to the Galactic equator by close to 90° (Courtesy of SAO and NASA)

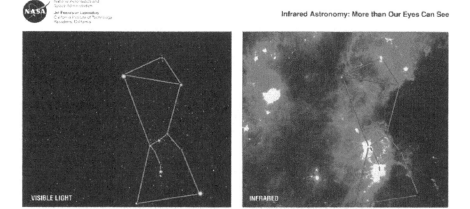

Fig. 10.4 Comparison of Orion constellation in visible (*left*) and infrared (*right*) wavelength revealing much more complicated configuration and giant molecular cloud inside. Electromagnetic spectrum is shown at the *bottom* (Courtesy of SAO and NASA)

(see Fig. 10.3). This is especially manifested when shifting from visible wavelengths to the far infrared, where dust emission significantly contributes to the brightness. Similar and even more spectacular differences are seen in the respective images of the constellation Orion (Fig. 10.4).

The luminosity of our galaxy is 6×10^{40} erg/s, ten millions times more than that of the Sun. We can say that the Milky Way has hundreds of billions of solar systems and its luminosity exceeds that of the Sun by seven orders of magnitude. The Sun is just one of the ordinary stars in the plane of the Galaxy located at about two-thirds (10 Kpc) of the way from its center, and revolving around it every 225–250 million years (see Fig. 5.1). Our solar system's plane is tilted by nearly 90° with respect to the equatorial plane of the Galaxy, and as the solar system orbits the galactic core, it wobbles or "bobs" up and down through the disk of the Galaxy, in a kind of sine-wave shaped path. This occurs because the mass of the Galaxy is actually spread out into a thin disk. This vertical oscillation happens every 30–33 million years while the system is orbiting the center of gravity of the Galaxy during the complete 225–250 million year orbit. We are currently "above" the plane (to the galactic North) by about 75–100 ly. We mentioned in a previous chapter that a passage through the galactic plane may trigger some extra bombardment of planets by comets from the Oort Cloud at the periphery of the solar system.

Fig. 10.5 Schematic view of quasar—the source of enormous energy release thought to be the bright nuclei of distant active galaxy (Adapted from Wikipedia)

Stars in galaxies have a nonuniform distribution with a maximum in the center. Stars are drawn by gravity to orbit a massive invisible object residing at the center. As we discussed in Chap. 6, in our Milky Way this object is associated with a supermassive black hole (SMBH) Sgr A* containing about four million solar mass stars. Other galaxies also exhibit structures similar to Sgr A*. The objects known as quasars are distinguished as sources of enormous energy release and are thought to be the bright nuclei of distant active galaxies (Fig. 10.5). Observations by the Chandra X-ray Space Observatory revealed a reservoir of hot gas around the Milky Way which may extend for a few hundred thousand light-years or even farther into the surrounding Local Group of galaxies. This means that the Milky Way and other galaxies are embedded in a halo of hot gas with temperatures of several hundred thousand degrees, which is much hotter than the surface of our Sun. The mass of the halo is estimated to be comparable to the mass of all the stars in the Galaxy.

In the structure of the Milky Way and other galaxies we first distinguish the vast gas-dust regions called molecular clouds, where spectacular star formation processes occur. The most well-known is a region in the constellation Orion (see Figs. 6.6 and 10.4). Let us also recall (see Chap. 6) that there are large multistar systems, both *open clusters* and the more massive *globular clusters* of spherical shape composed of 10^5–10^7 stars of relatively small (about solar) mass where processes of star formation have terminated. In distant galaxies such associations are observed as a single stellar source because the single stars within them are not resolved. Similarly, we also cannot resolve bright massive stars in the stellar complexes found in distant galaxies. Our observations are mostly prevented by the dust present in the interstellar medium, which is opaque to visual light, though it is more transparent to far infrared and some radio wavelengths. Infrared and radio astronomy are powerful tools to observe the interior

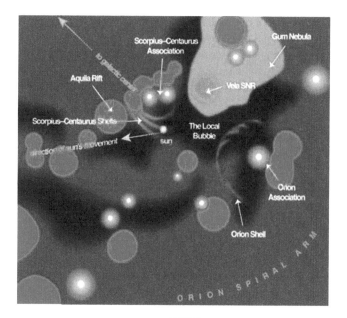

Fig. 10.6 Solar neighborhood within 1,500 ly: ▧—high density molecular clouds; ▧—hot ionized gas; ▧—low density hole in the interstellar medium; ▧—low density molecular gas in a spiral arm. Direction of the Sun's movement around Galactic center (toward apex) is shown (Courtesy of SAO)

regions of giant molecular clouds and galactic cores, particularly distant, red-shifted and active galaxies, in great detail. Water vapor and carbon dioxide in the Earth's atmosphere absorb a number of useful portions of the spectrum; therefore, high-altitude or space-based telescopes are used. Besides molecular clouds, the space within galaxies is filled with neutral and ionized gas in regions of different densities. The solar neighborhood within 1,500 ly, involving high-density molecular clouds, low-density molecular gas in a spiral arm, a low-density hole in the interstellar medium, and hot ionized gas among the familiar constellations, is shown in Fig. 10.6.

The "ambassadors" of deep space are galactic cosmic rays in a wide range of energies: from hundreds to billions of billions of electron volts. Cosmic rays communicate information about the most powerful processes occurring in the universe. They are composed mainly of protons (about three-quarters of the total number of nuclei) and electrons but include also nuclei of heavier chemical elements (nitrogen, oxygen, magnesium, silicon, iron, etc.), as well as photons and neutrinos. Enigmatic dark matter may possibly also be part of their composition. Cosmic rays of moderately high energy (up to hundreds of teraelectron volts)[2] are recorded by detectors on satellites and balloons. For particles of heavier energy, the Earth's atmosphere itself serves as a detector, and the entry of cosmic rays produces particles of secondary origin—broad atmospheric heavy showers recorded on the Earth's

---

[2] 1 TeV (one teraelectron volt) is equal to 10^{12} eV; 1 EeV is equal to 10^{18} eV.

surface, mostly at observatories located in the mountains. The amount of cosmic ray particles is reciprocal to their energy: the higher the energy, the fewer particles come in with a larger fraction of heavier nuclei in their composition.

The origin of cosmic rays in the range of energies from a few to millions of teraelectron volts (1 EeV) is related with processes in the Galaxy and/or in other galaxies (the metagalaxy), in particular, with the mechanisms of charged particle accelerations in electromagnetic fields near black holes, in active galactic nuclei, and in the powerful electric fields on relatively young supernova remnants, though none of these mechanisms has been confirmed experimentally. It is difficult to verify the above hypothesis and, in particular, to identify the real source of cosmic rays, because particles of high energy are retained by the Galaxy's magnetic fields, which distort their trajectories. Theory predicts a cutoff in the cosmic ray spectrum beyond the enormous energy of about 100 EeV (10^{20} eV). Sources of particles with ultrahigh energies as well as their distribution and composition (mass dependence on energy) remain even less clear. Of special interest is the investigation of photons and neutrinos of ultra-high energies that result, not only from mechanisms of acceleration, but also from the interaction of cosmic ray particles with the cosmic microwave background (CMB) radiation (see Chap. 11). Such neutrinos with energies up to 1,000 TeV were successfully detected most recently in the IceCube experiment in the Antarctic. Similar and more complex experiments are currently in preparation at some other observatories, in particular with the Telescope Array project in the USA. Undoubtedly, further study of galactic cosmic rays will promote a better understanding of the physical processes in space accelerators, the interaction of energetic particles with the interstellar medium, and magnetic fields and electron density distribution in different parts of the Galaxy. It may possibly even help us understand the nature of dark matter and the quantum theory of gravity.

Our view of the universe from the Earth reveals a hierarchical system of structures of ever-increasing size, as deep sky surveys have shown. Galaxies are in relatively close association with other galaxies and form clusters of progressively growing size that reveal an amazing self-organization of matter in the universe. A galaxy cluster—the rarest and largest of galaxy groupings—is a collection of up to thousands of galaxies bound together by gravity and experiencing rapid growth from a process called inflation (see Chap. 11). Following the virial theorem,[3] each member galaxy has a sufficiently low kinetic energy to prevent it from escaping. It was found that solitary galaxies that have not significantly interacted with another galaxy of comparable mass during the past billion years are relatively scarce. Only about 5% of the galaxies surveyed are thought to be truly isolated, and they may have interacted with other galaxies in the past, and may still be orbited by smaller, satellite galaxies. Note that isolated galaxies can produce stars at a higher rate than normal, as their gas and dust are not being stripped away by other nearby galaxies.

---

[3] The virial theorem relates the average over time of the total kinetic energy of a stable mechanical system consisting of N particles (in this case, galaxies), bound by potential forces, with the total potential energy of the system. Its technical definition was introduced by Rudolf Clausius in 1870.

Clustering ranges from pairs and loose groups to giant clusters forming a fractal-like hierarchy of clustered structures of thousands of galaxies. About 70–80 % of the mass in a cluster is assumed to be in the form of dark matter and 10–30 % consists of infalling very hot (hundreds of megakelvins) gas influx. This accompanies the ongoing processes of smaller galaxies merging to form larger scale clusters, resulting in the evolution of a smaller number of galaxies in a cluster. Only a few percent of the baryonic matter is in the form of galaxies. The average separation between galaxies within a cluster is of an order of magnitude larger than their diameter; hence, mutual attraction and tidal interactions between these galaxies as well as some exchange of gas and dust may play an important role in their evolution. Dramatic events like collisions occur when two galaxies pass directly through each other. Generally, they have sufficient relative momentum not to merge, and stars typically pass straight through without colliding. However, gas and dust interaction (both disruption and compression) in the interstellar medium can trigger bursts of star formation and severe distortion of the shape of one or both galaxies, forming bars, rings, or tail-like structures. In the extreme case, interactions take the form of galactic gradual mergers, resulting in the formation of a single, larger galaxy. Obviously, the new galactic morphology significantly changes compared to that of the original galaxies, and there are cases when one of the galaxies is much more massive and remains relatively undisturbed by the merger whereas the smaller galaxy is torn apart. This event is known as *galactic cannibalism*. Interactions and collisions significantly affect the evolution of galaxies.

Our galaxy, the Milky Way, is a member of the Local Group of nearly 20 other galaxies, including the famous bright Andromeda Galaxy, the Magellanic Clouds, and many dwarf companions. The Local Group is shaped like a disk of 100 million ly (approximately 30 megaparsec) in diameter. Another example is the group of galaxies known as Hickson Compact Group 44 (the NGC 3190 Group), located about 60 million ly away in the constellation Leo and containing several spiral and elliptical galaxies. It is shown in Fig. 10.7. Some clusters contain well over 1,000 galaxies and can therefore be hundreds of times more massive than the Milky Way. However, the Local Group of galaxies we belong to is relatively small in space scale. It is a part of a cloud-like structure within the Virgo Supercluster, a large, extended structure of groups and clusters of galaxies centered on the Virgo Cluster. The Virgo Supercluster is located about 64 million ly away and contains about 2,500 galaxies (Fig. 10.8). The clusters of galaxies around us within 450 Mpc are shown in Fig. 10.9. Clusters are about 3 Mpc in size, or 100 times the size of our Milky Way.

We see that clusters of galaxies form superclusters, and that some superclusters contain more than 100 clusters, or tens of thousands of galaxies altogether. The typical size of a supercluster is about 20–30 times the size of a cluster of galaxies or about 300 million ly (~100 Mpc). At the supercluster scale, galaxies are arranged into sheets and filaments surrounding vast empty voids. Above this scale, the universe appears to be the same in all directions (isotropic and homogeneous); see Chap. 11.

Fig. 10.7 The local group of galaxies Hickson 44 containing more than 30 galaxies (Courtesy of NASA)

Fig. 10.8 The supercluster of galaxies Virgo at about 64 million ly away containing about 2,500 galaxies (Courtesy of SAO and NASA)

Let us note that galaxy clusters and superclusters from the early universe, which are extremely far away, are hard to observe using ground-based visible-light telescopes because light emitted from these faraway structures has been stretched into longer, infrared wavelengths due to the expansion of space discussed in Chap. 11. NASA's Wide-field Infrared Survey Explorer (WISE), equipped with an infrared

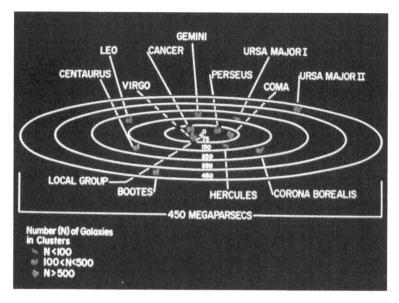

Fig. 10.9 The clusters of galaxies around us within 450 Mpc. The Local Group within ~50 Mpc is distinguished. The clusters are distinguished by the respective cancellations (Courtesy of SAO and NASA)

telescope, is well fitted to observe these structures. An example of a large galaxy cluster located 7.7 billion ly away that was discovered with the WISE in infrared wavelengths is shown in Fig. 10.10. Clusters and superclusters of galaxies may also be revealed from observations of the CMB (see Chap. 11), because galaxies containing hot gas look like specific holes in the CMB background. This is called the Sunyaev-Zel'dovich effect, and it is widely used in the study of the universe's structure. Very detailed information is also revealed from radio observations. A supercluster of galaxies as viewed in the radio wavelength is shown in Fig. 10.11.

A deep image of the universe as seen by the Hubble Space Telescope (HST) is shown in Fig. 10.12. The HST allows us to penetrate to a distance equivalent to about five billion ly, of the total 13.7 billion years since the universe's origin (see Chap. 11). The size of the observable universe containing more than 100 billion galaxies is about 30 times the distance between superclusters, or more than ten billion ly (~4 Gpc), close to that of the 13.7 billion year old universe. Basically, all bright spots in the image are galaxy clusters and superclusters. On the largest scale, one may assume that clumps of dark matter pulled the respective galaxies together in the continually expanding universe.

One of the most impressive discoveries in astrophysics is the fact that superclusters of galaxies are not uniformly distributed in the universe, which was shown by observations (Fig. 10.13) and supported by modeling (Fig. 10.14). Remarkably, this nonuniform distribution ultimately forms more or less organized structures. The result is that the large-scale structure of the universe (the *Cosmic Web*) consists of

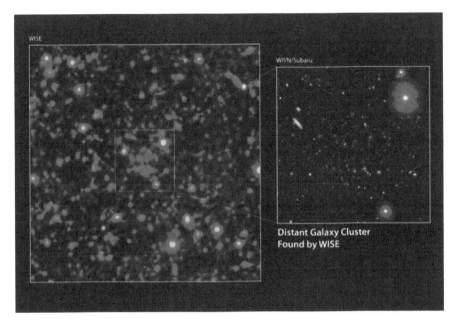

Fig. 10.10 A galaxy cluster 7.7 billion light-years away has been discovered using infrared data from NASA's Wide-field Infrared Survey Explorer (WISE). The discovery image is shown in the main panel. The inset shows a deeper, or more sensitive, optical and near-infrared composite constructed using data from the WIYN telescope at Kitt Peak in Arizona and Japan's Subaru Telescope on Mauna Kea in Hawaii. The *red* galaxies in the inset image are part of the cluster, while the *circles* highlight the galaxies seen by WISE that were used to detect the cluster. This galaxy cluster is the first of thousands expected to be discovered with WISE over the entire sky (Courtesy of NASA/ JPL-Caltech/UCLA/WIYN/Subaru)

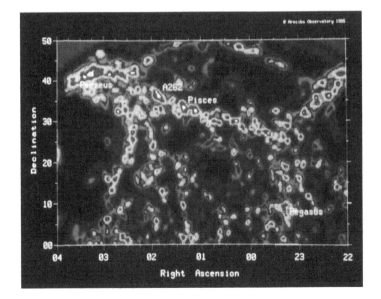

Fig. 10.11 Super cluster of galaxies as viewed in the radio wavelength (Courtesy of NASA)

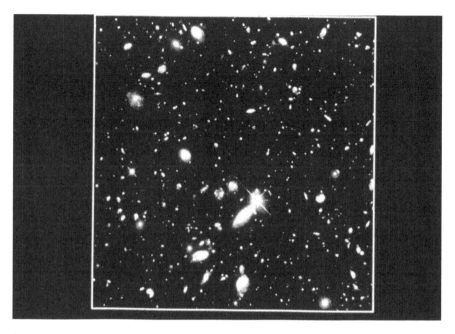

Fig. 10.12 The deep image of the universe as seen by the Hubble Space Telescope (HST) (Courtesy of HST Institute and NASA)

Fig. 10.13 Superclusters of galaxies. Their non-uniform distributed in the universe is evident (Courtesy of SAO and NASA)

Fig. 10.14 Simulation of the large-scale structure of the Cosmos. The image spans about 400 million light years across (Adapted from Wikipedia)

walls (strings of superclusters) separated by *voids* (Fig. 10.15). Walls with the clusters of galaxies located between them at the nodes are the largest observable structures. They are addressed as remnants of the structure of the early universe generated by the above-mentioned primordial quantum density fluctuations. These structures were revealed first by the COBE and WMAP satellites and then, more precisely, by the Planck satellite. We will discuss this in more detail in Chap. 11.

Galaxies: Formation and Dynamics

The formation of galaxies and galaxy clusters can be explained by invoking contemporary views on the expanding universe and Einstein's general relativity theory. The galaxies were probably formed within dark matter regions built by the aggregation of subgalactic mass halos and the collapse of self-gravitating objects during the development of gravitational instability that enhanced the quantum fluctuations emerging during the earliest inflation epoch. Clearly, internal sources of energy release through dissipative gas dynamic processes, shock heating during supernova explosions, and supermassive black hole accretion were involved in this process. Subsequently, the protogalaxies grew by accreting smaller clumps and/or combining with a halo of comparable size. Intense radiative cooling of gas accompanied by X-ray emission, recombination, and collisional excitation of emission lines should

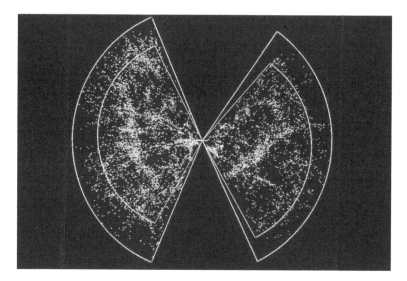

Fig. 10.15 The large scale structure of the universe (*Cosmic Web*) consisting of *walls* (strings of superclusters) and *voids* ("empty" space). Walls are the largest structures we observe. They are remnants of the early universe set up (Courtesy of SAO and NASA)

have taken place at this stage. The redistribution of the system's angular momentum, the formation of a gaseous galactic disk, and the onset of star formation should apparently be attributed to the same period of the early universe. Observations of galactic clusters and superclusters of hundreds of trillions of solar masses as well as their space correlation and red shifts are key in understanding how the structure of the universe evolved and thus in advancing the study of modern cosmology.

The combination of order and chaos is the most important mechanism for the formation of the structure and dynamics of galaxies. The main family of orbits in galaxies contains epicyclical, resonant, and nonresonant orbits inside and outside the corotation regions, as well as short- and long-period orbits, along with nonperiodic and vanishing (going to infinity) orbits. Planar rings, shock waves, and vortices (of cyclon-anticyclon type) are also distinguished in the symmetry plane of galaxies. At the same time, unstable chaos-generating orbits that usually do not combine with nonperiodic orbits can emerge from planar ones as a result of multiple bifurcations along the vertical axis, and they are directly related to various resonances. Interaction of the resonances produced by a large-amplitude perturbing force can lead to a chaotic motion. Note that the arms of spiral galaxies are characterized by the propagation of nonlinear density waves affecting periodic orbits, as follows from the theory of spiral density waves. Such interactions could apparently take place at the earliest formation stage of stellar systems, which could produce a certain order in the structure of matter in the universe in the course of evolution. Some galaxies were in close interaction and even experienced collisions, which led to new structures.

It is important to emphasize the difference between stellar dynamics and statistical physics and the inapplicability of the concept of thermodynamic equilibrium to any self-gravitating (stellar) system. This implies that whereas there is a tendency for equilibrium to be established in a thermodynamic system, no such process takes place in a self-gravitating system with a dense center. Thus, in contrast to a star with thermal equilibrium and a certain temperature at each radius, the stars in a stellar system move with a constant energy in a wide range of distances from the gravitating center, and the stellar velocity distribution at a given radius is not a temperature characteristic. Irrespective of the specific configuration of a stellar system, its entropy grows through the transfer of energy from the dense central region to the much more rarefied outer one.

Chapter 11
Cosmology: The Origin and Fate
of the Universe

Brief History

The origin and advancement of cosmology is intrinsically related with many great scientists. The first attempts to gain insight into the arrangements of the world go back to the Hellenistic school of the ancient Greeks with the most profound ideas of Aristotle. In his fundamental *Metaphysics*, he thoroughly developed a physical cosmology of rotating celestial spheres based on the planetary theory of Plato's student Eudoxus. Following this concept, he considered the fixed stars and the planets embedded in interconnected spheres, with the spherical Earth at the center of the universe and the planets moving along the seven concentric circles through mostly empty space. This concept was later developed in Ptolemy's epicycles and was basically preserved for fifteen centuries (although it was slightly modified by Christian, Muslim, and Islamic theological philosophers) until Copernicus introduced the heliocentric rather than the geocentric world theory.

A real breakthrough in cosmology was made by a key figure in basic science, Isaac Newton, who in his *Principia* formulated the laws of motion and the universal gravitation law which dominated scientists' view of the universe for the next three centuries. Newton's law of universal gravitation established originally presumed side-splitting ideas on strong mathematical ground and led the way to further studies. He first showed that the motion of celestial bodies and objects on Earth obey the same physical principles and, by deriving Kepler's laws of planetary motion from his mathematical description of gravity, he perfectly confirmed the heliocentric theory.

Newton's basic principles were greatly advanced by Albert Einstein, who revolutionized the former world attitude and founded the ideas of modern physics and cosmological concepts. He developed the special and general theories of relativity and applied the latter to model the structure of the universe as a whole. He showed that properties of the physical universe may change depending on motion, and then described gravity as an entity stipulated by time-space curvature created by both mass and energy. In an attempt to reconcile the general theory of relativity with his view of

© Springer Science+Business Media New York 2015
M.Ya. Marov, *The Fundamentals of Modern Astrophysics*,
DOI 10.1007/978-1-4614-8730-2_11

the universe as static, eternal, and unchanging, he introduced a hypothetical cosmological constant which practically turned out to be in accordance with the modern view of the expanding universe. Einstein's dream of unifying fundamental laws of physics with gravity motivates modern quests for mainstream physics, such as modern string theory, where geometrical fields emerge in a unified quantum-mechanical setting.

Alexander Friedman, in the 1920s, was the first to argue with Einstein that the universe is expanding rather than static. Based on his solution of Einstein's equations of general relativity theory, he predicted the red shift of galaxies. The prediction was perfectly confirmed soon after that by the thorough observations of Edwin Hubble, and his discovery of the receding of galaxies revealed that the universe is indeed dynamic. An idea of the universe's origin from an exploded point of immense density and temperature, later named the Big Bang, was suggested and theoretically grounded by Georges Lemaitre. The intriguing concept of the universe's original expansion was further supported by Georges Gamow, who predicted microwave radiation as a Big Bang remnant. In turn, Fred Hoyle developed the theory of nucleosynthesis as being responsible for the origin of chemical elements. As a key component, the modern concept of the Big Bang model incorporates superinflation quantum theory as first suggested by Alexei Starobinsky and refined by Alan Guth and Andrei Linde, who developed a simpler approach to the basic concept of the inflation model that is now well recognized. This allowed us to find a better approach to resolving many cosmological problems and, in particular, to refine the Big Bang standard model. Figure 11.1 shows a gallery of distinguished cosmologists.

Distinguished Cosmologists

Albert Einstein – General relativity theory

Alexander Fridman – Expanding Universe,
                              prediction of red shift of galaxies

Edwin Hubble – Finding receding of galaxies;
                          dynamical Universe

Georges Lemaître - Big Bang theory

Georges Gamov – Prediction of microwave
                            radiation as Big Bang remnant

Fred Hoyle – Theory of nucleosynthesis

Alex Starobinsky
Andre Linde   Superinflation quantum theory
Alan Guth

Fig. 11.1 Gallery of distinguished cosmologists (Credit: the Author)

Hubble's Law

In 1929 Edwin Hubble discovered that all galaxies are receding from us with a velocity which is proportional to their distance. This phenomenon is called, in his honor, Hubble's law. A linear expansion of the universe over its lifetime was found through the Doppler shift in spectral lines emitted by galaxies. The Doppler shift, which is well known from middle school physics courses, states that if a source of light or sound is moving relative to the observer, then the observer will see a change in the wavelength of light λ relative to the reference wavelength λ_0 (Fig. 11.2). Motion away from an observer causes an increase in the wavelength (red shift); motion toward an observer causes a decrease in the wavelength (blue shift). The amount of wavelength shift ($d\lambda$) at wavelength (λ) is proportional to the velocity (v) relative to the speed of light (c):

$$z = d\lambda / \lambda = v / c$$

In the nonrelativistic case ($v \ll c$):

$$z = d\lambda / \lambda = (\lambda - \lambda_0) / \lambda_0 = v / c.$$

In the relativistic case ($v \approx c$):

$$z = \left\{ \left[1 + (v/c) \right]^{1/2} / \left[1 - (v/c) \right]^{1/2} \right\} - 1.$$

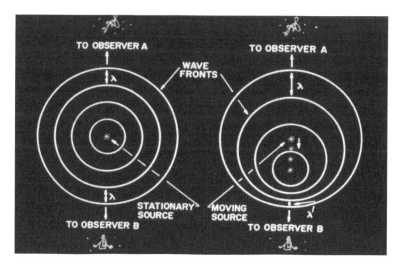

Fig. 11.2 Doppler shift. In the case of stationary light source (*left*) no shift of wave front (and therefore, no spectral lines λ) occurs; In the case of non-stationary light source (*right*) moving down in the plot observer A see *red* shift and observer B *blue* shift (Courtesy of G. Fazio)

From observations of spectral lines emitted by galaxies, Hubble found that distant galaxies show a greater red shift, and thus they are racing away (receding) faster than closer ones. It looks as if the wavelength of light emitted by distant galaxies is "stretching out," and the faster the galaxy moves, the larger the red shift. Similarly, the oldest galaxies exhibit the largest red shifts. Astronomers measure the receding of galaxies by the quantity z: the greater z is, the faster a galaxy moves away. Objects which are relatively close to us have $z \leq 10$.

The following simple relationship defines galaxy velocity v depending on distance L:

$$v = H_0 \times L,$$

meaning that the universe is expanding. This relationship includes the quantity H_0 normalized to 1 Mpc; H_0 is called the *Hubble constant* (Fig. 11.3). The most accurate value to fit observations is $H_0 = 72.8 \pm 2.4$ km/sec/Mpc. The inverse quantity $1/H_0$ is called the *Hubble time* and defines the age of the universe as $1/H_0 = 13.7$ billion years. Hubble's law underlies the *cosmological principle*, which states that an observer—wherever he or she may be in space—sees the universe as homogeneous (uniform) and isotropic (the same in all directions).

Hubble's discovery posed very important questions. First of all, because galaxies are running further apart, revealing the runaway expansion of the universe, what is it expanding from? Were galaxies closer together in the past? Therefore, running backward, was there an early converging of the whole universe in an incredibly compacted, compressed beginning? Such an extraordinary compression due to gravity would then mean that all the matter in the universe had been originally converted into a point of almost pure

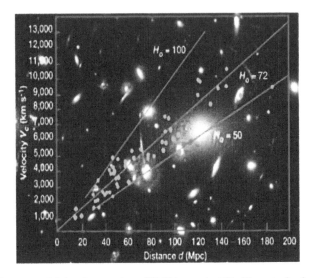

Fig. 11.3 Diagram explaining the meaning of Hubble constant H_0. The actual value according to modern data is $H_0 = 72.8 \pm 2.4$ km sec^{-1} Mpc^{-1} (Courtesy of G. Fazio)

energy and enormous temperature which progressively decreased with expansion. One could also suggest that the expansion left behind a kind of "afterglow" throughout space, and this was serendipitously discovered in 1965 by Arno Penzias and Robert Wilson, who were awarded the Nobel Prize in Physics for this discovery. This "relic" radiation, called the Cosmic Microwave Background (CMB) radiation, has a temperature of 2.725 K, $z \sim 1,000$, and fills the whole universe almost uniformly. It is therefore identified as the remnant of the original Big Bang.

The questions above are intimately related with the most fundamental problem of how the world came to exist as it is. The history of the universe is rooted in the Big Bang model, which is generally supported by the observational data now available. The Big Bang theory nicely explains different features of the origin, evolution, and current state of the universe. Indeed, according to the Big Bang theory, the universe began 13.7 billion years ago from a point of infinitely high temperature and density. After this original expansion, it continued to expand and cool. Evidence in support of the Big Bang theory includes the observed continuing expansion of the universe according to Hubble's law, whose echo is the CMB radiation and the present abundance of light elements (hydrogen, deuterium, helium, and lithium) which were synthesized soon after the explosion. However, because we lack knowledge about such a peculiar subatomic point, the universe pre-Big Bang is associated with a *singularity* where the known laws of physics break down. In other words, the physics of the processes is unclear until 10^{-43} sec (called the Planck time). Similarly, modern cosmology postulates that expansion of the youngest universe from a point involved an original instantaneous superinflation stage, the mechanism of which is also difficult to understand. We will return to this point later.

The CMB radiation was measured for the first time with the NASA Cosmic Background Explorer (COBE) satellite in 1992, then with NASA's Wilkinson Microwave Anisotropic Probe (WMAP) satellite launched in 2001, and with the best achieved resolution with ESA's Planck satellite launched in May 2009.[1] Their observations allowed us to penetrate to the earliest time of the young universe—about 380,000 years after the Big Bang (see Fig. 11.4). The accuracy of the measured uniformity and isotropy of the radiation was as high as 10^{-5} within only a factor of 2, to which the model of a homogeneous and flat universe corresponds. However, a small temperature anisotropy and inhomogeneities in the CMB (ripples in space) were also found. These were attributed to the original quantum fluctuations of density in the early universe after its exponential expansion, and its later density increase, leading to the formation of galaxies and galaxy clusters and creating the large-scale structure of the universe—the Cosmic Web that we discussed in Chap. 10. Essentially, the observed

---

[1] WMAP and Planck were launched and operated in the L2 Lagrange point; see Fig. 1.6.

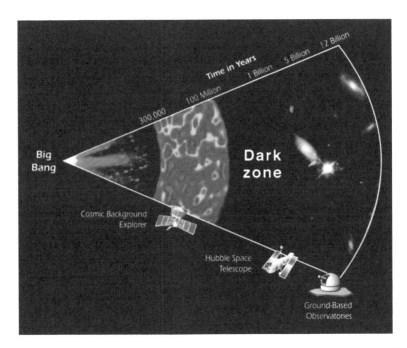

Fig. 11.4 Observable Universe. Diagram shows how deep back in time penetrate ground based telescopes, Hubble Space Telescope, and WMAP. The latter observed Universe in the state 380,000 years after Big Bang (Courtesy of NASA)

density fluctuations of baryonic matter represent the imprint of "acoustic waves" in the early universe (Fig. 11.5). Their measured spectrum perfectly confirms the prediction of the inflation model, which we will address in more detail below. Thus, these important measurements provided the image of a newborn universe that was in a nearly ideal state of thermodynamic equilibrium, and revealed primordial density perturbations that subsequently formed its large-scale structure.

The most detailed map of CMB temperature is now based on the results of Planck satellite precise cosmology measurements which allowed us to constrain the statistics of the CMB anisotropies to high accuracy. At the same time, Planck's CMB measurements have yielded improved evidence for anomalies in the CMB temperature field which are not easy to accommodate within the accepted paradigm. The Planck results also supported the standard cold dark matter model and basically confirmed the inflation theory with a high accuracy, although they set some limits on several inflationary parameters. Of great importance are the data measuring how photons polarized by the CMB are distributed in the universe that could be indicative of primary gravity waves remnants left behind the inflation stage though their traces difficult to identify unambiguously because they could be masked by an odd dust distribution in the metagalaxy. Generally, the Planck results greatly contributed to both standard and nonstandard cosmological models and fundamental physics.

Let us now address the problem of the universe's evolution, which is closely related to the intriguing question of its ultimate fate.

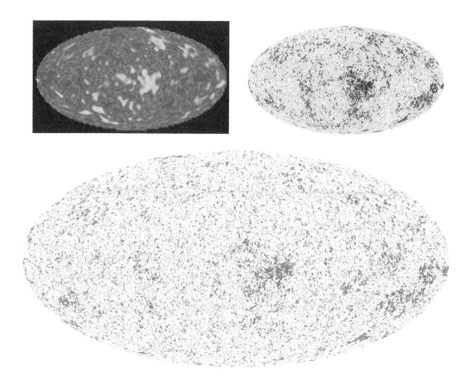

Fig. 11.5 Cosmic microwave background radiation according to PLANCK measurements. The temperature field inhomogeneities and anisotropy $\delta T/T \sim 10^{-5}$. CBM map according to COBE and WMAP measurements are sown in the upper *left* and *right* corners, respectively (Courtesy of ESA and NASA)

Evolution and Fate of the Universe

Evolution, involving self-organization processes against an originally chaotic background, shaped the universe and the properties of its objects at different evolutionary stages. These include the fundamental processes of the evolution of primary matter, the birth of stars and gas-dust clouds from a diffuse medium, and the formation of galaxies and galaxy clusters, as well as accretion disks and planetary systems. The basic ideas on the processes of matter transformation and stellar-galactic evolution are fundamental in the cosmology of the universe. Although great progress was achieved in the field in recent decades, many problems must still be resolved before we come to an unambiguous and internally coherent model of the universe's origin and evolution.

The scenario of the evolution and fate of the universe is based on an estimate of its total mass density ρ and how it is balanced with the critical mass value ρ_c. This value places constraints on the future expansion rate: if $\rho < \rho_c$ the universe will expand forever; if $\rho > \rho_c$ the universe will collapse on itself. The ratio of the present density ρ to the critical density ρ_c is called omega: $\Omega = \rho/\rho_c$. Historically, there were several models of the universe suggested to predict its fate:

Einstein-de Sitter model, $\Omega = 1$: The universe is assumed to be open, flat, infinite and static;

Friedmann-Lemaitre model, $\Omega < 1$: The universe is assumed to be open, infinite, hyperbolic, and dynamic, meaning it will expand forever;

Friedmann-Lemaitre model, $\Omega > 1$: The universe is assumed to be closed, finite, spherical, and dynamic, meaning it will expand and collapse on itself (the Big Crunch!);

Gold-Bondi-Hoyle model, $\Omega \sim 1$: The universe is assumed to be stationary; its expansion is continuously balanced by newborn matter generated in a "creation field" to maintain the mean density of the universe.

Basically, Einstein's general relativity theory supports the idea that the mean density of matter in the universe equals the critical mass value, which is expressed in the modern superinflation theory discussed below. The theory also favors the concept that the universe is open, extremely flat, and that space obeys Euclidean geometry. The critical density is estimated by the formula $\rho_c = 3H_0^2/8\rho G$, which includes the Hubble constant H_0 and the gravitational constant G. It equals $\rho_c = 5$ particles/m^3 or 1×10^{-29} g/cm^3. At the same time, the average density of the observed universe is only about $\rho_c = 0.2$ particle/m^3 or 4×10^{-31} g/cm^3. Therefore, for the visible baryonic matter (more precisely, what we can observe in the whole electromagnetic spectrum from radio waves to gamma rays) $\Omega_v = 0.045$, with stars accounting for only 0.005; the remaining is interstellar gas and dust. Another part of about 0.005 is due to small mass neutrinos. However, there also seems to exist invisible (dark) matter inferred from observations of galaxies—a very important but poorly understood component. The idea and term cold dark matter (CDM) was suggested by Fritz Zwicky back in 1933, but was underestimated for nearly half a century, when in-depth study of this phenomenon began. A bit earlier, in 1970, Vera Rubin, who studied the rotation speed of gas in galaxies, discovered that the total visible mass (from the stars and gas) does not properly account for the speed of the rotating gas. In other words, the observations showed that gravity to keep stars together should significantly exceed that caused by their visible mass; otherwise, the galaxies would disintegrate and the stars be scattered out in space. This galaxy rotation problem is thought to be explained by the presence of large quantities of unseen dark matter. This hidden mass was initially estimated to be $\Omega_{dm} = 0.23$.[2] The nature of dark matter is yet unknown and its behavior is different from the baryonic category of elementary particles as the main component of matter. Indeed, it shows no features related to electromagnetic radiation (e.g., such as light reflection) and exhibits itself only through gravitational forces. It is possibly composed of stable, massive, and electrically neutral particles of a nonbaryonic nature. Currently, weakly interacting massive particles (WIMPs) whose mass exceeds that of a proton by about two to three orders of magnitude ($m_X = 100$ GeV to 1 TeV) appear to be the most suitable candidates for CDM. They are assumed to have been born at the Big Bang under energies $E = kT > m_X$ ($k = 1.38 \times 10^{-16}$ erg/grad is the Boltzmann constant). Also recall (see Chap. 10) that the found "missing baryons" in an extended halo of hot gas around galaxies could be kept in mind to resolve the problem. Indeed, recent

---

[2] Planck observations showed that it is actually $\Omega_{dm} = 0.268$.

Fig. 11.6 Evolution of galaxies since ~ 2 billion years after origin of the universe until the contemporary epoch based on the HST wide field and planetary camera image (Courtesy of Space Telescope Science Institute)

observations with NASA's Chandra X-ray Observatory found evidence that the Milky Way is embedded in a halo of hot gas (1–2.5 million K) that extends for hundreds of thousands of light years and that the mass of the halo is estimated to be comparable to the mass of all the stars in the Galaxy. This means the galaxies' missing baryons have been hiding in the halo of gas.

Galaxies were probably formed within dark matter by the aggregation of subgalactic mass halos, the collapse of self-gravitating objects during the development of gravitational instability that enhanced the quantum fluctuations that emerged at the earliest inflation epoch. Subsequently, the protogalaxies grew by accreting smaller clumps and/or combining with a halo of comparable size (Fig. 11.6).

Redistribution of the system's angular momentum, the formation of a gaseous galactic disk, and the onset of star formation should apparently be attributed to the same period. The formation of galaxies and galaxy clusters in an expanding universe obeying modern gravitation theory and Einstein's general relativity theory can be associated with the presence of the adiabatic metric fluctuations in the inflation model, which is also consistent with the CMB observations. This makes it possible to refine the relative contents of visible and dark matter.

Visible and dark matter altogether result in $\Omega_m = \Omega_v + \Omega_{dm} < 1$; i.e., the total density of the universe is still less than the critical value of one. The balance is achieved, however, if in addition to the visible and dark matter, dark energy Ω_{de} is introduced. Indeed, recent observations indicated that the universe is expanding even faster than predicted by the known mass involving dark matter; in other words, it is accelerating. This was revealed from the observations of luminosity fading of distant and Ia type supernovae caused by the universe's expansion. Three scientists—Saul Perlmutter, Brian Schmidt, and Adam Riess—were awarded the Nobel Prize in Physics in 2011 for this work. Acceleration requires an additional energy density that is associated with an existence of an antigravity force in space. This is an analog of the cosmological constant Λ originally introduced by Einstein in order to resist the gravitational collapse in a static universe, the constant serving as an equivalent of a repulsing force. Later on the prominent physicist complained that, by introducing this constant, he "spoiled" his basic field equations of general relativity theory and referred it to as "the greatest blunder" in his scientific career. But his assumption turned out to be correct, and Einstein was completely right in his vision. Nowadays one can formally write the balance condition as follows: $\Omega_m = \Omega_v + \Omega_{dm} + \Omega_{de} = 1$ ($\Omega_{de} = \Lambda \sim 0.68$ is required).

According to the latest estimates based on the Planck measurements, dark energy amounts to 68.3 % of the total mass of the universe compared to 26.8 % of dark matter and 4.9 % of visible (baryonic) matter, although dark energy does not have to be associated with mass but rather with a positive cosmological constant of a yet unknown nature.[3] Of the 4.9 % baryonic matter, stars occupy only 0.5 % and 0.3 % is thought to belong to neutrinos. Note that heavy elements, which are crucial for the origin of planets, life and human existence, compose only 0.03 % of the overall matter of the universe (Fig. 11.7). Nowadays, the above balance of the composed matter/energy of the universe is confirmed by observations at ~1 % accuracy, which means that the mass total density is close to the critical one ($\Omega \sim 1$) indeed. We therefore think of the Big Bang model of the universe as the most probabilistic. In other words, in the framework of the contemporary model, the universe is considered as nonstatic, infinite, flat, and open (unbounded), and it will expand with acceleration forever.

All this means that the fate of our universe is fully defined by its dark energy. The comparative contribution of routine (baryonic) matter, dark matter, and dark energy is roughly 1:10:25. Dark energy is still an enigma and is probably related to the completely new understanding of a vacuum as a new form of energy density far exceeding everything that we know of. Fantastic vacuum properties have been found; it possesses negative pressure, and the density of its energy is independent of expansion, which it causes. Thus, a vacuum is the densest media, and its density is fully uniform over the whole universe. Its unit volume energy (probably caused by quantum fluctuations of some scalar field filling our universe) is incredible: it exceeds by a factor 10^{120} (!) that required by the observed velocity of the universe expansion, a discrepancy that remains hard to explain.

Many questions remain unanswered. In particular, we cannot answer the principal question of whether it changes with time and how the underlying cosmological

---

[3] Before the Planck measurements the figures were 72.8 % of dark energy, 22.7 % of dark matter, and 4.5 % of baryonic matter.

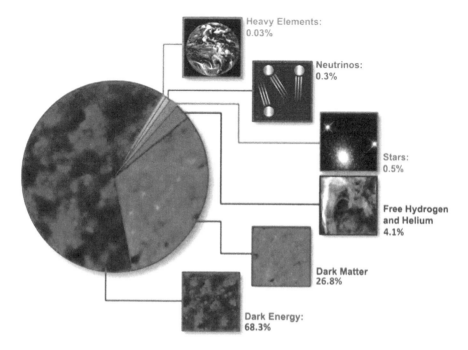

Heavy Elements:
0.03%

Neutrinos:
0.3%

Stars:
0.5%

Free Hydrogen
and Helium
4.1%

Dark Matter
26.8%

Dark Energy:
68.3%

Fig. 11.7 Bulk composition of the universe (Courtesy of NASA)

scalar fields and their evolution are related to the fundamental physical constants and/or their variations.[4] New efforts to obtain more insight into the properties of dark energy have been undertaken since 2013 in the framework of the Dark Energy Survey (DES) Science program using a gravity microlensing technique with USA Blanco 4-m telescope in Chile and the Japanese 8-m Subaru telescope in Hawaii. The goal is to observe hundreds of millions of galaxies and galactic clusters with very high resolution in order to map matter distribution in the universe in an attempt to detect the influence of dark energy on this distribution. Evidently, the light that left these faraway structures in visible wavelengths is stretched into longer, infrared wavelengths because of the accelerating expansion of space.

Nonetheless, the vacuum concept dramatically changed the former views on the earliest stages of the universe's origin. Note that if expanding matter performs work against gravity, its total energy E (and hence mass m according to the famous Einstein formula $E=mc^2$) decreases with time. Now the total mass of the universe is $\sim 10^{56}$ g (10^{50} tons), and originally at the Big Bang it was supposed to be nearly 40 orders of magnitude larger. Let us note that it corresponds to the theoretically maximum conceivable Planck density $\rho_g \sim h/2\rho c l_g = 10^{94}$ g/cm^3 composed of the dimension combination of minimum fundamental Planck length $l_g = 10^{-33}$ cm, Planck constant

---

[4] It was shown that the value of one of the fundamental constants $\mu = m_p/m_e$ (the ratio of proton and electron masses, $\mu = 1{,}836.15267247$) changed less than $\Delta\mu/\mu < 10^{-5}$ for 12–13 billion years.

$h = 6.67 \times 10^{-33}$ erg s, and speed of light $c = 3 \times 10^{10}$ cm/s. However, now it is thought that the original mass of the universe was much less ($\sim 10^{-3}$ g) because there was enormous energy (and hence, mass and density) of the vacuum scalar field. This follows from the previously discussed superinflation theory, which postulates a nearly instantaneous giant expansion of an original negligible piece of matter to the universe's enormous mass and size involving all its galaxies, stars, planets, and interstellar matter. Moreover, according to this model, the once born universe begins to reproduce itself, giving birth to numerous other universes having different properties. This idea underlies the multiverse concept that will be discussed in more detail below.

We can now briefly summarize the main sequence of the universe's origin and evolution scenario based on the Big Bang model. The universe started from a subatomic point of small mass at an infinitely high temperature and density. At the time moment $< 10^{-43}$ sec (referred to sometimes as a supergravity wall) at temperature $T \sim 10^{32}$ K it represented the Planck universe (a singularity). This moment is associated with the origin of space-time—this is why it is meaningless to ask what existed before that moment. Between 10^{-36}–10^{-34} sec ($T = 10^{28}$ K) at the stage of "impetuous boiling vacuum" when the universe was still far from thermal equilibrium, the above-mentioned superinflation occurred, which was responsible for the expansion of space by a factor $\sim 10^{50}$ and symmetry violation. This means that the expanding velocity grew instantaneously such that for 10^{-35} s the volume of space increased from the size of an atom to the size of an orange and then from an orange to a metagalaxy. Then expansion stopped, and the scalar field (energy) transformed to ordinary matter via Einstein's familiar equation, $E = mc^2$. At 10^{-6} sec ($T = 10^{13}$ K) at the state of thermal equilibrium, baryons (protons and neutrons) formed, then within 1–1,000 sec ($T = 10^{10}$–10^7 K) nucleosynthesis started, and nuclear reactions began to occur as the precursors of chemical elements. Currently, the earliest time visible in the universe's history is 380,000 years since its origin, when at $T \sim 10^3$ K atoms of H, D, He, and Li formed.

The principal (and the most enigmatic) phase in this scenario is superinflation, which occurred immediately after the born universe emerged from singularity. The inflationary theory allowed us to explain many features of our world and to make several important predictions, generally in accordance with the Einstein-de Sitter model. One of them is that the universe should be extremely flat (the curvature is less than 10^{-41}); all other models turned out to be much more complicated and hardly realistic. The theory predicts that inflation generates Gaussian adiabatic perturbations of the metric with a nearly scale-independent (flat) spectrum. Importantly, the flatness of the universe can be experimentally verified, since the density of a flat universe is related in a very simple way to the speed of its expansion given by the Hubble constant. Indeed, the observational data so far available are consistent with this concept: density and temperature perturbations produced during inflation made their imprint in the CMB and they comply with the distribution of matter in the universe (the Cosmic Web).

The earliest stage of the universe is hardly possible to imagine. An artist's attempt is reproduced in Fig. 11.8. That period is sometimes referred to as the dark ages with a hydrogen haze. In the following hundreds of millions of years to 1 billion years ($T = 100$ K), stars, galaxies, galaxy clusters, and the Cosmic Web formed. The haze scattered out, hydrogen became mostly ionized, and light radiated by stars/

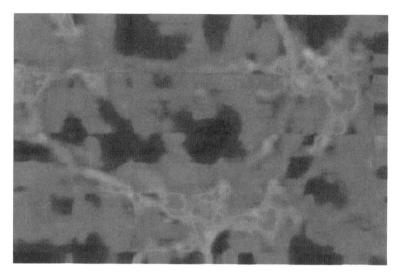

Fig. 11.8 Early stages of the universe. Artist's concept (Courtesy of SAO/G. Fazio)

Fig. 11.9 First stars formation. Artist's concept (Courtesy of SAO/G. Fazio)

galaxies appeared. Today, 13.7 billion years from the beginning, we have CMB radiation with $T = 2.735$ K as a remnant of the Big Bang. Observations of the brightness of distant galaxies showed that the process of star formation was most active in the early universe. Studies from the Hubble Space Telescope also lead to the conclusion that the universe made a significant portion of its first stars in a torrential firestorm of star birth just a few hundred million years after the Big Bang. Another artist's concept of that epoch is reproduced in Fig. 11.9.

We should be aware, however, that the Big Bang theory is still a theory, and that it still poses some unresolved problems. Similarly, the experimental search for dark matter and dark energy, as well as vacuum and antigravity properties, represents a potential Rosetta stone of modern astrophysics. It is not fully known how the universe became so isotropic/homogeneous, or what originated the density fluctuations from which the galaxies eventually formed. It is not clear what happened to the antimatter particles created in the Big Bang, though there is an attempt to explain it by incorporating elementary particle theory, which will be briefly discussed in the next section. We do not fully understand why the universe is flat and its total mass is close to the critical one ($\Omega \sim 1$), although, as we said above, it can be principally explained in the framework of the superinflation theory arguing for the formation of a smooth and identical universe. We try to test theories of the universe's early inflation period by uncovering the most massive of galaxy clusters billions of light-years away. However, the theory itself is far from complete. One of the key questions in cosmology remains open: How did the first bumps and wiggles in the distribution of matter in our universe rapidly evolve into the massive structures of galaxies we see today? Currently, we can only say that the universe is dominated by a cold dark matter and by dark energy associated with a cosmological constant and vacuum energy.

Moreover, modern theories pose the problem of how to compensate for the monstrous vacuum energy with a cosmological constant. In particular, one should explain why the value of Λ, in the case of the universe, turned out to be in the rather narrow range of time scales suitable for the formation of stars, galaxies, and planets, and ultimately, the development of life forms and intellectual beings. Under a different positive or negative value of Λ, the fate of the universe would be completely different, and this may be related with the paradigm of the anthropic principle, which states that the laws of nature are astonishingly well fitted to the existence of life.

This is especially important in light of the newest observations which showed that acceleration of the universe's expansion increases exponentially with time and, therefore, that its ultimate fate is defined by the concurrence of gravity and antigravity forces. Antigravity will dominate gravity, giving rise to repulsion rather than attraction, which means that the Λ term defines antigravity. Moreover, theory predicts that antigravity will grow and in billions of years the new process of a *Big Rip* of the universe will begin, resulting in the universe's destruction. In more detail, this implies the following scenario: because antigravity dominates, galaxy clusters, galaxies, stars, planetary systems, and then chemical elements, and even atoms will be eventually decomposed, leaving behind dark empty space.

Standard Model and Superstring (M) Theory

Main Interactions and Elementary Particles

Cosmology is viewed as the synergy of microphysics and macrophysics. It is intimately related with the physics of elementary particles bearing the most fundamental properties of matter and four fundamental interactions (forces) in nature: the

strong force, involving *u* and *d quarks* building protons and neutrons (nucleons) composed of three quarks each, and *gluons* keeping quarks together in the nucleus; the **weak** (or electroweak) force responsible for radioactivity and involving *leptons*—electrons and neutrinos interacting with W^+, W^-, and Z^0 bosons, and photons; the **electromagnetic** force involving *photons*; and **gravity** involving *gravitons*. Bosons are weakly interacting positively or negatively charged particles and zero-charged neutral particles which obey Bose-Einstein statistics, which are widely used in physics and give the boson its name. While the mass of a gluon, photon, or graviton is zero, weak calibrating bosons are quite heavy particles with a mass nearly two orders of magnitude (86.97) more than the mass of a proton (which is, in turn, 1,852 times heavier than an electron). Besides these, there are bosons of strong interactions called mesons composed of even numbers of quarks and antiquarks. Note that in galactic cosmic rays another particle called a muon is distinguished, which belongs to the lepton family, and can be regarded as a heavy (207 times more massive) electron with a lifetime of only 2.2 µs, but with a velocity close to the speed of light. Besides electron neutrinos there are also tau and muon neutrinos and their energy/mass transformations—neutrino oscillations—from one mode to another were discovered. These transformations (called different "aromatics") are probably related with the basic principles of matter structure and the universe as a whole. Let us note that, since Einstein, the unification of all four interactions has been considered a fundamental principle of nature, and many have searched for it.

All elementary particles have antiparticles of opposite charge but equal mass and spin. *Spin* (also called spin quantum number) is one of the fundamental characteristics in quantum mechanics, defining the internal moment of elementary particle momentum. It is an important property pertinent to elementary particles themselves, rather than to their spacious motion. In contrast to classic mechanics, spin characterizes exchanges through interactions within a quantum-mechanical system. Spin can acquire either integer or semi-integer positive values, in particular, 0 for scalar particles like π and K mesons (bosons of strong interactions made of a quark and an antiquark); 1/2 for fermions (electrons, quarks, neutrinos, protons, neutrons, and muons); 1 for vector particles like photons, gluons, and W and Z bosons; and 2 for tensor particles like gravitons. To be familiar with the terminology, note also that a group of heavy elementary particles—protons, neutrons, and hyperions—is distinguished, called baryons (from here we get the term baryonic matter that we used earlier). In turn, baryons and mesons constitute the group of particles participating in strong interactions, which are called hadrons. For example, one can describe a meson as a strong interaction particle (hadron) composed of one quark and one antiquark and having an integer spin.

Standard Model and Symmetry Principles

The prerequisites for unification of the above interactions are rooted in what is called the standard model of particle physics: the model of strong and electroweak forces acting between elementary particles and jointly with electromagnetic forces underlying all matter and forces acting in nature. Similar to general relativity theory, the standard model obeys symmetry (invariant) principles reflecting the universal character of the laws of nature in terms of their form preservation, regardless of the position of an observer in space. Principles of symmetry are important not only in time-space transformations but also in the physics of elementary particles, which are generally based on quantum mechanics with a caveat that will be briefly discussed below. Here we deal with internal (or local) symmetry, historically also called calibration symmetry. As we said earlier, gluon fields create strong forces which "glue" quarks into protons and neutrons. There is, however, another local symmetry intrinsically related to the inner quark properties called (conditionally) color. This means that each u and d quark can exist in three different states and a special kind of symmetry between them exists. However, quarks of different colors cannot exist independently. Instead, they are unified by gluons to form colorless combinations of three quarks (like mesons) forming a proton or neutron. This property of strong interaction is known as confinement. In quantum field theory such symmetry/interactions are studied by quantum chromodynamics.

An even more unusual symmetry is pertinent to electroweak interactions. Here, because of internal symmetry, weak nuclear forces caused by an exchange of W and Z bosons are coupled with electromagnetic forces caused by photon exchange. In electroweak theory, symmetry is responsible for photons, and W and Z particles originate as clots of energy of four fields. From the local symmetry, the forms of physical laws remain unchanged if, instead of a field of electrons or neutrinos, mixed fields of these particles in a certain proportion are taken; simultaneously, other particle families (e.g., u and d quarks) are mixed in the same proportion. Interestingly, because weak nuclear forces result from W and Z boson exchange, electron-neutrino symmetry also means symmetry between electromagnetic and weak nuclear forces. This sort of symmetry is analogous to the symmetry of different coordinate systems in a gravitational field, following from the symmetry of general relativity theory, but it is completely different from what we are used to in our everyday environment.

We therefore see that a significantly different approach compared to our standard views of the world around us must be applied to understand microphysics (also known as elementary particle or high energy physics). The powerful tool that allows us to deal with intimate processes in atoms and subatomic structures is quantum mechanics. Instead of material particle motion as in Newtonian mechanics, quantum mechanics deals with wave functions and their transformations, which serve as a convenient mathematical formalism to trace the probable position and velocity of particles in a system under consideration, and eventually, to describe the state and structure of matter. Quantum mechanics also grounds different types of interactions and physical entities such as common and local principles of symmetry. As renowned physicist and Nobel prize winner Steven Weinberg said, the principles of symmetry and different modes of wave function transformations using these principles rather than matter itself dominate in physics. Therefore, the principles of symmetry define the very existence of all known forces in nature.

Symmetry Breaking and the Higgs Boson

Nowadays unification of three of the four (except gravity) interactions has been accomplished in the framework of the advanced standard model, though some differences between them (specifically, difficulties in unifying the weak and electromagnetic forces with the strong nuclear ones) still remain. This is a key step of the Unified Field Theory, which is also called the Theory Of Everything (TOE) because it underlies the nature of our world, and it is still a work in progress. The development of the Unified Field Theory was Albert Einstein's dream for almost the last 30 years of his life, as he unsuccessfully tried to combine gravity and electromagnetic interactions. The modern standard model of strong, weak, and electromagnetic interactions is basically the field theory including time-space symmetries of special relativity theory and internal symmetries of electromagnetic and other fields responsible for interaction transfer. There are, as we now know, photon, electron, quark, and other fields. In turn, electrons and quarks, for example, are clots of energy of different fields which, jointly with momentum, determine the properties of these particles in the quantum field theory approach.

Unfortunately, quantum field theory, which should also include symmetries of general relativity theory, fails to describe the physics of gravity. This is probably caused by the extremely low force of graviton interaction, which is less than the electrostatic Coulomb force interaction of two electrons by a factor of 10^{42}. We are perfectly familiar with the opposite situation in which electromagnetic force yields to the gravitational interaction between bodies of large mass. This is just what we feel every day living on our massive planet and what we observe in space around us. In essence, the TOE is addressed as an ultimate goal to unify these fields under very high energy on the principles of symmetry as the fundamental basis of both matter and the universe's structure, specifically at very early stages of its existence. In other words, the goal is to construct the most comprehensive theory with no requirements to reduce to (or deduce) other, more fundamental principles.

In the framework of the standard model, the phenomenon of spontaneous symmetry breaking was also suggested. It appears in some solutions of the basic model equations, although the symmetry of the equations themselves defining particle properties (which probably have a deeper and more general sense) is preserved. Symmetry breaking was discovered in electroweak interactions and, as a common effect of the model, chiral symmetry breaking is found in strong interactions as well. The importance of this phenomenon is difficult to exaggerate, because it is intrinsically related with the origin of the mass of elementary particles. As a matter of fact, the symmetries of the standard model prohibit particles to have mass, but they have it. This means that in the standard model all particles acquire their masses due to symmetry breaking between the weak and electromagnetic forces. Otherwise, all quarks, electrons, and W and Z bosons would have zero mass similar to photons and neutrinos; only the mechanism of spontaneous breaking of electroweak-electromagnetic symmetry allows us to avoid such a scenario. Moreover, in the framework of symmetry breaking it is possible to resolve the key problem of the predominance of matter over antimatter, which is referred to as *CP*-invariance breaking. Note that the above-mentioned neutrino transformations could play a part in the attempts to explain the asymmetry of the universe.

In order to resolve the problem and explain how such a mechanism of symmetry breaking takes place in nature, the existence of a hypothetical quantum field of non-zero value filling all of space (the Higgs field) was introduced mathematically in the standard model equations. This involved a new short-lived subatomic particle, the Higgs boson, as an excitation (quantum) of this field above its ground state. The Higgs mechanism theory, which endowed mass to elementary particles through interaction with the Higgs field, was postulated independently about 50 years ago, in 1964, by Peter Higgs from the University of Edinburgh, Great Britain, and Francois Englert and Robert Brout from Université Libre de Bruxelles, Belgium.

Great efforts to experimentally confirm the existence of this missing particle which would help to keep the standard model in order were undertaken, most recently with the Large Hadron Collider (LHC) at CERN. It was announced recently that the birth of a new particle having an energy of 125 GeV (that is, only slightly more than $1,000\,m_p$ with an accuracy of 1–2 %) was indeed found. With a high probability it is associated with the Higgs boson, though some additional evidence (in particular, more statistics and the proof that it has zero spin) is still required. Francois Englert and Peter Higgs became Nobel prize winners in 2013 (Robert Brout died in 2011).

The importance of the Higgs boson discovery is difficult to overestimate. This is exactly the particle which presumably gives mass to all atomic matter, including electrons, quarks, leptons, and W and Z bosons. It is assumed to interact with other elementary particles through the weak interaction; it has zero spin quantum number, no electric or color charge, and it is easily decomposed through different combinations of particles or channels. Its mass turned out within the earlier estimated 50–1,000 proton masses (m_p). The masses of all other fundamental particles such as quarks and electrons are supposed to be proportional to that of the Higgs boson. Not accidentally, it has been called the "divine particle" because it appears to hold the key to basic concepts of the matter in the universe. In other words, everything we know about the universe goes through the Higgs field; otherwise, elementary particles would not have mass, like photons. The formation of atoms and molecules composing the universe would then be impossible, and the world would be completely different. It is worth mentioning that the theory of electroweak unification of elementary particles that predicted the Higgs boson was developed earlier independently by the three Nobel Prize winners Steve Weinberg, Abdus Salam, and Sheldon Glashow.

One may reasonably assume that discovery of the Higgs boson will open new horizons in physics and that symmetry breaking in weak and strong interactions is just a part of a more fundamental principle underlying the overall structure of matter (including the relationship of the Higgs boson and dark matter) and unifying gravity with all other forces in nature. Remarkably, CP-invariance breaking was observed in LHC experiments as well with the decay of neutral Bs mesons. These mesons are produced in large amounts from proton collisions. We may therefore conclude that the discovery of the Higgs boson and CP-invariance breaking, which are basically in accordance with the standard model, opens windows to the physical composition, the origin, and probably unknown parts of the universe.

Grand Unification

Now addressing the problem of grand unification, we may assume that this question will presumably be resolved in the framework of quantum gravitation theory. It is known that the intensity of interactions depends on the mass of particles transferring the interactions (such as W and Z bosons) and on some numbers characterizing the probability of emission and absorption of particles (photons, gluons, and bosons). These numbers entering the basic equations of theory are called constants of interaction, or coupling constants. Theory predicts that under extremely high energy the constant values become equal. This implies that negligible gravity forces greatly increase while other forces of particle interaction decrease simultaneously, and, therefore, all forces acting between elementary particles become comparable. The enormous energy at which such a situation is attained (i.e., all forces in nature unify) is called the Planck energy (10^{19} GeV). In other words, at this energy all forces become comparable by their value and are assumed to represent the variation of a single force. Let us note that this value is only three orders of magnitude lower than the energy when symmetry violation of electroweak and strong interaction occurs, and their respective constants become equal. This indicates that all interactions in nature may indeed be unified in the framework of a more fundamental theory such as the TOE. However, only a theoretical approach is really conceivable, because the Planck energy exceeds by a hundred thousand billions times ($\sim 10^{14}$) what was accomplished with the most powerful particle accelerators on Earth (the energy of an elementary collision in the LHC is ~ 2.5 TeV). The famous cosmologist Stephen Hawking figured out that, in order to reach the energy under which grand unification would occur, one would require an accelerator the size of the solar system! This means that the theory could hardly ever be verified experimentally.

Despite the great progress in our knowledge about matter's properties, many extremely complicated problems remain to be solved, including the problem of hierarchy connected with the fantastic difference of fundamental energies in elementary particle physics (for example, the mass of the Higgs boson is only about 10^2 GeV, that is, 10^{17} times less than the Planck energy!). Attempts to find the problem's solution based on the idea of supersymmetry unifying the constants of strong, weak, and electromagnetic interactions under very high energy ($E \sim 10^{16}$ GeV), to confirm possible breaking of conservation laws for lepton and quark numbers and stability of matter, to find a candidate for a dark matter particle, etc., are all outstanding issues. It has been found that the laws of physics are not invariant relative to three existing symmetry transformations: C, P, and T. This means that the laws will operate differently for particles and antiparticles (C); there will be different paths of evolution of the universe and its mirror reflection (P); and the universe will behave differently with time inversion (T)—not the case for the classical mechanics laws. An important consequence that follows is related to the early universe. Indeed, one may perceive that if an equal amount of matter and antimatter were born at the Big Bang, it would result in the annihilation of quark-antiquark pairs with an additional enormous energy release. This did not happen, however, because of the non-invariance of symmetry transformations that allowed an excess of quarks over antiquarks (in other universes an opposite situation could be expected!). As a result, existence is based on matter rather than on antimatter. The fact serves as another brilliant evidence to confirm the theory of grand unification.

Although some limitations and enigmatic problems are left, the advanced standard model, which proved its validity in numerous experiments dealing with strong and electroweak interactions and conservation law verification and led to the discovery of the Higgs boson, serves as an important clue for understanding the very early history of the universe. It argues that under the enormous temperatures at the first few instants after the Big Bang when all particles had yet no mass, there was symmetry between the weak and electromagnetic forces. Baryonic asymmetry, it seems, occurred only after $\sim 10^{-10}$ s when the temperature dropped down to 10^{15} K, and then the first massive particles (quarks, electrons, and W and Z bosons) appeared, followed by the generation of strong and electroweak interactions in nucleus building.

Superstring Theory

In support of the new concept of matter organization, relativistic quantum mechanics string theory was proposed. It is better known as *superstring theory* and is also called M-theory (standing for the Mother (M) of all theories). This mathematical theory (involving string-based multidimensional space) integrates everything: it underlies the diversity of elementary particles, quantum mechanics, gravity, and the nature of the universe (cosmology). Mathematically, it is harmonious, slim, and compatible with quantum mechanics, and it satisfies the principles of fundamental space-time symmetry (known as conformal symmetry) as well as internal symmetries underlying the contemporary standard model of strong, weak, and electromagnetic interactions.

String theory is referred to as a theory of fundamental particles and forces in which the basic entity is an exceedingly short one-dimensional structure (a "string") rather than a point-like particle. It postulates that all elementary particles of different energies and masses result from oscillations of ultramicroscopic loops—tiny vibrating strings of minimal possible Planck length (10^{-33} cm) and of enormous tension. Steven Weinberg associates them with "tiny one-dimensional cuts on the smooth tissue of space." They can be either open or closed. Every particle exists like a wave produced by a string vibration similar to overtones (analogous with musical tone/pitch). Thus, string vibrations (resonance modes in the infinite wavelength) are responsible for the nature and mass of the whole variety of elementary particles and all four interactions in nature. String theory predicts the existence of a particle having zero mass and spin twofold that of a photon, which is associated with a graviton: a quantum of gravity radiation.

When the ideas of supersymmetry are applied to string theory, the outcome is superstring theory. Superstring theory requires a ten-dimensional space-time; however, all but four of these dimensions (length, breadth, height, and time) are hidden in the present-day universe. In other words, string-based space is assumed to be a multidimensional entity (strings exist in a ten-dimensional space) and leads to a new view of the microworld and the universe. Time and space are deduced from this theory and are limited in terms of having no sense before Planck time (10^{-42} s after the Big Bang). This is why when curious people ask the question about a time when the Big Bang occurred or what existed beforehand, the answer is: there was as yet no time, since space and time appeared just at the instant of the Big Bang.

Therefore, superstring theory, developing in conjunction with modern quantum electrodynamics, quantum field, and calibration theories (all rooted in quantum mechanics), is one of the prime candidates for a Theory of Everything, embracing all of the forces and particles of nature. Superstring theory is called on to explain all principal properties of the matter in the universe and the universe itself and it appears to be a real step toward a TOE.

We can see that, similar to elementary particle theory, modern cosmology deals with categories that are very far away from our daily life perceptions. Indeed, it is impossible to imagine Planck's length, time, density, and energy as key parts of the original explosion, or a "boiling vacuum" followed by superinflation of enormous scale for vanishingly small time lapses. All these concepts are well beyond our ideas of routine thinking. Moreover, the events and their sequence at the very beginning are impossible to test or verify experimentally. Therefore, in our basic understanding we entirely rely on the theory, which does not contradict the known fundamental physical laws and allows development of a rather coherent and comprehensive approach to the concept of how the universe formed.

Modern View of the Universe: The Multiverse and Wormholes

Modern cosmology assumes that an infinite multitude of other (parallel) universes exist, our own rather young universe being just one of them. Parallel universes form the multiverse (Fig. 11.10). As we noted above, superinflation theory assumes self-production of universes, in other words, universes are born and decay in different regions and at different times. Expansion of an original piece of a former universe occurs instantaneously, each piece having different "colors" pertinent to different properties of each particular universe. The origin of a universe and its specific "color" might be the result of quantum oscillations in vacuum (*creatio ex nihilo*). Universes floating in space are infinitely (exponentially) far away from us and hence cannot be observed. They are associated with invisible "vesicles" which form "temporal-spacious foam" and they may experience collisions. Our universe's origin might be the result of such a collision that led to the Big Bang and its effects. It is further assumed that, because space is filled up with quantum fluctuations, time-space tunnels resembling black holes (Fig. 11.11) could exist in a multiverse allowing an escape to hyperspace, and that distant (in both time and space) universes may be coupled through wormholes along a hidden dimension (Fig. 11.12). The processes may resemble the chaotic inflation in the Big Bang. Accretion of (phantom) matter on a worm hole with a regular magnetic field may result in black hole appearance, although a difference would be hard to reveal. A wormhole could be assumed theoretically as the case of a charged or rotating (Kerr) black hole,

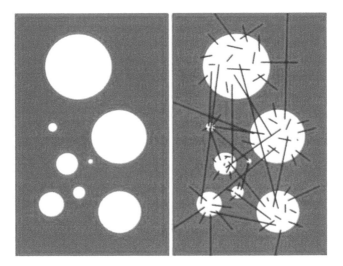

Fig. 11.10 The concept of multiverse. Universes floating in space are associated with invisible "vesiculars" which form "temporal-spacious foam" and may experience collisions (Courtesy of I.D. Novikov, N.S. Kardashev, A.A. Shatskii)

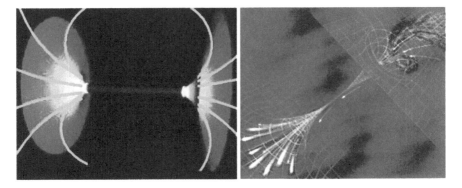

Fig. 11.11 The concept of worm holes as time-space tunnels resembling *black* holes (*left*). Accretion of (phantom) matter on a worm hole with a regular magnetic field may result in *black* hole appearance (*right*) (Courtesy of I.D. Novikov, N.S. Kardashev, A.A. Shatskii)

giving us a theoretical chance to avoid the singularity, which basically signals the breakdown of any theory. Extending the solutions of general relativity equations to some extremes reveals a hypothetical possibility of exiting the black hole into a different space-time with the black hole acting as a wormhole, though any perturbation may destroy this possibility. Thus, the possibility of traveling to another universe remains only theoretical. A wormhole configuration might be imagined as two black holes of a cone shape connected by a tie-plate with $r = r_g$ in the middle (see Fig. 11.12). Here exotic matter could exist having energy density $\rho < p_{ii}$, where p_{ii} is the pressure of the exotic (probably dark) matter.

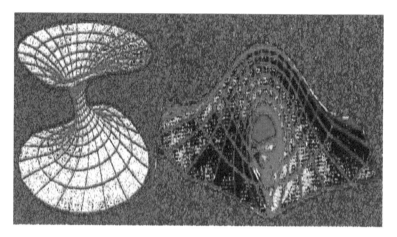

Fig. 11.12 Time-space tunnels are assumed to exist in multiverse to escape to hyperspace. Distant (in both time and space) universities may be coupled through the worm holes along a hidden dimension (Courtesy of I.D. Novikov, N.S. Kardashev, A.A. Shatskii)

We may conclude that our universe as a part of a multiverse is a more complicated entity than was earlier assumed, representing a chain of mysterious events that no scientist has yet fathomed. The multiverse is permanently changing, and the Big Bang can be viewed as a routine event of parallel universe interactions, and the process of gradual formation of order out of chaos. Likewise, the evolution of the universe is a continuous self-organization process that has led to its currently observed structure with its host of galaxies, galaxy clusters, and planetary systems. As Andrei Linde observed, the universe as a whole is immortal. Each of its parts could be separated from a singularity somewhere in the past, and it could be terminated in a singularity somewhere in the future. However, there is no end to the evolution of the whole universe. Moreover, proceeding from the postulated theory of modern cosmology, the initial quantum fluctuations and the birth of post-inflation domains (probably with different space-time dimensions and different physical constants) may be considered as a universal paradigm of an infinite process of ordering of an infinite number of initially chaotic structures filled with fields and matter. The theoretical and experimental physics underlying cosmology and astronomical observations are progressing so quickly that new breakthroughs may be expected in the near future, which will significantly complement and possibly even modify our current understanding of the world we live in.

Conclusions

Hopefully, the universe has been presented here not just in its complexity and infinity but also with the beauty of the world where we, human beings, live on the planet Earth – a Pale Blue Dot in the vastness space, as Carl Sagan called it. As natural sciences are the powerful tool to understand our environment, astrophysics is the fascinating area to gain knowledge about outer space – its structure, origin and evolution, stretching from the home planet well beyond to the conceivable edge of the universe's space frontiers. The field is enormous in scope, having many specific branches that are not easy to comprehend and even more difficult to describe using quite simple language as attempted here in the brief sketches covering the most important topics of the field.

Certainly, this is not an easy read for fun – it is aimed towards the goal of serious thinking about the nature of our world. Your response whether the goal was reached or not is just what the author would like to know in order to improve the contents and/or account it for whenever necessary.

Since the ancients' time throughout human history, curiosity together with fear of the unknown and the desire to ensure security from natural phenomena (coming from the heavens first!), was perhaps the main driving forces in the hope of understanding the overall scientific context in which people live and the motivation for inventions and primitive technology development. It was also the desire for wealth and power in spite of the obvious costs and risks. The fundamental sciences – first of all physics, chemistry, biology – were at the frontiers of disclosing unknown and mysteries in nature. This impetus motivated the first yet timid steps beyond the home planet to space that mankind undertook in the middle of last century and has become a worthy imperative (both scientific and cultural) in the twenty-first century.

The progress in the study of the outer space started from intense competition between the former USSR and USA for dominance in military rocket technology and highlighted most of the history of space exploration for a few decades. Ironically, we gained our knowledge on the space environment and benefited the basic sciences owing to the technology development motivated by the rivalry of superpowers for

© Springer Science+Business Media New York 2015
M.Ya. Marov, *The Fundamentals of Modern Astrophysics*,
DOI 10.1007/978-1-4614-8730-2

world domination, when the wish to invent the most effective means of killing people which brought us to threshold of the civilization destruction in the 1960s. Another incentive was national pride, like in the case of the Moon race. Historical lessons learned are that scientific progress runs jointly with political ambitions and social-economic concerns, leaving basic science secondary or tertiary consideration at best; we are not, however, going to touch upon and discuss these paradoxical issues being focused in the book on the natural science.

One should be aware that the challenge and the allure of vigorous space exploration and utilization are undeniable and instead of competition, pooling international efforts would come as a new paradigm. It seems unbelievable that only slightly more than a half of century passed since the beginning of space exploration the great breakthroughs in this new human enterprise with the well-articulated goals and optimum mix of robotic and human explorers have been accomplished. Owing to the remarkable confluence of scientific goals and technological progress the results obtained are really impressive. They broadened our horizons tremendously in understanding immense physical-chemical phenomena and matter/energy transformation in space, but also impacted on our mind in terms of philosophical concepts and world attitude. Astrophysics including in the broad sense planetary science, space science, solar and plasma physics, cosmogony, cosmology, and astrobiology manifested the fabulous progress in our civilization development. These areas pushed back frontiers of the unknown and the future progress is difficult to conceive.

Humanity would strive to explore and develop capability to expand its presence throughout the solar system. While more difficult and costly to travel with humans to other worlds, it may offer unique opportunities for our civilization further evolution and unprecedented perspectives. Certainly, we should address carefully first the feasibility of sending humans to planets, the Moon and Mars being the first on the agenda, on technical, scientific, policy-related and economic levels. It is also important to bear in mind the capabilities of robotic spacecraft are yet far to be exhausted and they are progressively growing to gain as much as possible knowledge about nature of solar system bodies before astronauts will enter there. No doubts, however, that this time will come and lunar infrastructure deployment and the first human missions to Mars will be undertaken in this century. As Johann Wolfgang Goethe noticed, "Knowing is not enough; we must apply. Willing is not enough; we must do."

The challenge for astrophysics dealing with the key concepts of the origin of the world and its evolution is to answer fundamental questions about the universe, our past and future. We need to understand as much as we can: where are we in terms of structure of the Universe and our location in space; where did we come from in terms of origin of the Universe and its elements (galaxies, stars, planets); and where are we going in terms of evolution and ultimate fate of the Universe and in particular, destiny of the home planet within the frontiers of the solar system? In other words, what will happen to us in the future, bearing in mind that humans are the product of a long trail of cosmic evolution, and we are fully dependent on our home planet on space environment? We articulated these demanding goals as viable importance for human beings and our civilization concern and tried to clarify the relevant issues.

Which discoveries can we expect which further advance astrophysics in the coming years? It is not an easy task to compile a short list. Finding dark matter properties may bring revolution in physics. The nature of dark energy is associated with the problem how we understand the Universe and its discovery will bring new breakthroughs in cosmology. Detection of extrasolar Earth-like planets with favorable natural conditions and possibly signs of life will advance astrobiology and extend the views about habitability in the Galaxy. Possibly it will also answer intriguing question: are we alone in the universe? The new challenges are related with the further advancement in modern cosmology considering our universe as a part of Multiverse represented as "temporal-spacious foam" in infinite space coupled through Worm holes along a hidden dimension and "allowing" to escape to hyperspace. This is a guess for the new physics underlying cosmology (involving grand unification, postulated theory of the initial quantum fluctuations and the birth of post-inflation domains) which may be addressed as a universal paradigm of an infinite process of ordering of an infinite number of initially chaotic structures in space filled with fields and matter. The progress expected in cosmology in the near future may significantly complement and possibly even modify our current understanding of the World where we live.

We may conclude with quotation from Albert Einstein: "The most incomprehensible fact about nature is that it is comprehensible."

Additional Reading

Barucci MA, Boenhardt H, Cruikshank DP, Morbidelli A (eds) (2008) The solar system beyond Neptune. University of Arizona Press, Tucson, 592 pp

Beaty JK, Peterson CC, Chaikin A (eds) (1999) The new solar system, 4th edn. Sky Publishing/Cambridge University Press, Cambridge, 421 pp

Boss A (2009) The crowded universe. The search for living planets. Basic Books, New York, 227 pp

Boyd TJM, Sanderson JJ (2003) The physics of plasmas. Cambridge University Press, Cambridge, 532 pp

Cole GHA, Woolfson MM (2002) Planetary science: the science of planets around stars. Institute of Physics Publishing, Bristol/Philadelphia, 508 pp

de Pater I, Lissauer J (2010) Planetary sciences, 2nd edn. Cambridge University Press, New York, 547 pp

Encrenaz TJ, Bibring P, Blanc M, Barucci MA, Roques F, Zarka PH (2004) The solar system, 3rd edn. Springer, Berlin, 512 pp

Enciclopedia of Astronomy and Astrophysics, (EAA), 2001. Nature Publishing Group and Institute of Physics Publishing

Fowler CMR (2005) The solid earth: introduction to global geophysics, 2nd edn. Cambridge University Press, New York, 685 pp

Fridman AM, Marov MY, Kovalenko IG (eds) (2006) Astrophysical discs, vol 337, Astrophysics and space science library. Springer, Dordrecht

Galimov EM, Krivtsov AM (2012) Origin of the moon. New concept. Geochemistry and dynamics. De Gruyter, Berlin, 168 pp

Gargaud M, Martin H, López-García P, Montmerle T, Pascal R (2012) Young Sun, early Earth and the origins of life, Lessons for Astrobiology. Springer, Berlin, 301 pp

Green B (1999) The elegant universe. Superstrings, hidden dimensions and the quest for the ultimate theory. Vantage Books, A Division of Random House, Inc., New York

Green B (2004) The fabric of the cosmos. Space, time, and the texture of reality. Alfed A. Knopf, New York

Grotzinger J, Jordan T, Press F, Siever R (2006) Understanding earth, 5th edn. W.H. Freeman, New York, 236 pp

Harland DM (2003) The big bang. A view from the 21st century. Springer Praxis Publishing, Chichester, 262 pp

Harrison ER (2000) Cosmology: the science of the universe, 2nd edn. Cambridge University Press, Cambridge, 567 pp

© Springer Science+Business Media New York 2015
M.Ya. Marov, *The Fundamentals of Modern Astrophysics*,
DOI 10.1007/978-1-4614-8730-2

Hartmann WK (2003) A traveller's guide to mars. The misterious landscapes of the red planet. Workman Publishing, New York, 468 pp

Hartmann WK (2005) Moons and planets, 5th edn. Brooks/Cole, Thomson Learning, Belmont, 428 pp

Heiken GH, Vaniman DT, French BM (1991) Lunar sourcebook. A user's guide to the moon. Cambridge University Press, Cambridge, 736 pp

Huebner WP (ed) (1990) Physics and chemistry of comets. Springer, Berlin, 376 pp

Huntress WT, Marov MY (2011) Soviet robots in the solar system. Mission technologies and discoveries. Springer-Praxis, New York, 453 pp

Kaku M (2005) Parallel worlds. A journey through creation, higher dimensions, and the future of the cosmos. Doubleday, New York

Keeton C (2014) Principles of astrophysics, vol XXI. Springer, New York, 434 pp

Kippenhahn R, Weigert A, Weiss A (2012) Stellar structure and evolution, 2nd edn. Springer, New York, 604 pp

Kitchin CR (2012) Exoplanets. Finding, Exploring and Understanding alien Worlds. Springer, New York, 281 pp

Kivelson MG, Russel CT (eds) (1995) Introduction to space physics. Cambridge University Press, Cambridge, 568 pp

Lang KR (2003) The Cambridge guide to the solar system. Cambridge University Press, Cambridge, 452 pp

Lang KR, Whitney CA (1991) Wanderers in space. Exploration and discoveryin the solar system. Cambridge University Press, Cambridge, 316 pp

Lewis JS (2004) Physics and chemistry of the solar system, 2nd edn. Elsevier, Academic, San Diego, 684 pp

Lissauer JJ (1993) Planet formation. Ann Rev Astron Astrophys 31:129–174

Lissauer JJ (1995) Urey prize lecture: on the diversity of plausible planetary systems. Icarus 114:217–236

Marov MY (1985) Planetas del Sistema solar. Editorial MIR, Moscu, 292 pp

Marov MY, Grinspoon DH (1998) The planet venus. Yale University Press, New Heaven/London, 442 pp

Marov MY, Kolesnichenko AV (2013) Turbulence and self-organization. Modeling astrophysical objests, vol 389, Astrophysics and space science library. Springer, New York, 651 pp

Marov MY, Rickman H (eds) (2001) Collisional processes in the solar system, vol 261, Astrophysics and space science library. Kluwer, Dordrecht/Boston/London, 357 pp

McFadden L, Weissman PR, Johnson TV (eds) (2007) Encyclopedia of the solar system, 2nd edn. Academic, San Diego, 982 pp

McSween HY Jr (1999) Meteorites and their parent planets, 2nd edn. Cambridge University Press, Cambridge, 322 pp

Miller R, Hartmann WK (2005) The grand tour: a traveller's guide to the solar system, 3rd edn. Workman Publishing, New York, 208 pp

Moche DL (2004) Astronomy. A self-teaching guide, 6th edn. John Wiley and Sons, Hoboken, 343 pp

Morrison D, Owen T (2003) The planetary system, 3rd edn. Addison-Wesley, New York, 531 pp

Narlikar JV (2002) An introduction to cosmology, 3rd edn. Cambridge University Press, Cambridge, 541 pp

Phillips AC (1999) The physics of stars, 2nd edn. Wiley, Chichester, 262 pp

Reipurth B, Jewitt D, Keil K (eds) (2007) Protostars and planets V. The University of Arizona Press, Tucson, 951 pp

Sagan C (1983) Cosmos. Abasus, London, 413 pp

Sagan C (1994) Pale blue dot. A vision of the human future in space. Random House, New York

Satz H (2013) Ultimate horizons, Probing the Limits of the Universe. Springer, Heidelberg, 172 pp

Seager S, Dotson R (eds) (2010) Exoplanets. The University of Arizona Press, Tucson, and Lunar and Planetary Institute, Houston, 526 pp

Stahler SW, Palla F (2005) The formation of stars. Wiley-VCH, Weinheim, 865 pp

Tayler RJ (1995) The hidden universe. Wiley and Praxis Publishing, Chichester, 213 pp

Taylor SR (2001) Solar system evolution, 2nd edn. Cambridge University Press, Cambridge, 484 pp

Wasson JT (1985) Meteorites: their record of early solar system history. W.H. Frieman, New York, 274 pp

Weinberg S (1988) The first three minutes. Basic Books, New York, 198 pp

Whipple FL (1981) Orbiting the Sun. Planets and satellites of the solar system. Harvard University Press, Cambridge

Wilkinson J (2009) Probing the new solar system. CSIRO Publishing, Canberra, 310 pp

Zelik M, Gregory SA, Smith EP (1992) Introductory astronomy and astrophysics, 3rd edn. Saunders College Publishing, Philadelphia

Zombeck MV (2007) Handbook of space astronomy and astrophysics, 3rd edn. Cambridge University Press, Cambridge, 767 pp

Index

© Springer Science+Business Media New York 2015
M.Ya. Marov, *The Fundamentals of Modern Astrophysics*,
DOI 10.1007/978-1-4614-8730-2

Lightning Source UK Ltd.
Milton Keynes UK
UKOW06n1244040516

273539UK00003B/23/P